Applied Microbiology

Sanjai Saxena

Applied Microbiology

Springer

Sanjai Saxena
Department of Biotechnology
Thapar University
Patiala, Punjab, India

ISBN 978-81-322-2258-3 ISBN 978-81-322-2259-0 (eBook)
DOI 10.1007/978-81-322-2259-0

Library of Congress Control Number: 2015931517

Springer New Delhi Heidelberg New York Dordrecht London

Printed on acid-free paper

Springer (India) Pvt. Ltd. is part of Springer Science+Business Media (www.springer.com)

Preface

Microorganisms appeared on the face of the earth around 3.5 billion years ago and evolved in due course of time in two clearly distinguishable forms – prokaryotes and eukaryotes. Eukaryotic microorganisms possess membrane bound cell organelles and comprise of fungi and protists, while prokaryotes lack membrane bound cell organelles and include eubacteria and the archaebacteria. Initially prokaryotic microorganisms dominated the earth, but during the course of their evolution they transformed the earth's anaerobic environment into aerobic and simultaneously generated massive amounts of organic compounds. Thus these evolved forms of prokaryotes created an environment suited for the evolution and maintenance of more complex life forms.

Microorganisms exhibit metabolic plasticity i.e. they adapt and survive changes which occur in the biosphere and therefore are ubiquitous in their existence as compared to the complex life forms. Advances in biochemistry, molecular biology and physiology have provided us tools to understand the genetic and metabolic makeup of microorganisms which have evolved in due course of time for their successful exploitation. Thousands of microorganisms have been recovered from different niche and are available as pure cultures in different culture collections across the globe while thousands are still to be explored or cultured.

Applied microbiology is primarily associated with exploitation of these microorganisms directly or indirectly in processes and products that are of economic, environmental and social importance throughout the world. Knowledge related to genetic engineering has revolutionized applied microbiology by enhancing the desired traits and removing the undesired traits from microorganisms thereby enhancing their commercial applicability.

Today microorganisms are playing a key role in the production of a variety of products via fermentation processes which include production of enzymes for use in commercial products like detergents, medicines, personal care products etc., chemical feedstocks, foods and pharmaceuticals. Microorganisms also play an important role in agricultural practices as well as remediating the environment. The purpose of this book is to provide a consolidated resource on practical exploitation of microorganisms in different fields like agriculture, environment, food, chemical and pharmaceuticals. The 12 chapters provide an in-depth understanding of the knowledge related to the specified field with practical approach. This book is specifically targeted for undergraduates and postgraduates who take up practical research in their degree programs. It will also prove to be a useful resource book for research institutes.

The inspiration to write this book is primarily students who have been striving hard to find consolidated information accompanied with industrial aspects. There has been a long-felt need for a comprehensive book on the application of biotechnological processes by exploitation of microorganisms by undergraduate and postgraduate students and researchers. This book is first of its kind meeting this requirement. I have attempted to prepare a suitable textbook by using a direct approach that should be very useful for students.

Patiala, India Sanjai Saxena

Acknowledgments

I would like to thank Dr. Devendra Kumar, my father, who has provided immense help in careful reading and editing of the manuscript. I am also thankful to Dr. Mamta Kapila, Springer for her interest in my proposal for this book and for introducing it to Springer..

I also acknowledge constant encouragement of my mother for writing this book. My wife and my daughter have been a constant inspiration and great support to make this project possible.

Contents

About the Author

Sanjai Saxena currently is an Associate Professor in the Department of Biotechnology, Thapar University, Patiala, India. Concurrently Dr. Saxena is also Coordinator of TIFAC-Centre of Relevance & Excellence (CORE) in Agro & Industrial Biotechnology, in the University. Dr. Saxena possess over 15 years of research and teaching experience in microbial secondary metabolites, microbial biochemistry, microbial diversity, biological control, mycology and drug discovery. His research work has resulted in significant extramural funding, over 35 refereed journal articles, 25 abstracts, 5 book chapters, and 1 US and 2 Indian patents. Recognizing his work, the Association of Advancement of Biodiversity Science, Karnataka, has conferred upon him Eminent Microbiologist Award in the year 2014 and inducted him as a fellow in the society.

1 Diversity of Industrially Relevant Microbes

1.1 Introduction

The discovery of the microbial world is much attributed to van Leeuwenhoeck (1677) who observed and described single-celled organisms which he originally referred to as *animalcules* with his handcrafted microscopes. However, the different activities and functions of these organisms were identified after approximately 200 years later while performing fermentations, understanding diseases in humans and animals and in agriculture. Incidentally the term microbe was given by Prof. Charles E. Sedillot (1804–1833) who is one of the pioneers of modern medicine, surgery, anaesthesiology, histopathology and infectiology. Sedillot understood the existence and action of microorganisms which he termed as *microbes* while studying the development of post-operative infections. Pasteur demonstrated that there are specific activities of yeasts and bacteria which are responsible for specific fermentations which he published in papers between 1857 and 1860. He was able to demonstrate the development of wine diseases and role of pasteurisation to preserve wine storage. Martinus W. Beijerinck was one of the great general microbiologists who made fundamental contributions to microbial ecology by highlighting microbial association with plants for fixing the atmospheric nitrogen. He isolated the aerobic nitrogen fixing microorganism *Azotobacter* as well as root nodule organism *Rhizobium*. The first microbe (prokaryote) evolved around 3.6 billion years ago and since then has undergone a process of evolution by exploiting a vast range of energy sources and thriving in different habitats which existed during the course of evolution. All the basic biochemical processes of the life evolved and developed from their microbial ancestors. Microbes are considered to be the common ancestors of all organisms which not only grow everywhere but are present in abundance. Microorganisms correspond to the richest collection of molecular and chemical diversity. They drive the ecosystem processes by maintaining the nutrient cycles as well as maintain elegant relationships between themselves and higher organisms. Microbial diversity is a great resource of biotechnological exploration of novel microorganisms and their products for exploitation in different processes. Microbial diversity has been explored using several approaches like phylogeny, physiology, metabolism and genomics.

1.2 Realm of Microbial Existence

Microbial diversity from six different environments has been studied using culture-dependent and culture-independent methods. Microbial communities in coastal subsurface sediments are scarcely investigated and have escaped attention so far (Kopke et al. 2005). The study of marine microbial biodiversity is vital to the understanding of the different processes of the ocean, which may present potent novel microorganisms for

S. Saxena, *Applied Microbiology*,
DOI 10.1007/978-81-322-2259-0_1, © Springer India 2015

screening of bioactive compounds. As the microbial communities have a complex ecosystem process, biodiversity study explores their distribution and roles in the habitat. Endophytic existence of microbes has become a new resource for their exploitation in new processes and development of novel products. Extremophilic environments are also new niches for exploration and identification of new microbes on the basis of their physiological and phylogenetic uniqueness. These have been exploited in screening of enzymes, biopolymers and antibiotics for industrial applications.

1.2.1 Diversity of Soil Microbes

Soil is a very complex habitat dominated by microbes which exist on solid phase. It is estimated that a gram of undisturbed soil may contain ten billion microorganisms representing 6,000–10,000 different genomes. Associating microbial diversity with soil functions is a very complex situation since it is not possible to assess the microbial diversity despite using molecular tools and techniques which have been lately used for viable but not cultivable microorganisms. Natural products isolated from soil samples play a significant role in discovery and development of new drugs and biocatalysts. The novel cultivation technologies like gene mining by direct cloning of soil DNA and screening of the resulting complex metagenomic libraries have lead to increase the discovery rate of new biomolecules and minimise the re-evaluation of already known natural products. The use of different isolation parameters, viz. pH, salt, temperature and metal concentration, has enhanced the identification of distinct microbial type which have an altered physiology and genetic mechanisms of overcoming or tolerating different physical as well as chemical stresses. These microbes can be explored for their abilities to produce novel enzymes as well as clinically important drugs. Plants also have a significant impact on the microbial composition of soil in the rhizosphere due to rhizodeposition and decay of litter roots. Microbes especially bacteria are attracted to aromatic pollutants like naphthalene, benzene and chlorinated herbicides as a precedent to their degradation. The terms biodegradation, biocatalysis and biotransformations have been used interchangeably depending on the aspect of chemical transformations being carried out by these microorganisms. Biocementation is a process of enhancement of the strength and stiffness of properties of soil and rocks through microbial activity or products. Chemical grouting is a process to fill soil voids with fluid grouts. Water-insoluble gel-forming biopolymers of microbial origin such as xanthan, chitosan, polyglutamic acid, sodium alginate and polyhydroxybutyrate can also be used as grouts for soil erosion control, enclosing of bioremediation zone and mitigating soil liquefaction (Momemi et al. 1999; Etemadi et al. 2003; Gioia and Ciriello 2006).

A variety of molecular methods like ARDRA, DGGE, TGGE or RISA have been developed to assess the bacterial diversity in the soil. However, fungal diversity in the soil could not be studied using the same techniques since the concentration of fungal DNA is much less than that of bacterial DNA. Another hindrance has been the use of specific primers without co-amplification of DNA from other eukaryotic organisms such as plants, algae and nematodes. Soil is a vast playground of interactions within different types of microbes which lead to microbial community development. The various interactions are (1) neutral associations, (2) positive associations and (3) negative associations.

1.2.1.1 Neutral Associations

Neutralism or neutral association between microbes refers to the occupation of two different species of microbes in the same environment without affecting each other. This type of association is generally transitory in nature.

1.2.1.2 Positive Associations

Positive associations comprise of mutualism, syntrophism and commensalism. Mutualism is essentially a relationship in which each organism is benefitted from the association. Syntrophism is a mutualistic association which involves the exchange of nutrition between two species.

Lichen, an association between fungus and blue-green algae, is an example of syntrophism. Mutualistic interaction between *Thiobacillus ferrooxidans* and *Beijerinckia lacticogenes* helps in ore leaching. Leaching is the process of recovering metal from the ore, where microorganisms play the important role of oxidising insoluble metal sulphides to soluble sulphates. Microorganisms may also form mutualistic relationships with plants in soil, an example of which is nitrogen-fixing bacteria, i.e. *Rhizobium*, growing in the roots of legumes (plants of the family Leguminosae). In this *Rhizobium*–legume association, *Rhizobium* bacteria are benefited by protection from the environmental stresses while in turn the plant is benefited by getting readily available nitrate nitrogen released by the bacterial partner.

Commensalism refers to a relationship between organisms in which one species of a pair benefits whereas the other is not affected. This happens commonly in soil with respect to degradation of complex molecules like cellulose and lignin. For example, many fungi can degrade cellulose to glucose, which is utilised by many bacteria. Many bacteria are unable to utilise cellulose, but they can utilise the fungal breakdown products of cellulose, e.g. glucose and organic acids.

1.2.1.3 Negative Associations

The negative associations comprise of antagonism, competition, parasitism and predation. Antagonistic relationship between microorganisms generally results in inhibition or adversely affects the growth and survival of other species. This is generally mediated by signal molecules which induce the inhibition or adverse effects and have been referred to as antibiotics, and this phenomenon is known as antibiosis. This interaction has a great importance in the discovery and development of a variety of antimicrobial drugs. Soil-dwelling *Streptomyces* species has fuelled in the development of an array of antibacterial and antifungal antibiotics like streptomycin, griseofulvin, neomycin, cycloheximide, etc.

Competition refers to the interaction between microorganisms for limited nutrients and space. In this interaction, a microorganism which outcompetes or eliminates other for the limited resource in lesser time is found to predominate. Microorganisms exhibiting adaptability and faster growth rate are better competitors. A relationship in which one organism lives in or on other organism is referred to as parasitism. The parasite lives in intimate physical contact with the host and forms metabolic association with the host. Mycoparasites and presumptive mycoparasites have biocontrol potential; some are responsible for the natural suppressiveness of soils to certain plant pathogens. *Trichoderma* and *Gliocladium virens* have been successfully used as commercial biofungicides to control a range of economically important soil borne fungal plant pathogens.

Several species of *Trichoderma* were used successfully against certain pathogenic fungi. *Trichoderma* sp. was used as commercial biofungicides to control a range of economically important soilborne fungal plant pathogens.

1.2.2 Marine Microbial Diversity

Oceans and sea contribute to the largest ecosystem on the Earth which has a profound effect on the world's climate. Microorganisms have been evolving since the last 3.8 Ga (10^9 years) of their 4.6 Ga existence and have made Earth habitable for all other species. Hence, the interactions of these organisms with other life forms and their diversity have been frequently questioned.

As compared to the plant and animal diversity, microbial diversity research is difficult, and the role of microbiologist is very crucial in understanding their diverse assemblages. Microbial diversity in marine ecosystem has been ventured into a couple of decades ago, and hitherto unknown groups of microorganisms have been identified (Zobell 1946; Wood 1959). Marine microorganisms occur in vast number (Box 1.1). Ocean water contains up to 10^6–10^9 microorganisms per ml of several thousand different types. The direct interactions of marine microorganisms with other organisms have been distributed in two broad classes, viz. pathogenic and symbiotic.

Box 1.1: Marine Microorganism

Any microorganism that grows in marine environment is known as marine microorganism, independent of the fact that whether they are abundant in other aquatic or terrestrial environment

Box 1.2: Facts About Marine Microbes

Marine microbes occur in vast numbers with huge genetic diversity

These microbes are key to all biogeochemical cycles and therefore crucial for the functioning of marine ecosystem

Marine microbes degrade organic matter in the ocean, thereby playing a key role in the maintenance of fixed carbon dioxide

Cyanobacteria, diatoms, picophytoplanktons and nanophytoplanktons (marine phototropic microorganisms) are responsible for more than 50 % of the oxygen produced on the Earth

Marine microbes represent largely untapped source of novel bioactive compounds and metabolic pathways which could be exploited for new biotechnological applications and products

Marine microbes occupy critical bottom trophic levels; in marine food webs they play an indispensable role in ensuing supply of sea food products

As life evolved after the formation of water on Earth, marine microorganisms are considered as the foundation of the life and therefore have a critical role in habitability and sustainability (Box 1.2). A wealth of knowledge on dominant types of microorganisms existing in oceans (Box 1.3) have been collected using technological improvements in biological sciences. There exist several questions that remain unsolved since either appropriate methodologies have not been developed or applied or else the amount of work to answer these questions is beyond the scope and resources of most labs individually.

Box 1.3: Significant Achievements in Marine Microbiology

1997: The development of techniques for the enumeration of microbes in oceans for the first time

1979: Microbes discovered in hydrothermal vents

1980: Bacteria in the ocean are found to be actively synthesising DNA and RNA

1982: The discovery of marine bacteria which is predated by a group of highly specialised small protists (heterotrophic non-flagellates)

1983: It was established that primary production in the ocean is carried out by microbes smaller than 2 μm

1989: Flow cytometry allows the discovery of picocyanobacteria – *Prochlorococcus*, the most abundant photosynthetic microorganisms on the earth.

1990: The first culture-independent assessment of bacterial diversity through rRNA analysis

1989–1990: The discovery of a vast number of viruses, i.e. 10 million/ml of ocean water, and their role in nutrient cycling

1994: High abundance of Archaea in marine plankton even in cold as well as oxygenated waters

2000: The discovery of photoheterotrophy in the sea by metagenomic techniques

2002: The discovery and isolation of new marine microorganisms and new metabolic pathways – *Pelagibacter* and Thaumarchaeota

2006–2007: High-throughput sequencing introduced in marine microbial ecology reveals that bacterial diversity is larger than expected

Several studies have indicated that marine microbial species are not cosmopolitan in their existence but they are restricted to specific habitat types and geographical regions.

1.2.2.1 Symbiotic Interactions with Marine Invertebrates

Marine invertebrates comprising of corals, sponges, squids and shipworms are associated with unique species of bacterial or archaeal symbionants. The basis of symbiosis is shelter, dispersal of nutrients and possibly a route for reproduction. The marine invertebrates find that their association with microorganisms makes them adaptable to survive in inhospitable conditions in the marine environment. Hydrothermal vents in marine environment is inhospitable for unprepared tube worms *Piftia*, but these survive in sulphur-rich, oligotrophic zones by harbouring chemoautotrophic bacteria which synthesised organic carbon using energy from the respiring reduced inorganic sulphur compounds. The bacteria provide organic compounds to their hosts, allowing worms to live on inorganic sources. Similarly luminescent bacterial symbionants in squids enable it to hunt in the moonlit water without casting a shadow, thereby remaining undetected by the predators.

Marine environment today is considered as an 'emerging gold mine' for novel bioactive compounds which possess antibacterial, antifungal, antiviral and anticancer properties/activities. Antifouling and anti-biofilm-forming properties have been reported from marine microbes.

Natural products derived from marine microorganisms exhibit enormous range of novel chemistries and provide challenging template for new entities via synthetic chemistry. However, only 1 % of the microorganisms can be isolated using traditional culturing techniques which is a major bottleneck. Sorbicillactone A is a novel alkaloid which is reported from *Penicillium chrysogenum* associated with sponge *Ircinia fasciculata*. The compound has shown promising activities in several mammalian and viral systems qualified for therapeutic human trails (Bringmann et al. 2003).

Marine surface-associated microorganisms are considered as a novel source/unique matrix for novel bioactive metabolites as there exists a necessity to evolve allelochemicals which would serve as defence for protecting them from the fierce competition that exists between the microorganisms on the surface marine eukaryote (Penesyan et al. 2010).

The microorganisms which thrive on the surface of eukaryotic marine organisms have been referred as epibionts. During the interactions between the host and epibiotic microorganisms, it is thought that these microbes acquire nutrients from the eukaryotic host, while the host benefits from the wide range of the bioactives produced by its associated microorganisms that appear to be widespread in marine environment (Harder 2009). Gamma-proteobacterium *Pseudoalteromonas tunicata* is known to produce several bioactive compounds which play a role in defending the host against surface colonisation by producing antimicrobial, anti-larval and anti-protozoan compounds (Holmström and Kjelleberg 1998; Egan et al. 2001; Franks et al. 2006).

Marinophilus is a new genus of actinomycetes which produces a set of a series of structurally unique antitumor antibiotics which have been named as marinomycins A and B. These compounds also exhibit antibacterial activities with a MIC range of 0.125–0.625 μg/ml against vancomycin-resistant *Enterococcus* (VRE) and methicillin-resistant *Staphylococcus aureus* (MRSA). Aplasmomycin is a new antibiotic which selectively inhibits gram-positive bacteria during in vitro assay and plasmodia under in vivo conditions. It is produced by *Streptomyces griseus* ss-20 isolated from shallow sea sediment in Sagami Bay, Japan (Okami et al. 1976). Another marine actinomycetes *Micromonospora marina* isolated from Mozambique strait produces thiocoraline, a novel bioactive depsipeptide which inhibits RNA synthesis and is cytotoxic to lung and colon cancer cell lines and melanoma (Erba et al. 1999). Lomaiviticins A and B have been isolated from *Micromonospora lomaivitiensis* which was isolated from an ascidian. These exhibit a potential antitumor activity (He et al. 2001). These also exhibit potential antimicrobial activity against the gram-positive microorganisms *S. aureus* and *E. faecium* with a MIC ranging from 6 to 25 ng per spot in the assay.

Cladosporium herbarum isolated from marine sponge *Callyspongia aerizusa* produced acetyl Sumiki's acid using sea water in the medium.

The acid and its derivative exhibit antibacterial activity against *Bacillus subtilis* and *S. aureus* at a concentration of 5 μg/disc (Jadulco et al. 2001). Ulocladol, a p56 tyrosine kinase inhibitor, was isolated from *Ulocladium botrytis* inhabiting *Myxilla incrustans* (Holler et al. 1999).

1.2.3 Halophilic Environment

Phylogenetic diversity of microorganisms existing in hypersaline environment is surprising and classified in three domains – Archaea, Bacteria and Eucarya. They exhibit metabolic diversity and include oxygenic and anoxygenic phototrophs, aerobic heterotrophs, fermenters, denitrifiers, sulphate reducers and methanogens. *Halothermothrix orenii*, an anaerobe, has been isolated from the Tunisian salt lake which can withstand a salt concentration of 200 g/l and a temperature of 68 °C. *Dunaliella* is a green alga which exists in Dead Sea and has variants which produce large quantities of β-carotene, which is a precursor of vitamin A. Ectoine and hydroxyectoine have been isolated from *Halomonas elongata*, and *Marinococcus* M52 have been used as enzyme stabilisers as well as moisturisers in cosmetics (Oren 2002). *Haloferax mediterranei* is a halophilic archaeon which produces a co-polymer of β-hydroxybutyrate and β-hydroxyvalerate which can be used for the production of thermoplastics. *Ralstonia eutropha*, non-halophile, has been used for the production of bioplastics by ICI used for the production of bottles for the cosmetics. *Haloferax mediterranei* accumulates as much as β-hydroxyalkanoate using starch as cheap source of carbon and energy. It also produces copious amounts of anionic polysaccharides which can be used as stabilisers, thickeners and gelling agents.

1.2.4 Plant: Microbe Interaction

Microbial diversity plays a significant role in agroecosystems as well as in forest ecosystems. The microorganisms have been found to be associated to plants as pathogens, epiphytes and endophytes. Microbes play an important functional role in the growth and yield of crop plants directly or indirectly. The beneficial interactions between the plants and microbes have resulted in applications of microbes directly as microbial inoculants in agricultural biotechnology. Further based on their effects on the physiology and metabolism of plants, these have been referred to as biofertilisers, plant strengtheners, phytostimulators and biopesticides (Lugtenberg et al. 2002). Microorganisms which have been exploited in industry for use as biopesticides and biofertilisers have been discussed in detail in Chap. 4. It has been reported that many plant-associated microorganisms have inherent ability to produce phytohormones like indole-3-acetic acid (IAA),

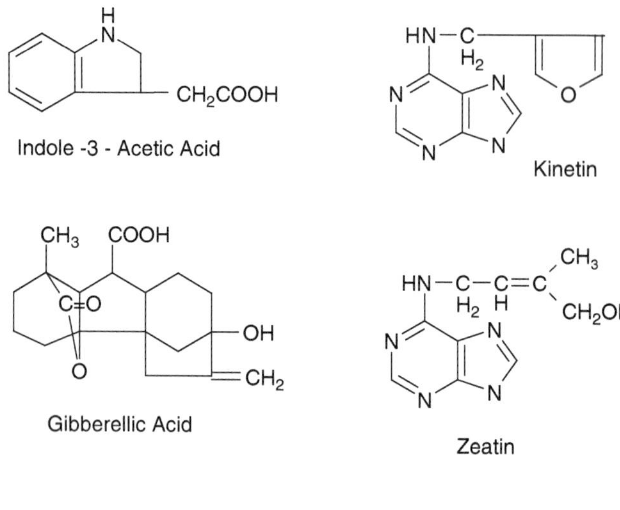

Fig. 1.1 Phytohormones produced by plant-associated microbes

gibberellins and cytokinins (Fig. 1.1). The ability of microorganisms to produce plant hormones has been displayed with rhizospheric, epiphytic and symbiotic bacteria which stimulate and facilitate the growth of plants, which are therefore referred to as PGP (plant growth promoters). Free living microorganisms also possess the ability to produce phytohormones. Many rhizospheric as well as epiphytic bacteria have been reported to produce IAA, the prominent one being *Azospirillum* spp., *Azotobacter* spp., *Alcaligenes* spp., *Erwinia* spp., etc. Apart from bacteria, fungi, viz. *Phoma* spp., *Aspergillus* spp., *Trichoderma* spp. and *Fusarium* spp., also possess the capability to synthesise IAA. Gibbere-llins are diterpenes which have isoprene units fused into a four-ring structure. Gibberellins primarily influence the cell division and cell elongation primarily in cells constituting the intercalary cell division. Phytopathogenic fungi such has *Ustilago maydis*, *Botryodiplodia theobromae*, *Fusarium semitectum*, *Fusarium equiseti*, *F. oxysporum* and *F. moniliforme* have been found to be active producers of gibberellins. Among the bacteria, *Azospirillum*, *Pseudomonas*, *Bacillus*, *Acinetobacter*, *Flavobacterium*, *Micrococcus*, *Agrobacterium*, *Clostridium*, *Rhizobium* and *Xanthomonas* have been found to produce gibberellins. Rhizobacteria belonging to the genera *Azotobacter*, *Azospirillum*, *Rhizobium* and *Pseudomonas* species produce cytokinins. *Rhizopogon*, *Suillus* and *Paxillus* are the genera of mycorrhizal fungi which produce cytokinins. Zeatin has been reported from *Phoma* species, while kinetin is produced by *Fusarium solani* and *Trichoderma viride*. It can thus be concluded that plant-associated microflora are the richest source of microorganisms which produce phytohormones and thus could be exploited for commercial production of phytohormones through fermentative route for agricultural and research applications.

Apart from producing phytohormones, plant-associated microbes have been recently found to the producers of pharmaceutically active compounds. Historically major drug molecules have been isolated from microbes that have been predominantly inhabiting the soil. Among soil microorganisms, actinomycetes and fungi have been the prolific producers of the pharmaceutically active compounds.

In the last decade, considerable amount of knowledge has been generated on the biology of endophytic microorganisms. These microorganisms actually colonise the plant tissue and develop symbiotic association with their host in which the host (macrophyte) protects and feeds the microorganism which in turn produces signal molecules/bioactive compounds which promotes growth and increases competitiveness of the host. They also help the plant in combating with different types of abiotic and biotic stresses, for example, drought, salinity, plant pathogens and attack by insects.

Endophytic fungi have been explored and exploited for their ability to produce putative phytochemicals of their host plant which may also possess medicinal properties. Much attention on exploitation of endophytic fungi was drawn with the detection of paclitaxel (Taxol®) in the endophytic fungus *Taxomyces andreanae* which was isolated from *Taxus brevifolia*, the latter being the source of important anticancer drug (Stierle et al. 1993, 1995). Several questions were raised subsequent to this discovery with regard to horizontal gene transfer between the host and the endophyte or vice versa. More recently it has been experimentally demonstrated that the production of putative medicinal compounds of plant origin by endophytes extends to other pharmaceutically important natural products such as camptothecin (Amna et al. 2006) and podophyllotoxin (Kour et al. 2008). Another compound a dimeric indole alkaloid of plant origin exploited as an anticancer drug is vincristine, reports exist wherein an endophytic *Fusarium oxysporum* has been isolated from the plant *Catharanthus roseus* from where vincristine was originally isolated (Lingqi et al. 2000).

Endophytes are also fountainheads of unique chemical structures which have been modified through evolution and probably involved in host plant protection and communication (Gunatilaka 2006). L-783,281 is an antidiabetic compound which has been isolated from the endophytic fungus *Pseudomassaria* species originally isolated from African rainforest. It acts as an insulin mimetic which does not get destroyed in the digestive tract (Zhang et al. 1999). Examinations of endophytic fungal cultures have yielded a wide variety of bioactive natural products and potential 'leads' possessing antiparasitic, antimicrobial, cytotoxic, neuroprotective or immunomodulatory properties (Staniek et al. 2008; Verma et al. 2009; Aly et al. 2010). MK-3118 is an orally available semisynthetic β-1,3-glucan synthase inhibitor derived from enfumafungin initially isolated from *Hormonema* sp., an endophyte living in the leaves of *J. communis*. MK-3118 exhibited potential inhibitory activity against *Candida albicans* and *Aspergillus fumigatus* glucanase with EC_{50} of 0.6 ng/ml and 1.7 ng/ml, respectively. This drug has entered phase I clinical trials for the treatment of fungal infections (Motyl et al. 2010). Tauranin is a new metabolite reported from *Phyllosticta spinarum* isolated from *Platycladus orientalis* collected from Arizona, USA, and exhibits potential activity against non-small cell lung cancer (NCI-H460), breast cancer (MCF-7), CNS cancer (SF-268) and metastatic prostate cancer (PC-314) (Wijeratne et al. 2008).

Nodulisporic acid A and its congeners are structurally complex fungal metabolites having insecticidal activity, isolated from *Nodulisporium* species, endophytic in Hawaiian plant *Bontia daphnoides* (Ondeyka et al. 1997). *Rhizopus oryzae*, an endophyte on *Foeniculum vulgare* (fennel) in the Mediterranean plant, is found to produce mycelium-bound lipase which is active over a pH of 3–8 and thermostable in nature with maximum activity at 160 °C (Torres et al. 2003). Similarly an endophyte *Acremonium* species also produces a glucoamylase with strong amylopectin-hydrolysing activity with a pH of 3–7.2 and temperature up to 60 °C (Marlida et al. 2000).

Bacterial endophytes also produce bioactive compounds which could be exploited by the pharmaceutical as well as agrochemical industries. Pseudomycins are a group of peptide antifungal compounds produced by plant-associated bacterium *Pseudomonas syringae*. Pseudomycin A has an impressive activity against the human pathogens *Candida albicans* and *Cryptococcus neoformans* (Harrison et al. 1991). Similarly kakadumycin A was reported from endophytic *Streptomyces* (NRRL 30566) which is related to echinomycin and possesses antibacterial as well as antimalarial activity (Castillo et al. 2003).

As endophytic microorganisms are present in tropical and temperate rainforest plants which occupy more than 7 % of the Earth's land surface, they exhibit a huge and untapped biodiversity for harnessing novel chemical entities for pharmaceutical as well as agricultural applications. Apart from that, they also appear as unique sources of novel enzymes which could be used for industrial applications. Endophytic microflora of plants involved in phytoremediation of toxic metallic and non-metallic pollutants could also be exploited for developing efficient in situ and ex situ bioremediation systems.

1.2.5 Microbe–Microbe Interactions

The microbe–microbe interactions are generally found in the rhizospheric zone while studying the soil microbial community dynamics. The plant-associated microbes existing in the root region modulate the occurrence of other microorganisms in and around their vicinity for their own benefit as well as for the survival of the plant. Similarly there is a plethora of interactions, both positive and negative, which occur during the food fermentations between different microorganisms. New avenues of microbe–microbe interactions are being explored by associating the microbial community dynamics of the gut and skin with different human diseases such as irritable bowel syndrome, diabetes, obesity, etc. The roles of prebiotics and probiotics are also being explored in the development of new therapeutic interventions (Wallace et al. 2011). Similarly skin microbiome has also become a new facet of study for understanding different diseases.

1.2.6 Animal–Microbe Interactions

The symbiotic associations can be commensal, mutualistic and parasitic depending on the effect on the partners in the relationship. Like plant–

Table 1.1 Interactions between animals and microbes

Microorganism	Animal host	Interaction
Termitomyces sp.	*Macrotermes* (termite)	Fungi produce celluloses to degrade plant matter, and termite consumes both plant matter and fungi
Xenorhabdus nematophilus	*Steinernema carpocapsae (soil nematode)*	Digestive tract symbiosis
Buchnera sp.	Aphids (*Aphidoidea*)	Nutritional symbiosis for amino acid synthesis; intracellular location of symbiont
Vibrio fischeri	*Euprymna scolopes* (Hawaiian squid)	Light organ symbiosis and functions to produce light to camouflage squid
Aeromonas veronii	*Hirudo medicinalis* (medicinal leech)	Digestive tract symbiosis; extracellular location of symbiont
Amylostereum sp.	*Sirex cyaneus (wood wasp)*	Fungus produces cellulases and xylanases to degrade plant matter to feed the offspring from the wasp eggs
Oceanospirillales	*Osedax* sp. (marine polychaete)	As endosymbionts, bacteria digest organic matter

microbe interactions, animal–microbe interactions could be endosymbiotic or ectosymbiotic depending upon their location. At time the symbionants live on the surface of the animals and are referred to as episymbionts. The symbionts are broadly classified as primary and secondary symbionts. Primary symbionts are associated with their host through long evolutionary history and primarily provide their host with vitamins and other nutrients. On the other hand, secondary symbionts may positively or negatively affect the host; they generally are facultative in nature.

Wolbachia pipientis is the most ubiquitous endosymbiont on the plant. This alphaproteobacterium is a gram-negative, bacillus- or coccoid-shaped bacterium of Rickettsiaceae family which is maternally as well as horizontally transmitted and is obligately intracellular. *Wolbachia* has been found to infect insects in a range of 20–75 %. Insects are also being explored for their microbial gut communities, such as those associated with termites, cockroaches and fruit Flies. These existing gut microbes produce vitamins which are needed by the hosts, help in the efficient digestion of food when it is of lower quality and control the invasion of pathogenic bacteria.

The two best studied bioluminescent symbioses are the Hawaiian bobtail squid *Euprymna scolopes* and *Vibrio fischeri*. The bacterial luminescence helps the squid to escape predation through counter-illumination or helps the fish hosts to attract mates; in turn they obtain shelter and nutrients from the host. Some animal microbe interactions have been summarised in Table 1.1. The endosymbionts of animals are recently being explored for new therapeutic molecules; this is particularly important in the case of marine invertebrate animals. The diversity of microflora in guts of ruminants and humans is another exciting area, and only surface has been scratched for their possible exploitation for the development of industrial products and processes.

1.3 Summary

The diversity of microorganisms is tremendous as they are associated with every life form existing on the earth. In fact it is the microorganisms from which the multicellular life forms have evolved. However, only a few of them could be grown under laboratory conditions and could be explored for development of products and processes.

Selected Reading

Aly AH, Debbab A, Kjer J, Proksch P (2010) Fungal endophytes from higher plants: a prolific source of phytochemicals and other bioactive natural products. Fungal Divers 41:1–16

Amna T, Puri SC, Verma V, Sharma JP, Khajuria RK et al (2006) Bioreactor studies on the endophytic fungus *Entrophospora infrequens* for the production of an anticancer alkaloid camptothecin. Can J Microbiol 52:189–196

Bringmann G, Lang G, Mühlbacher J, Schaumann K, Steffens S, Rytik PG, Hentschel U, Morschhäuser J, Müller WE (2003) Sorbicillactone A: a structurally unprecedented bioactive novel-type alkaloid from a sponge-derived fungus. Prog Mol Subcell Biol 37:231–253

Castillo U, Harper JK, Strobel GA, Sears J, Alesi K, Ford E, Lin J, Hunter M, Maranta M, Ge H, Yaver D, Jensen JB, Porter H, Robison R, Millar D, Hess WM, Condron M, Teplow D (2003) Kakadumycins, novel antibiotics from *Streptomyces* sp NRRL 30566, an endophyte of *Grevillea pteridifolia*. FEMS Microbiol Lett 224(2):183–190

Egan S, James S, Holmström C, Kjelleberg S (2001) Inhibition of algal spore germination by the marine bacterium Pseudoalteromonas tunicata. FEMS Microbiol Ecol 35:67–73

Erba E, Bergamaschi D, Ronzoni S, Faretta M, Taverna S, Bonfanti M, Catapano CV, Faircloth G, Jimeno J, D'Incalci M (1999) Mode of action of thiocoraline, a natural marine compound with anti-tumour activity. Br J Cancer 80:971–980

Etemadi O, Petrisor IG, Kim D, Wan MW, Yen TF (2003) Stabilization of metals in subsurface by biopolymers: laboratory drainage flow studies. Soil Sediment Contam 12:647–661

Franks A, Egan S, Holmstrom C, James S, Lappin-Scott H, Kjelleberg S (2006) Inhibition of fungal colonization by *Pseudoalteromonas tunicata* provides a competitive advantage during surface colonization. Appl Environ Microbiol 72:6079–6087

Gioia F, Ciriello PP (2006) The containment of oil spills in porous media using xanthan/aluminium solutions, gelled by gaseous CO_2 or by $AlCl_3$ solutions. J Hazard Mater 138:500–506

Gunatilaka AAL (2006) Natural products from plant-associated microorganisms: distribution, structural diversity, bioactivity, and implications of their occurrence. J Nat Prod 69:509–526

Harrison LH, Teplow DB, Rinaldi M, Strobel G (1991) Pseudomycins, a family of novel peptides from Pseudomonas syringae possessing broad-spectrum antifungal activity. J Gen Microbiol 137:2857–2865

He H, Ding WD, Bernan VS, Richardson AD, Ireland CM, Greenstein M, Ellestad GA, Carter GT (2001) Lomaiviticins A and B, potent antitumor antibiotics from Micromonospora lomaivitiensis. J Am Chem Soc 123:5362–5363

Holler U, Konig G, Wright AD (1999) A new tyrosine kinase inhibitor from a marine isolate of *Ulocladium botrytis* and new metabolites from the marine fungi *Asteromyces cruciatus* and *Varicosporina ramulosa*. Eur J Org Chem 11:2949–2955

Holmström C, Kjelleberg S (1998) Marine Pseudoalteromonas species are associated with higher organisms and produce biologically active extracellular agents. FEMS Microbiol Ecol 30:285–293

Jadulco R, Proksch P, Wray V, Sudarsono, Berg A, Berg A (2001) New macrolides and furan carboxylic acid derivative from the sponge-derived fungus Cladosporium herbarum. J Nat Prod 64:527–530

Kopke B, WIlms R, Engelen B, Cypionka H, Sass H (2005) Microbial diversity in coastal subsurface sediments: a cultivation approach using various electron acceptors and substrate gradients. Appl Environ Microbiol 71(12):7819–7830

Kour A, Shawl AS, Rehman S, Sultan P, Qazi PH, Suden P, Khajuria RK, Verma V (2008) Isolation and identification of an endophytic strain of *Fusarium oxysporum* producing podophyllotoxin from *Juniperus recurva*. World J Microbiol Biotechnol 24:1115–1121

Lingqi Z, Guo B, Li H, Zeng S, Shao H, Gu S, Wei R (2000) Preliminary study on the isolation of the endophytic fungus of Catharanthus roseus and its fermentation to produce products of therapeutic value. Chin. Trad. Chin Tradit Herb Drugs 31:805–807

Lugtenberg B, Chin-A-Woeng T, Bloemberg G (2002) Microbe-plant interactions: principles and mechanisms. Antonie Van Leeuwenhoek 81:373–383

Marlida Y, Saari N, Hassan Z, Radu S, Bakar J (2000) Purification and characterization of sago starch-degrading glucoamylase from Acremonium sp. endophytic fungus. Food Chem 71:221–227

Momemi D, Kamel R, Martin R, Yen TF (1999) Potential use of biopolymer grouts for liquefaction mitigation. In: Leeson A, Alleman BC (eds) Phytoremediation and innovation strategies for specialized remedial applications. Batelle Press, Columbus, pp 175–18

Motyl MR, Tan C, Liberator P et al (2010) MK-3118, an oral enfumafungin with potent in vitro activity against Candida and *Aspergillus* spp. In: Abstracts of the fiftieth interscience conference on antimicrobial agents and chemotherapy, Boston, MA, 2010. Abstract F1-847. American Society for Microbiology, Washington, DC, USA

Okami Y, Okazaki T, Kitahara T, Umezawa H (1976) Studies on marine microorganisms. V. A new antibiotic, aplasmomycin produced by a streptomycete isolated from shallow sea mud. J Antibiot 29(10):1019–1025

Ondeyka JG, Helms GL, Hensens OD, Goetz MA, Zink DL, Tsipouras A, Shoop WL, Slayton L, Dombrowski AW, Polishook JD, Ostlind DA, Tsou NN, Ball RG, Singh SB (1997) Nodulisporic acid A, a novel and potent insecticide from a Nodulisporium sp. isolation, structure determination, and chemical transformations. J Am Chem Soc 119:8809–8816

Oren A (2002) Diversity of halophilic microorganisms: environments, phylogeny, physiology, and applications. J Ind Microbiol Biotechnol 28:56–63

Penesyan A, Kjelleberg S, Egan S (2010) Development of novel drugs from marine surface associated microorganisms. Mar Drugs 8:438–459

Staniek A, Woerdenbag HJ, Kayser O (2008) Endophytes: exploiting biodiversity for the improvement of natural product-based drug discovery. J Plant Interact 3:75–93

Stierle A, Strobel G, Stierle D (1993) Taxol and taxane production by *Taxomyces andreanae,* an endophytic fungus of Pacific yew. Science 260:214–216

Stierle A, Strobel G, Stierle D, Grothaus P, Bignami G (1995) The search for a taxol-producing microorganism among the endophytic fungi of the Pacific yew, Taxus brevifolia. J Nat Prod 58:1315–1324

Torres M, Dolcet MM, Sala N, Canela R (2003) Endophytic fungi associated with Mediterranean plants as a source of mycelium-bound lipases. J Agric Food Chem 51:3328–3333

Verma VC, Gond SK, Mishra A, Kumar A, Kharwar RN, Gange AC (2009) Endophytic actinomycetes from *Azadirachta indica* A. Juss.: isolation, diversity and anti-microbial activity. Microb Ecol 57:749–756

Wallace TC, Guarner F, Madsen K, Cabana MD, Gibson GR, Hentges E, Sanders ME (2011) Human gut microbiota and its relationship to health and disease. Nutr Rev 69(7):392–403

Wijeratne EMK, Paranagama PA, Marron MT, Gunatilaka MK, Arnold AE, Gunatilaka AAL (2008) Sesquiterpene quinines and related metabolites from *Phyllosticta spinarum*, a fungal strain endophytic in *Platycladus orientalis* of the sonorant desert. J Nat Prod 71:218–222

Wood EJF (1959) Some aspects of marine microbiology. J Mar Biol Assoc India 1:26–32

Zhang B, Salituro G, Szalkowski D, Li Z, Zhang Y et al (1999) Discovery of small molecule insulin mimetic with antidiabetic activity in mice. Science 284:974–977

Zobell CE (1946) Marine microbiology. Chronica Botanica, Waltham

Microbial Technology and Biotechnology

2

2.1 Introduction

Microorganisms since time immemorial have been exploited for the process of baking, brewing and food preservation. However, they also produce things which have immense value and are usually classified as primary metabolites that are essential for their own growth and secondary metabolites which are non-essential. Both primary and secondary metabolites have played a tremendous role in the development of chemical, pharmaceutical and food industries wherein they have been exploited for novel products and process development.

The advent of modern biotechnology began with the birth of recombinant DNA technology in 1972. Biotechnology currently refers to an array of enabling technologies which have applications in different industrial sectors. Genetic engineering, protein engineering and metabolic engineering are the three major disciplines which comprise biotechnology. The fourth discipline is biochemical process engineering which encompasses commercial production of biotechnological products. Modern microbial technology relies on the metabolic potential of microorganisms and the varieties of methods by which they have been harnessed. Genetically modified microorganisms (GMMOs) today find applications in the areas of human health, agriculture, environment, food, chemicals, paper and textile industries. By genetic engineering the molecular diversity and chemical selectivity of the microorganisms help in the production of desired products as well as cheaper and ecofriendly process development.

The different molecular methods which contribute to the development of GMMOs are (1) gene transfer methods to deliver specific genes in the desired host, (2) cloning vectors, (3) selectable marker gene for identification of recombinant microorganisms and (4) promoters to control the expression of desired genes. The most common recombinant microorganisms which express the genes of other organisms are *E. coli*, *Bacillus subtilis*, *Saccharomyces cerevisiae*, *Pichia pastoris*, *Hansenula polymorpha* and *Aspergillus niger*. A variety of products and processes have been developed through these recombinant microorganisms.

The strategies of creating a GMMO are based on (a) overexpression of the target gene in the native host or heterologous host, (b) alteration of gene sequence resulting in a change in the target protein sequence and (c) disruption or complete removal of the target gene or pathway.

In recombinant strains of *Corynebacterium glutamicum*, the increased number of *dapA* gene for dihydrodipicolinate synthase was introduced, and it enhanced the lysine titre when compared to the wild type (Eggeling et al. 1998). Site-directed mutagenesis and DNA shuffling have been

The chapter entails the application of microbial interventions in different industrial sectors. The role of GMMOs would also be discussed in light of their application for higher productivity or bioprocess development for green technologies.

S. Saxena, *Applied Microbiology*,
DOI 10.1007/978-81-322-2259-0_2,

effectively used in improving the enzymes from bacteria and fungi for application in the detergent industry. Feedback regulation has also been manipulated at the molecular level and has been effectively applied in *Corynebacterium glutamicum* for isoleucine production. Several examples exist wherein recombinant microorganisms play an important role in the industrial sector for the production of specific/desired products as well as for the development of green and energy-efficient processes.

2.2 Healthcare Industry and GMMOs

Human insulin was the first recombinant protein which was produced through a genetically modified *E. coli* containing human insulin genes. This work was done at Genentech Inc, USA, and the USFDA approved the clinical use of this product in 1982. Recombinant interferon gamma-1b (Actimmune) has also been developed by Genentech with Boehringer Ingelheim for the treatment of chronic granulomatous disease in 1990. IFN-γ 1b was produced by expressing the genes encoding the interferon in *E. coli*. Similarly the first recombinant vaccine was Engerix-B® produced by the gene encoding for hepatitis B surface antigen expressed in *Saccharomyces cerevisiae*, the common baker's yeast. Till date 151 protein-based recombinant pharmaceuticals have been licensed up til 2009, by the FDA and EMEA, of which 29.8 % are being produced in *E. coli*, 18.5 % in *Saccharomyces cerevisiae* and 11.2 % in hybridoma cells while 39 % in mammalian cell lines. Some recombinant products produced by *Saccharomyces cerevisiae* and *E. coli* have been listed in Table 2.1.

In animals bovine somatotropin hormone regulates animal growth as well as milk production. The bovine somatotropin (bST) gene has been expressed in *E. coli* and was approved by the USFDA in 1994 under the commercial name Posilac™. Studies have indicated that bST-supplemented cows produced 10–15 % more

Table 2.1 Therapeutic interventions developed using recombinant GMMOs

Recombinant product	Application	Expression system	Approving authority, year
Dukoral	Oral cholera vaccine	*E. coli*	USFDA/EMEA, 2004
Actimmune	Interferon gamma-1b	*E. coli*	USFDA, 1990
Viraferon	Interferon alfa-2b	*E. coli*	EMEA, 2000
IPLEX	Mecasermin rinfabate recombinant (insulinlike growth factor)	*E. coli*	USFDA, 2005
Kepivance	Palifermin (keratinocyte-like growth factor)	*E. coli*	EMEA, 2004; USFDA, 2005
Humulin	Human insulin	*E. coli*	USFDA, 1982
Preotact	Human parathyroid hormone	*E. coli*	EMEA, 2006
Nesiritide	Recombinant natriuretic peptide B	*E. coli*	USFDA, 2001
Engerix-B	Hepatitis B vaccine	*S. cerevisiae*	USFDA, 1998
Gardasil	Human papillomavirus vaccine [types 6,11,16,18]	*S. cerevisiae*	USFDA/EMEA, 2006
Regranex	Human platelet-derived growth factor	*S. cerevisiae*	USFDA, 1997; EMEA, 1999
Valtropin	Recombinant somatropin	*S. cerevisiae*	EMEA, 2006
Regranex	Human platelet-derived growth factor for wound healing	*S. cerevisiae*	USFDA 1997; EMEA, 1999
Iprivask	Recombinant-specific inhibitor of human thrombin isolated from *Hirudo medicinalis*	*S. cerevisiae*	EMEA, 1997
Fasturect	Recombinant rasburicase (urate oxidase)	*S. cerevisiae*	EMEA, 2001

milk. The enzyme phytase helps in the release of phosphate from phytate which is a primary storage form in plants. It has been observed that ruminants possess the capacity to obtain phosphorus, an essential element from their feed; however, nonruminants like pigs and chicken are unable to obtain phosphorus. The gene of this enzyme has been isolated from *Aspergillus niger* and constitutively expressed in the industrial strain of *Aspergillus niger* for commercial production of phytase for treatment of commercial feed.

Apart from recombinant biopharmaceuticals for human and veterinary purposes, the genetically modified microorganisms also find applications in the development of diagnostic kits. The antigenic coat protein of HIV has been cloned and expressed in *E. coli* for the large-scale production of protein for the development of ELISA-based diagnostic kit. Similarly tau proteins have been implicated in Alzheimer's disease. To diagnose Alzheimer's disease non-invasively, antigens against tau proteins can be tested in human cerebrospinal fluid. The antigens against tau proteins have been produced in large scale using recombinant *E. coli* expressing Alzheimer's antigen for the development of enzyme-linked immunosorbent assay, INNOTEST h TAU.

2.3 GMMOs in Agriculture

Genetically modified organisms find limited applications in agriculture. *Bacillus thuringiensis* var. *kurstaki* is a soilborne bacterium which is exploited as a biopesticide since it produces unique crystalline proteins (δ-endotoxins or *cry* proteins) which possess larvicidal activities against different insect species. These toxins broadly are non-toxigenic to mammals, birds and fishes and thus can be effectively used as bioinsecticides.

Bt toxins have been introduced successfully into several plant-associated bacteria like *Pseudomonas* sp. and *Azospirillum* sp. for delivering Bt toxins inside the plants without altering their genomes. One important gram-positive bacterium is *Clavibacter xyli* ssp. *cyanodontis* which resides in the xylem of *Cynodon dactylon* (Bermuda grass). This bacterium also colonises *Zea mays* (corn) when introduced artificially, and thus, recombinant *Clavibacter xyli* expressing Bt toxins have been used as a bio-inoculant to provide corn resistance against insect attacks. It has experimentally shown moderate control of the European corn borer. Nitrogen assimilation is important for plant metabolism, and it has been reported that genetically modified rhizobacteria *Sinorhizobium meliloti* expressing the *nifA* gene of *Klebsiella pneumonia* exhibited better root nodulation in alfalfa (*Medicago sativa*) plants. Further it was also found that the recombinant *S. meliloti* significantly increased the plant biomass as compared to the wild type. Thus, possibilities exist wherein genetically modified microorganisms living endophytically or colonising in the rhizosphere could be suitably exploited for improving plant growth and yield without genetically modifying them.

2.4 Role of GMMOs in Chemical Industry

With the increasing societal concern about the environment, climate change and limited natural resources, there has recently been considerable effort exerted to produce chemicals and materials from renewable biomass. The common method of production of these polymers is directly through fermentation or polymerisation of the monomers produced by fermentation. Advances in metabolic engineering and molecular biology have helped in the systematic development of superior strains and processes for the production of polymers and monomers.

Polymers are macromolecules which are composed of a series of low molecular weight monomers. Microorganisms naturally produce polyhydroxyalkanoates (PHAs) in the form of granules that the organisms use as an energy storage material. They are genuine polyester molecules with properties similar to petroleum-derived polymers. Further they are biodegradable; the enzyme which depolymerises them is widely present in fungi and bacteria. *E. coli* has been transformed by transferring the PHA genes which the microorganism is lacking for the commercial production of PHA since *E. coli* exhibits a robust growth and its metabolism is well characterised and apart from it lacks PHA depolymerase which is responsible for its degradation.

Similarly *E. coli* knockout mutant strain has also been developed for the commercial production of lactic acid exhibiting a production of 138 g/L of lactic acid with a yield of 0.99 g lactic acid/g glucose and an overall productivity of 3.54 g/L/h. Adipic acid is the building block of nylon-4, 6 and nylon-6, 6. Using a recombinant *E. coli* strain, cis, cis-muconic acid has been produced at a concentration of 36.8 g/L which is subsequently hydrogenated to adipic acid with 0.97 g/g conversion yield at room temperature. Similarly glucaric acid production has been carried out through *E. coli* by metabolically constructing a direct synthesis pathway by introducing myo-inositol-1-phosphate synthase from *Saccharomyces cerevisiae*, myo-inositol oxygenase from mouse and urinate dehydrogenase from *Pseudomonas syringae*.

Antibiotic production is significantly enhanced using GMMO's. Cephamycin path genes from *Streptomyces cattleya* was cloned in *Streptomyces lactamgems* led to 2.3 fold increase in Cephamycin C production. The precursors of semi-synthetic cephalosporins are 7-aminocephalosporanic acid (7-ACA) or 7-amino-deacetoxycepalosporanic acid (7-ADCA). *P. chrysogenum* was transformed with *Streptomyces lipmanii* cefD and *S. clavuligerus* cefE genes which allowed the production of intermediate deacetoxycephalosporin C (DAOC) at titers of 2.5 g/L, along with penicillin V. A recombinant *E. coli* was constructed containing a D-amino acid oxidase gene from *Trigonopsis variabilis* and the glutaryl-7-aminocephalosporanic acid acylase gene from *Pseudomonas* species which was capable of converting cephalosporin C to 7-ACA directly.

For the production of alcohol, a recombinant strain of E. coli was developed which expressed alcohol dehydrogenase II and pyruvate decarboxylase genes from *Zymomonas mobilis*. 43 g/L of ethanol production was achieved by this recombinant strain. *Lactobacillus plantarum* has been genetically engineered overexpressing the two sorbitol-6-phosphate dehydrogenase genes (srlD1 and srlD2) for high sorbitol production. It is used in food and pharmaceutical industries, and its annual requirement is approximately 500,000 tonnes. Thus, recombinant microorganisms play a critical role in commercial production of primary and secondary metabolites apart from the development of novel ecofriendly processes.

2.5 GMMOs in Textile Industry

The exploitation of microbial enzymes in the textile industry began in early 1900. To commercially use them in the industry on a large scale, these enzymes have to be produced at high levels. Heterologous hosts have been used for the production of enzymes to overcome the inherent limitations of the naturally producing microorganism. *Bacillus stearothermophilus* produces a heat-stable and broad-pH active α-amylase. The gene of this enzyme was cloned and expressed in *Bacillus licheniformis* for commercial production of the enzyme. A variety of recombinant enzymes such as cellulase, pectinases, proteases, etc., have been used in the textile industry which have been discussed in Chap. 9 of this book. Recombinant enzymes produced by GMMOs not only find applications in the textile industry but also other industrial sectors such as paper, environment, food, feed, biosensors, pharmaceuticals and fine chemicals.

2.6 Environmental Applications of GMMOs

Microbial bioremediation essentially refers to the use of microorganisms to detoxify the polluted environment which is due to heavy metals such as mercury and lead and organic compounds such as petroleum hydrocarbons, radionuclides such as uranium and plutonium and other compounds such as explosives, pesticides and plastics. There are two basic strategies of microbial bioremediation, bioaugmentation and biostimulation for the remediation of polluted sites. The pioneering work in this field was carried out by Gunsalus and A. M Chakraborty by developing a recombinant *Pseudomonas*, *Pseudomonas fluorescens* HK44, for degradation of camphor, octane, salicylate and naphthalene. It is the first life form to be the subject of an intellectual property case. As soon as the prospect of releasing GMMOs for bioremediation became a reality, much of the research effort in the field was aimed at biosafety and risk assessment. There are many issues surrounding the use of GMMOs for use in bioremediation such as (1) their effectiveness as compared to their counterparts in the nature, (2) their influence on indigenous microbial community, (3) their fitness in nature and (4) their containment. There are only a few cases where the use of GMMOs turned out to be much better in performance than its non-manipulated counterpart. Commercial bioremediation largely relies on the naturally occurring microbes identified from the contaminated sites.

2.7 Food Industry and the Role of GMMOs

The enzymes produced by genetically modified microorganisms have been effectively used in the food industry for over a decade. The gene encoding for calf stomach chymosin was cloned and expressed in an industrial strain of *Kluyveromyces lactis*, a yeast that had been used for many years in the safe production of food ingredients. Chymosin produced through the recombinant yeast strains possesses the same chemical and biological properties as that of calf rennet. The preparation has been registered under the brand name Maxiren® and has been commercially available since 1988. Currently there are approximately 30 different enzymes, many of them in food use, that are produced by GMMOs. The recombinant brewer's strain having an amylase gene from *Saccharomyces diastaticus* together with a gene for copper resistance was approved in 1993; however, it could not go commercial because of the unwillingness of the industries to face a negative consumer reaction. Similarly recombinant strains of *Saccharomyces* for wine production have been developed but have not become a commercial reality till date.

Despite much research in the genetic and molecular studies on probiotic bacteria, *Lactobacillus* species, and *Saccharomyces* species, and modifications carried out, the applications of genetically modified microorganisms have not been realised in food processes.

2.8 GMMOs for Bioethanol Production

The major technical roadblock for the development of bioethanol industry is the lack of suitable microorganisms which can convert the biomass into fuel ethanol. However, in the last two decades, several microorganisms have been genetically engineered exploiting them as bugs which enhanced capacity to transform the lignocellulose biomass. Lignocellulose biomass basically comprises complex carbohydrates that cannot be fermented by brewer's yeast *Saccharomyces cerevisiae*. The success has been met with gram-negative microorganisms like *Escherichia coli*, *Zymomonas mobilis* and *Klebsiella oxytoca* which have been genetically engineered to metabolise a wide spectrum of sugars and produce ethanol. *Zymomonas mobilis* is homoethanol fermentative because it converts pyruvate into ethanol utilising pyruvate decarboxylase (PDC), which only consumes one NADH, H+ for each ethanol produced. The recombinant *E. coli* expressing the pdc and adhII genes were co-expressed under the control of native lac promoter, and the resulting construct was named as PET (production of ethanol) operon. This recombinant *E. coli* strain became homoethanol fermentative thereby producing ethanol exclusively. *Zymomonas mobilis* has high ethanol yield and productivity since it metabolises glucose anaerobically using the Entner–Doudoroff (ED) pathway as compared to the EMP pathway or glycolysis. However, the limitation of *Zymomonas mobilis* is that it ferments only glucose, sucrose and fructose. The first recombinant strain was engineered to ferment xylose

The transformed strain CP4 (pZB5) grew on xylose, and the ethanol yield was 86 %. The recombinant microorganisms thus have played a significant role in the development of the bioethanol industry.

2.9 Summary

Thus, one can conclude that genetically modified microorganisms have a tremendous impact in the human healthcare sector and chemical industry as well as for bioethanol production. Advances in functional genomics and bioinformatics tools, combined with existing recombinant DNA technologies, will help us better understand the physiology and metabolic potential of the organisms we study and, in turn, will lead to the development of GMMs best suited to our needs.

Selected Reading

Amarger N (2002) Genetically modified bacteria in agriculture. Biochimie 84:1061–1072

Demain AL (2000) Microbial biotechnology. TIBTECH 18:26–31

Demain AL, Vaishnav P (2009) Production of recombinant proteins by microbes and higher organisms. Biotechnol Adv 27:297–306

Eggeling L, Oberle S, Sahm H (1998) Improved L-lysine yield with *Corynebacterium glutamicum*: use of dapA resulting in increased flux combined with growth limitation. Appl Microbiol Biotechnol 49:24–30

Garvilescu M, Chisti Y (2005) Biotechnology—a sustainable alternative for chemical industry. Biotechnol Adv 23:471–499

Lee JW, Kim HU, Choi S, Yi J, Lee SY (2011) Microbial production of building block chemicals and polymers. Curr Opin Biotechnol 22:758–767

Wandrey C, Liese A, Kihumbu D (2000) Industrial biocatalysis: past, present, and future. Org Process Res Dev 4:286–290

3 Fermentation Technology

3.1 Introduction

The knowledge of fermentation is ancient perhaps prehistorical. Sumerians and Egyptians possessed the knowledge of technique by which they could convert the starchy grains into alcohol. In due course of time, fermentation processes have been developed to manufacture a vast range of materials from chemically simple feedstocks such as ethanol to high complex protein structures. Fermentation process comprises of a variety of chemical reactions such as oxidations, reductions, polymerisation and hydrolysis as well as biosynthesis and formation of cells. The term fermentation has been used variably by different groups of individuals.

BOX 3.1 Microbial Fermentation: Multiple Denotations

- *Any process involving the mass culture of microorganisms, aerobically or anaerobically*
- *Any biological process that occurs in the absence of oxygen*
- *Spoilage of food*
- *The production of alcoholic beverages*

As the spectrum of products produced through fermentation enhanced, viz., from ethanol to antibiotics to recombinant protein, it brought in technological changes in the fermentation industry which pertained to novel methods of bioprocess operations. The most common strategy of fermentation was a batch process, while the fed-batch process was commonly used for the production of antibiotics and baker's yeast. Till today, a majority of industrial fermentation operations are carried as a batch or fed-batch process. Commercially, the fermentations are classified as submerged (SmF) and Solid Substrate Fermentation (SSF). Solid Substrate Fermentation (SSF) is generally carried out on a solid substrate. Submerged fermentations employ substrate in a dissolved state or solid substrate suspended in large amount of water. These could be further classified as aerobic and anaerobic. Penicillin production by *Penicillium chrysogenum* is carried out by submerged aerobic fermentation.

The fermentation process is influenced by a variety of factors inclusive of temperature, pH, medium composition, dissolved oxygen, precursors and mode of operation such as batch, fed-batch or continuous process. Microbial fermentations are generally oriented to produce (1) microbial cells or biomass, (2) microbial metabolites, (3) microbial enzymes and (4) recombinant proteins and carry out (5) biotransformations, i.e. modifications of compounds.

3.2 Batch Fermentation

In batch fermentation process, a batch of culture medium is inoculated aseptically with the desired microorganism which is generally referred as the

S. Saxena, *Applied Microbiology*,
DOI 10.1007/978-81-322-2259-0_3, © Springer India 2015

seed or starter culture under optimal fermentation conditions which are maintained for a defined duration for the maximum product formation in a closed vessel referred to as fermenter/bioreactor. The duration for maximum product formation in the fermenter is referred to as the 'batch time' or 'fermentation time' which extends from a few hours to 6 days. In some cases, more specifically traditional food fermentations, the fermentation time can last for a month or so.

Generally, the fermentation process begins with pure starter culture which is grown in Petri dishes or in liquid medium in shake flask. The inoculum is built through successive stages to 5–10 % of the total working volume of the production fermenter. This helps in the reduction of the batch time (Fig. 3.1). After the fermentation time/ batch time is over, the bioreactor or fermenter is harvested, cleaned and prepared for the next batch. The harvested contents are then subjected to different processes such as biomass separation, broth concentration and subsequently solvent extraction and purification to obtain the final product.

After completion of the fermentation time, the spent medium is removed and subjected to different processes for product recovery. Meanwhile, the production fermenter is cleaned, sterilised and prepared for the next batch of fermentation. Hence, during batch fermentation, there is a downtime between two batches.

All the processes beginning from activation to fermentation that start in the production fermenter are referred to as upstream processes while those which are employed after the fermentation is over for product recovery are referred to as downstream processes (Fig. 3.2).

The typical phases of microbial growth observed in a batch fermentation process are (1) lag phase, (2) log or exponential phase, (3) stationary phase or idiophase and (4) death phase (Fig. 3.3). Thus, during optimisation studies, the fermentation time is adjusted based on the association of the product with either the growth phase or the stationary phase. The lag phase is the first major phase in a batch fermentation process which denotes the time taken by the cells to adapt to the new environment. This may be absent in some fermentative processes. The second major phase is referred to as the log phase or exponential phase. This phase is marked by exponential increase in the number of cells present. Bu'Lock et al. (1965) have referred this phase as 'tropophase'. In this phase, the sole products of metabolism are essential to growth, such as amino acids, proteins, nucleotides, nucleic acids, lipids, carbohydrates, etc., or produce ethanol, acetone and butanol which are by-products of energy-yielding metabolism. They are also referred to as primary metabolites being characteristically associated with the growth phase. Mathematically, microbial growth in exponential

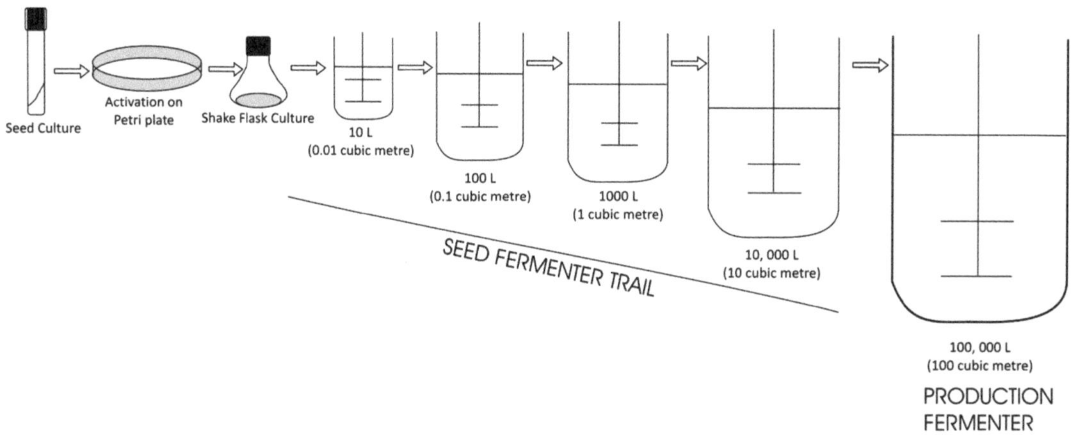

Fig. 3.1 The process of inoculum development for submerged batch fermentation processes

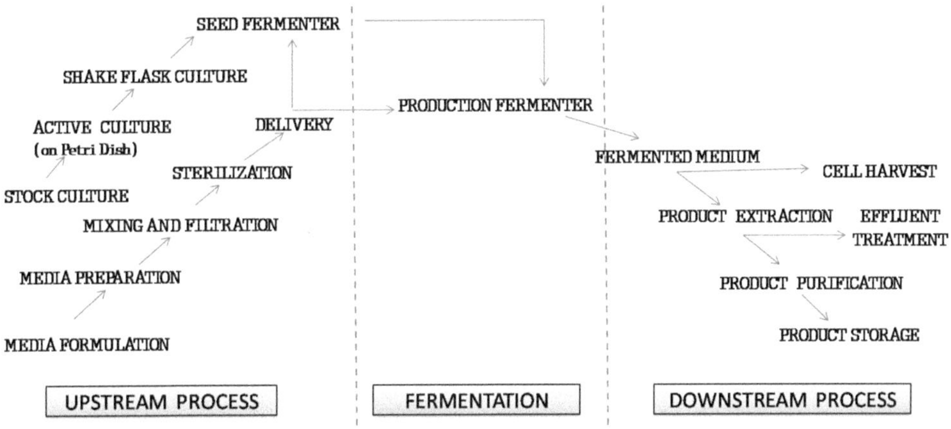

Fig. 3.2 Different phases of fermentative production process

Fig. 3.3 A typical growth curve of a microorganism in a batch process

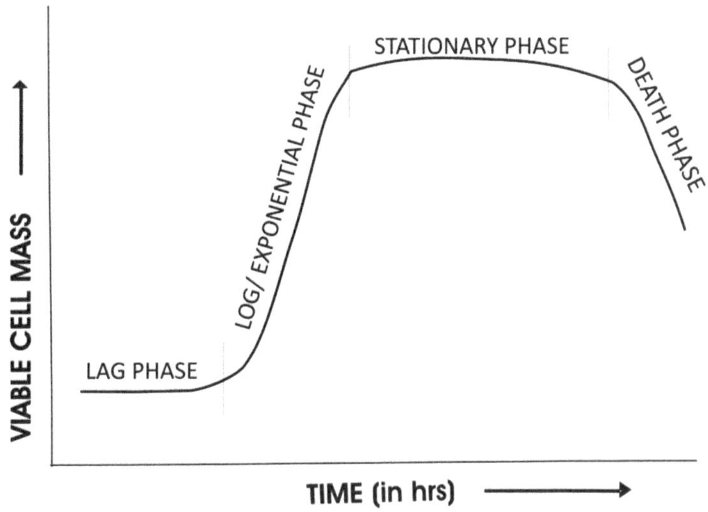

phase can be defined in terms like *mean generation time* and *mean growth rate*. Mean generation time is the time required for microbial population to double and is also referred to as doubling time. Mean growth rate is defined as the number of generations per unit time often expressed as generations per hour. The rate at which the microbial cells increase exponentially is referred to as specific growth rate and is given by the equation

$$\frac{dX}{dt} = (\mu - k_d)$$

where X is the cell concentration, μ is the specific growth rate and k_d is the cell death rate. However, during exponential growth, there is negligible death and hence is omitted. The specific growth is often limited by substrate S until it attains a non-limiting level where μ attains its maximum value, i.e. μ_{max}. The dependence of the growth rate on substrate concentration typically follows the Monod growth kinetics given by the following equation:

$$\mu = \mu_{max}\frac{S}{k_s + S},$$ where k_s is the saturation constant.

Numerically, k_s is the concentration of a growth-limiting substrate when the specific growth rate is half of the maximal value. Stationary phase is the third major phase of microbial growth in a batch fermentation, where there is an equilibrium achieved between the number of cells dividing and dying. This phase generally results in depletion of one or more essential growth nutrients as well as association of toxic growth together with by-products. Stationary phase is also known as "idiophase" wherein the products synthesised do not have any role in cell metabolism and hence are referred to as secondary metabolites. The last phase of batch fermentation process is the death phase which is also referred to as the decline phase. The phase is characterised by higher rate of cell death than cell division. It is also represented by first-order kinetics similar to the exponential phase. The equation for death kinetics is given under.

$$\frac{\mathrm{d}X}{\mathrm{d}t} = (-k_{\mathrm{d}})X$$

3.3 Continuous Fermentation

In continuous fermentation process, a part of the medium is replaced by fresh nutrient medium at more or less regular intervals when the fermentation process is ongoing, i.e. in exponential growth phase. Thus, steady state is achieved, and the fermentation continues non-stop. Steady-state formation of new biomass in the vessel is equivalent to the loss of cells from the vessel. In steady-state condition, the specific growth rate is a function of dilution rate which is a controllable function. Thus, the primary objective of continuous culture is to control cell growth at an optimum level of productivity. This can be achieved by two systems, chemostat and turbidostat. The chemostat mode of operation of continuous culture involves the maintenance of steady cell growth by a constant inflow of fresh medium consisting of nutrients (phosphorus, nitrogen, glucose) at a concentration that is growth limiting. Increase or decrease in the concentration of the growth-limiting factor is correspondingly expressed by increase or decrease in the growth rate of cells.

Turbidostat is a continuous culturing method developed by Bryson and Szybalski (1952) in which microorganisms can be grown at $\mu_{\max}$ without the risk of small fluctuations in culture volume causing wash out. If the turbidity tends to increase, the feed rate is increased to dilute the turbidity back to its set point. When the turbidity tends to fall, the feed rate is lowered so that growth can restore the turbidity to its set point.

3.4 Fed-Batch Fermentation

Fed-batch fermentation is an intermediate between batch and continuous fermentation processes in which a sterile culture medium is either periodically or continuously added to the inoculated fermentation batch. A very prominent example of fed-batch fermentation process is penicillin fermentation. The production of penicillin is a two-stage fermentation: an initial growth phase which is followed by the production phase (idiophase). During the production phase, the biomass should be maintained at a relatively low growth rate by feeding glucose at a low dilution rate along with the precursor phenylacetic acid which also is toxic to *Penicillium chrysogenum* above threshold concentration. There are two basic approaches in fed-batch fermentation – fixed-volume fed-batch and variable-volume fed-batch fermentation.

3.4.1 Fixed-Volume Fed-Batch

In fixed-volume fed-batch process, the limiting substrate is fed without diluting the culture. In this method, a very concentrated feed solution or a powdered form of nutrient is supplied to the culture to minimise the volume increase. An extended version of this method is the cyclic fed-batch culture for fixed-volume systems which involves a periodic withdrawal of a defined portion of culture and uses the residual culture as a starting point for further fed-batch process with sterile water or medium containing the feed substrate. As a result of dilution, the biomass concentration is decreased and results in an increase in specific growth rate. As the feeding continues, there is a gradual decrease in the growth rate with increase in biomass and it reaches the maximum sustainable concentration in the vessel once again.

3.4.2 Variable-Volume Fed-Batch

A variable-volume fed-batch culture process essentially involves volume changes with the fermentation time due to substrate feed. Hence, the volume changes are dependent on the requirements, limitations and objectives of the operator. As the name implies, a variable-volume fed-batch is one in which the volume changes with the fermentation time due to the substrate feed. The way this volume changes is dependent on the requirements, limitations and objectives of the operator.

The feed can be provided according to one of the following options:

(i) The medium used in the batch mode is added.
(ii) The solution is of the same concentration of a limiting substrate as that used in the initial medium.
(iii) A very concentrated solution of the limiting substrate is added at a rate less than (i) and (ii).

This could be further classified into repeated fed-batch (RFB) and single fed-batch (SFB). In RFB, the fermentation has reached to a stage after which it was not further effective and therefore a quantity of culture was removed and refilled by fresh medium. This decrease in volume increases the specific growth rate followed by a gradual decrease as the quasi-steady state is established. In SFB, the supplementary growth medium is added during the fermentation, but no culture is removed until the end of the batch. The disadvantage of this process is that much of the fermenter volume remains unutilised until the end of the batch and the batch time is limited by the fermenter volume. The fed-batch mode ensures the presence of antibiotic throughout the course of the fermentation with the intention of keeping the antibiotic-marked plasmid while working with recombinant strains.

3.5 Components in a Typical Bioreactor

A bioreactor or a fermenter is a vessel designed to cultivate and grow microorganisms in large quantity under defined conditions to form the by-product (Fig. 3.4). The laboratory fermenter has a capacity from 2 to 100 L, but in commercial/industrial settings, the operation is large scale between 100 and 250 m^3.

The body of a bioreactor is made up of glass or steel depending upon its volume. The small-scale fermenters are generally made up of glass or steel. For pilot and large-scale process, stainless steel (>4 % chromium) and mild steel (coated with glass or epoxy material) are generally used for making the bioreactor vessel. Inclusion of nickel improves engineering, while presence of molybdenum enhances resistance to halogen salts, brine and seawater. Side plates have lower thickness than top and bottom plates and are hemispherical to withstand pressures. The top plate and vessel are sealed to maintain airtight aseptic containment. Three types of surfaces are sealed, viz., glass–glass, glass–metal and metal–metal. They basically are gasket, lipseal and 'O' ring. The seals should be changed after a defined time. Metal strips are attached radially from the walls at every one tenth of vessel diameter to prevent vortex formation. These are known as baffles. This movement minimises microbial growth on baffles and walls of the fermenter. If needed, cooling coils may be attached to baffles. Aeration provides sufficient oxygen to the organism in the fermenter. Sparger is a device to introduce air into a fermenter. Generally, fine bubble aerators are used as they facilitate the oxygen transfer to a greater extent. Air supply to a sparger is supplied through an air filter. There are three types of sparger, viz., porous, orifice and nozzle. Microbial activity as well as mechanical agitation generates heat during the fermentation process; however, this should not have drastic effects on the fermentation efficacy, and hence there is a need to maintain the heat being generated. Removal of excess heat can be achieved through the internal coils present in the fermenter by using cooling jackets.

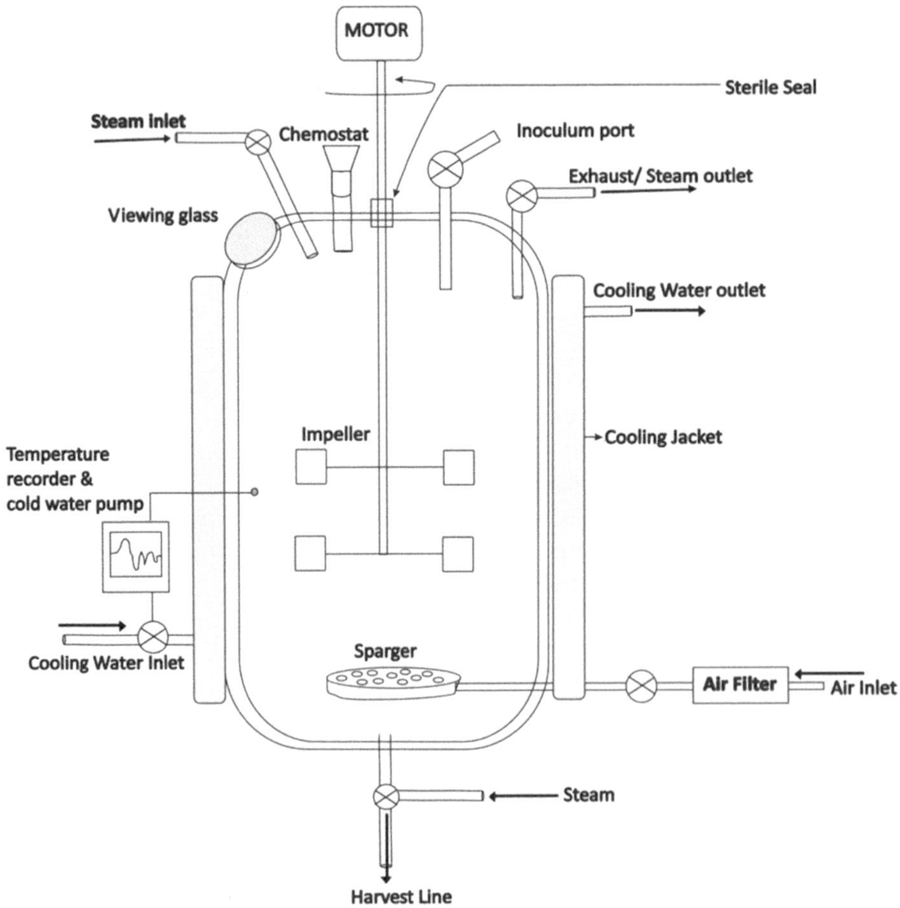

Fig. 3.4 Components of a typical bioreactor

Agitation is required to ensure the uniform suspension of microbial cells is achieved in a homogeneous nutrient medium for which we use an impeller. Thus, the impeller achieves a number of mixing objectives like suspension of solid particles, bulk fluid, gas-phase mixing, etc. The impellers basically are of two types – disc turbine and variable pitch open turbine. Silicone tubes are generally used as feed ports and to connect the nutrient reservoir with the fermenter. They are sterilised in situ using steam after the connection has been made prior to feeding the fermenter.

Foam formation during the fermentation process needs to be controlled or else it will lead to wetting of filters thereby causing contamination. Hence, a foam-sensing and control unit is preferably installed. The unit consists of a probe which is inserted through the top plate of the fermenter and set at a defined level above the broth surface. When the foam rises and touches the probe tip, a current is passed through the circuit of the probe. The current actuates the pump and antifoam is released within seconds. Different types of valves are used for the control of flow of gases and liquids in and out of the bioreactor. Globe valves are used for opening and closing the steam or waterline; they do not regulate the flow. Ball valves are compatible for aseptic operations as they can handle mycelial broths as well as can be operated at high temperature. Diaphragm valves are used for flow regulation and for stem services within pressure limits. Safety valves are installed in every air or steam vessel and pipe layout which is subjected to work under pressure. These valves ensure that the pressure never exceeds the safe upper limit of the specified value.

3.6 Types of Submerged Bioreactors

3.6.1 Stirred Tank Fermenter (STF)

Microbial submerged fermentation received prominence during the antibiotic era which began in the early 1940s marked by the discovery of penicillin. Thus, stirred tank fermenter (STF) became a choice for the production of antibiotics. The height-to-diameter ratio (aspect ratio) of this fermenter is 3–4. It has a central shaft which supports three to four impellers (Fig. 3.5a). The typical decision variables are type, size, location and the number of impellers; sparger size and location. The vessel consists of four baffles spaced vertically. Typically, the baffle width is 8–10 % of the vessel diameter. Impellers and baffles determine the hydrodynamic pattern in the reactor, which in turn influence mixing times, mass and heat transfer coefficients, shear rates, etc. The fermenter performs multiple functions like homogenisation, suspension of solids, dispersion of gas–liquid mixtures, aeration of liquid and heat exchange. Conventional fermentation is carried out in a batch mode.

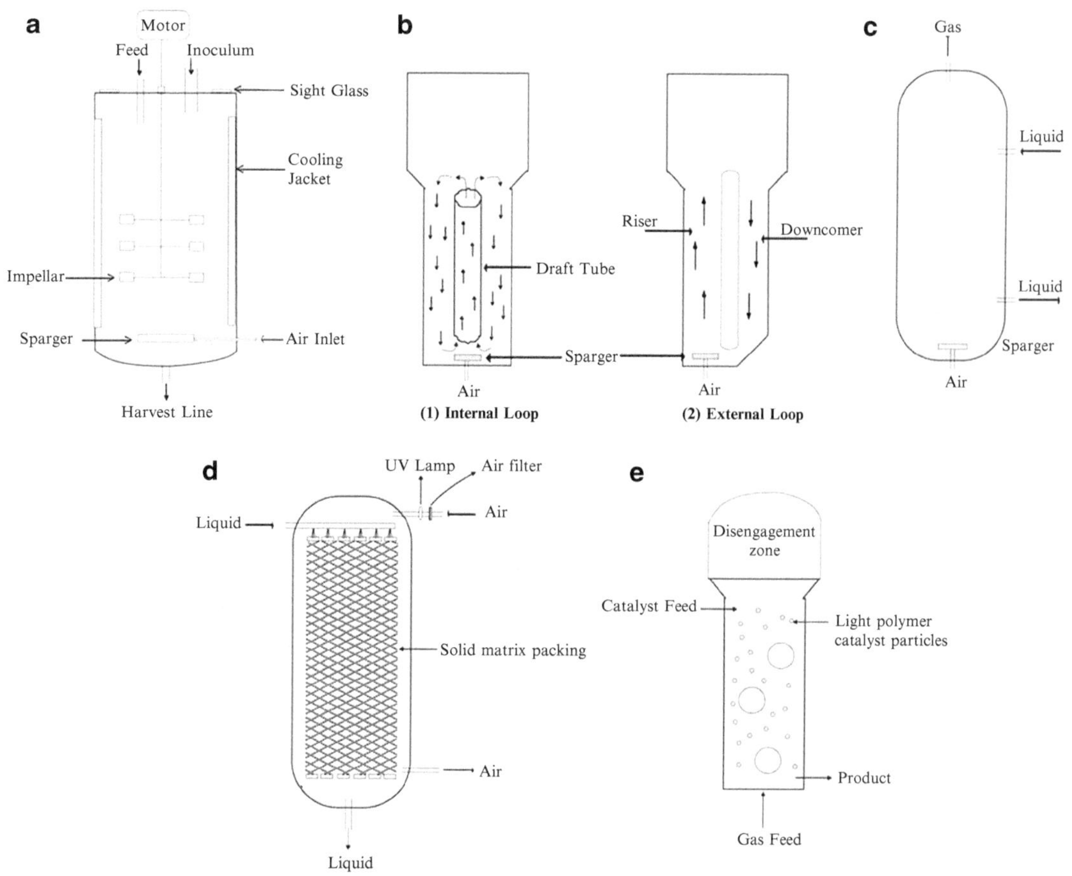

Fig. 3.5 Different types of fermenters employed in submerged fermentation processes: (**a**) stirred tank fermenter, (**b**) airlift fermenter, (**c**) bubble column fermenter, (**d**) trickle bed fermenter, (**e**) fluidised bed fermenter

3.6.2 Airlift Fermenter (ALF)

Airlift fermenter (ALF) is a pneumatic fermenter which does not have any mechanically controlled stirrer for homogeneous mixing of the fermentation medium (Fig. 3.5b). There are two designs of ALF, viz., internal loop design and external loop design. The internal loop-designed ALF has a draft tube which is provided in the central section of the reactor. Air is forced through a porous plate at the bottom of the draught tube. Air bubbles flow up in the central draught tube, some coalesce and exit at the top of the column while other bubbles follow degassed liquid and circulate down from the area outside the draught tube. In the external loop design, the riser and the downcomer are separate tubes joined at the top and at the bottom. Airlift fermenters are preferably easier to operate and adaptable to cultures which are shear sensitive compared to mechanically agitated stirred tank fermenters.

3.6.3 Bubble Column Fermenter (BCF)

It is a cylindrical vessel with an aspect ratio of 4–6. The fermenter is sparged from the bottom into a liquid phase or liquid–solid suspension. Industrial bubble columns usually operate with an aspect ratio of at least 5 (Fig. 3.5c). These fermenters have two types of mode, viz., semi-batch mode and continuous mode. In continuous operation, the gas and the suspension flow move concurrently upward into the column, and the suspension that leaves the column is recycled to the feed tank. Bubble columns exhibit homogeneous (bubbly flow) regime, the churn turbulent (heterogeneous regime) and slug flow regime. Overall, bubble column fermenters have excellent heat and mass transfer characteristics, meaning high heat and mass transfer coefficients, little maintenance and low operating costs. The bubble column fermenters/rectors are of three types, viz., trickle bed reactors (with fixed or packed beds), fluidised bed reactors and bubble column reactors.

3.6.3.1 Fluidised Bed Fermenter (FBF)

Fluidised bed fermenters (FBFs) have received considerable attention in recent years primarily in the area of biocatalysis and biotransformations which have immense demand in the chemical industry (Fig. 3.5e). Majority of the FBFs developed for biological systems involving cells as biocatalysts are three-phase systems (solid, liquid and gas). FBFs are essentially immobilised cell reactors. Basically, the particles used in FBBs can be of three different types: (1) inert core on which the biomass is created by cell attachment, (2) porous particles in which the biocatalyst is entrapped and (3) cell aggregates/flocs (self-immobilisation). Usually, fluidisation is obtained either by external liquid recirculation or by gas fed to the reactor.

3.6.3.2 Trickle Bed Fermenter (TBF)

These fermenters are cylindrical vessels with solid supports like wood chips, rocks and plastic structure. These support materials offer spaces for flow of liquid, gas and microorganisms. The nutrient broth is sprayed on the top of the support material and it slowly trickles down the solid support (Fig. 3.5d). The air flows in countercurrent to liquid flow till the bed. These reactors are suitable for liquids with low viscosity and a few suspended solids.

3.7 Solid Substrate Fermentation

Solid Substrate Fermentation (SSF) is generally carried out on a solid support with little or no free water under non-septic conditions and in natural state. This process requires less energy as compared to submerged fermentation. The water is generally present as thin film between the solid substrate particles. The water film is discontinuous and is filled in intervening gas phase. The SSF processes use filamentous fungi predominantly followed by yeasts and bacteria under aerobic conditions. At times, the organisms are used as consortium of pure cultures. The solid substrates commonly used in solid substrate fermentation are the by-products of food processing, forestry and agriculture. At times, artificial and inert supports are also used with nutrient solution absorbed within the matrix.

Solid substrate fermentation is generally preferred in processes where the end product is preferred in the solid form (e.g. fermented foods), or when the product yield is much higher in SSF when compared to SmF. At times, SSF is adopted as a strategy due to socio-economic reasons, and the fermentation process is carried out by relatively unskilled workers as some processes are resistant enough to be relatively taken up by contaminants.

Basic steps of solid substrate fermentation are generally similar to SmF and include:

(i) Inoculum preparation
(ii) Preparation of substrate
(iii) Bioreactor preparation
(iv) Inoculation and loading
(v) Bioreactor operation
(vi) Unloading
(vii) Downstream processing
(viii) Disposal of waste

During the process development, adequate attention has to be given to all the above steps; however, some specific issues need to be addressed during different steps.

Preparation of Substrate: The substrate need to be brought to appropriate particle size for efficient fermentation for which it need to be cut, milled, cracked or granulated. There could be a requirement of preprocessing by addition of minimal water and nutritional supplements or cooking or pretreatment with enzymes to enhance the bioavailability of the nutrients. Further, the substrate may be sterilised or pasteurised outside the bioreactor, or alternatively, this step may be carried out inside the bioreactor.

Inoculum Preparation: SSF processes which involve the use of filamentous fungi use spore-based inocula. Thus, sufficient concentration of spore with high viability is desired for use as inocula. Mycelium inocula can be prepared by submerged fermentation which could be dried on solid substrate and subsequently ground into a fine powder.

Inoculation and Loading: The inoculation step either prior to loading or after loading. In case the inocula can be mixed in the bioreactor, then the best method would be spraying it in the form of a mist on the substrate bed. If the substrate is sterilised or pasteurised and inoculated outside, then loading step is carried out quite carefully in order to prevent or at least minimise the entry of contaminants. At large scale, loading will need to be mechanically assisted.

Bioreactor Preparation: It essentially involves cleaning and sterilisation before the addition of the substrate. At times, sterilisation in place is adopted for the substrate as well as the bioreactor.

Bioreactor Operation: The operation of the bioreactor will depend upon the specific bioreactor design However, the general operating variables which need to be manipulated would be flow rate and temperature of the inlet air, the bed mixing speed and the cooling water temperature, in order to control key fermentation parameters, such as bed temperature and water activity, at the optimum values for growth and product formation.

Unloading: After fermentation, the steps of leaching/drying are undertaken in the bioreactor while the product recovery steps are carried outside the bioreactor. Hence, solids have to be unloaded from the reactor for downstream processing. In large-scale operations, unloading is done mechanically.

Downstream Processing: The specific products from the solids are recovered and then extracted for product recovery. Thus, extraction from the solids represents a significant step in the SSF downstream process which is not similar to SmF. However, after extraction, the general principles of downstream processing are similar for both SSF and SmF.

3.8 Role of Bioreactor in Solid Substrate Fermentation

Bioreactor design is an important facet of solid substrate fermentation since it is the site of bioconversion and product formation. It has two major functions in SSF:

(i) It primarily holds the substrate bed as well as serves as a barrier for release of the inoculum into the surroundings and simultaneously prevents it from getting contaminated by microorganisms thriving in the surroundings.

(ii) It controls the key environmental conditions such as bed temperature and water activity at values which are optimal for growth and product formation by the microorganism.

3.9 Types of Solid Substrate Bioreactors

Many different bioreactors have been used in SSF processes and have been given different names by different authors. However, based on similarities in design and operation, SSF bioreactors can be divided into different groups on the basis of how the solid substrate is mixed and aerated (Fig. 3.6):

Tray Bioreactors: These bioreactors constitute the first group. In these bioreactors, the bed is static or mixed very infrequently, i.e. once or twice per day. The air is circulated around the bed but not blown forcefully.

Packed Bed Bioreactors: In these bioreactors, also the bed is static and mixed very

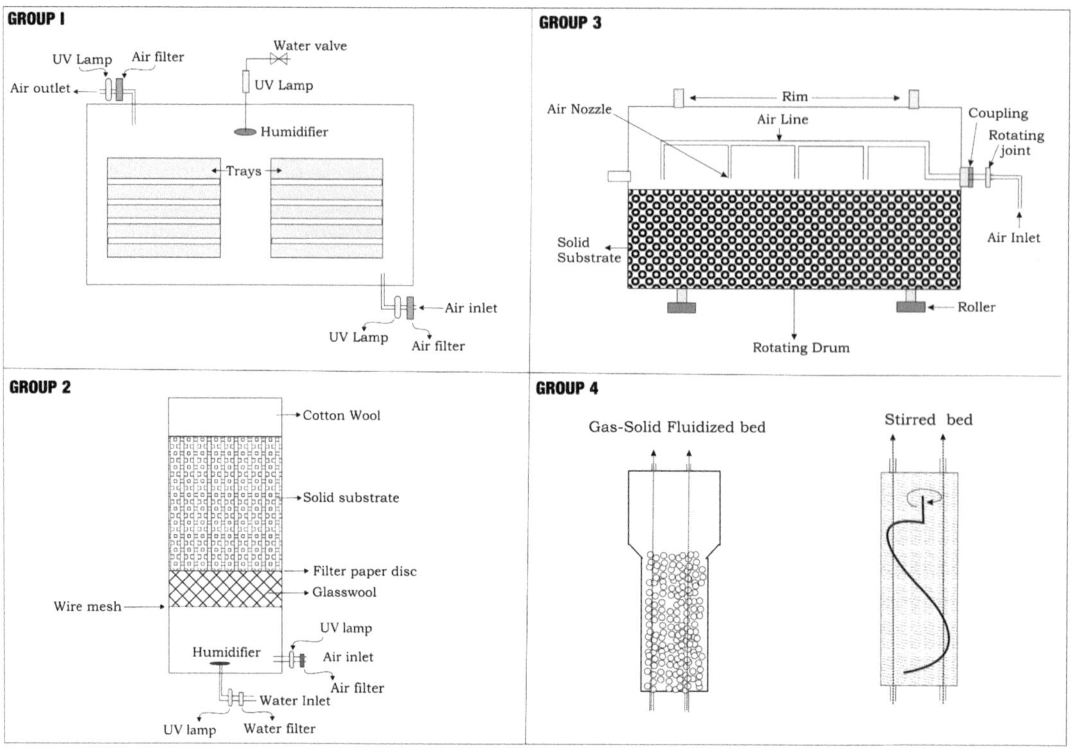

Fig. 3.6 Different groups of solid substrate bioreactors: Group 1, tray fermenter; Group 2, packed bed reactor; Group 3, stirred drum bioreactors; Group 4, gas–solid and stirred bed reactors

infrequently, i.e. only once per day; however, the air is blown forcefully through the bed. These constitute the group two bioreactors.

Group III Bioreactors: In these bioreactors, the bed is continuously mixed or intermittently mixed with a frequency from minutes to hours and air is circulated around the bed. The intermittent mixing of the substrate is carried out by bioreactors in two modes, viz., stirred drum bioreactors and rotating drum bioreactors.

Group IV Bioreactors: The characteristic feature of these bioreactors is that the bed is agitated and air is blown forcefully through the bed. It operates in two modes, those are mixed continuously and others mixed intermittently with intervals from minutes to hours between the mixing events. The bioreactors which fulfil this criterion are gas–solid fluidised beds, rocking drum and stirred-aerated bioreactors.

Thus, the design of the bioreactor plays a significant role in solid substrate fermentation since the selection of the reactor is based on the morphology and growth characteristics of the microorganism, aeration of the substrate and the type of the product being produced.

Both solid and submerged fermentations have their respective advantages and disadvantages and are therefore used in the industry as per the quality and quantity requirements of the end product (Table 3.1). For example, the production of citric acid is carried out by both SmF and SSF. However, SSF is generally being used by manufacturers who are unable to invest in higher infrastructural setup and operational cost.

Some traditional applications of SSF are:

(i) Tempe, in which cooked soya beans are fermented by the fungus *Rhizopus oligosporus*, subsequently fried and eaten as a substitute to meat.

(ii) Soy sauce using koji fermentation process in which the cooked soya beans are inoculated with the fungus *Aspergillus oryzae*, which subsequently covers them along with secretion of enzymes; subsequently, the fermented beans are transferred to brine where they are slowly degraded into a dark brown sauce.

(iii) Ang kak or the red rice is prepared by the cultivation of the fungus *Monascus ruber* which is also a source of lovastatin, the anti-hypercholestrolaemic drug. The fermented red rice is ground into a fine powder and used as a colouring agent in cooking.

SSF technology is currently being explored for the production of technical enzymes, pigments, organic acids, growth hormones, animal feeds and biopesticides (Table 3.2).This technology is also being adopted for biobleaching, bio-pulping and bioremediation.

Table 3.1 Comparison between submerged and solid substrate fermentation

Submerged fermentation	Solid substrate fermentation
Substrates are soluble sugars	Substrate is polymeric and insoluble
High volume of water consumption and effluent release	Very limited water consumption and no release of effluent
In-depth fermentation process	Shallow fermentation process
Uniform temperature, pH, C and N source distribution	Gradient of temperature, pH, C and N source distribution
2-phase system (liquid and gas)	3-phase system (solid, liquid and gas)
Temperature and oxygen transfer are controlled	Temperature, oxygen transfer and water content are controlled
Inoculum ratio is low	Inoculum ratio is large
No intra-particle resistance	Intra-particle resistance
Fermenting organism is evenly distributed throughout the medium	Fermenting organism sticks to the solid surface and grows
High-energy consuming process	Low-energy consuming process
High-cost technology	Low-cost technology
Low concentration of the end product	High concentration of the end product

Adapted from Raimbault (1998), Prabhakar et al. (2005)

Table 3.2 Products manufactured by solid substrate fermentation

Process/product	Substrate	Microorganism	Reference
Citric acid	Cassava bagasse	*Aspergillus niger*	Vandenberghe et al. (2000)
Lactic acid	Sugarcane bagasse	*Rhizopus oryzae*	Soccol et al. (1994)
Mushroom	Cassava bagasse	*Pleurotus ostreatus* *Lentinus edodes*	Barbosa et al. (1997)
Biopesticide	Potato waste	*Beauveria bassiana*	Soccol et al. (1997)
Gibberellic acid	Coffee husk	*Gibberella fujikuroi*	Machado et al. (2000)
Amylase	Cassava bagasse	*Rhizopus arrhizus*	Pandey et al. (2000)
Mycophenolic acid	Wheat bran	*Penicillium brevicompactum*	Sadhukhan et al. (1999)
Cephalosporin	Sugarcane bagasse	*Acremonium chrysogenum*	Candra et al. (2004)

3.10 Media for Industrial Fermentations

The nutritional requirements of microorganisms used in industrial fermentation processes have complex and varied requirements for growth and product formation. This variation is not only encountered in the broader sense, i.e. microorganism type (bacteria, fungi, yeast, etc.), but strictly at species and strain levels.

Microbial environment is largely driven by the composition of the growth medium. When pure compounds in precisely defined concentrations are used, then the medium is referred to as chemically defined or synthetic medium. This is generally preferred for research purposes to establish specific nutritional requirements for growth and product formation by systematically eliminating or adding a chemical species in the medium formulation. Chemically defined media are reproducible and exhibit fewer tendencies to foam during the fermentation process. Moreover, they offer easy product recovery and purification; however, they are unsuitable for industrial fermentations since they are expensive and the process becomes cost intensive.

Natural or complex substrates such as corn steep liquor, molasses and stick liquor which are not chemically defined completely are preferred for media development and formulation in industrial fermentations. Sometimes inorganic nutrients and vitamins are blended to satisfy the metabolic requirements of the fermenting microorganism (Table 3.3). During industrial fermentation processes, there are several stages in which media are required, viz., inoculum propagation step, pilot-scale production and main production. The media are formulated to keep into the technical objective of the fermentation process. For instance, the media designed for inoculum propagation would be entirely different from the main production medium. Similarly, if the objective of the fermentation process is production of biomass or primary metabolite, then the medium would be formulated which shall optimally promote the growth of the microorganism. Media formulation is essentially blending complex nutrient resources in a manner to achieve proper chemical balances desired by the organism in a

Table 3.3 Comparison of a chemically defined medium and natural complex medium used for production of Penicillin

Chemically defined/ synthetic medium (g/L)		Complex/natural medium (g/L)	
Sucrose	20	Corn steep liquor (CSL)	50
Lactose	10	Glucose/sucrose	30
Peptone	5	$(NH_4)_2$ SO_4	10
$(NH_4)_2$ SO_4	13	KH_2PO_4	2
KH_2PO_4	3	$CaCl_2$. $2H_2O$	0.06
Na_2SO_4	0.5		
EDTA	0.25		
$MgSO_4$.7 H_2O	0.05		
$CaCl_2$. $2H_2O$	0.25		
$FeSO_4$. $7H_2O$	0.02		
$MnSO_4$.$4H_2O$	0.02		
ZnSO4.$7H_2O$	0.01		
Na_2 MoO_4.$2H_2O$	0.01		
$CuSO_4$.$5H_2O$	0.005		

cost-effective manner. The nutrients which are primarily used for media development in fermentation are carbon, nitrogen, sulphur, minerals and vitamins. Hence, media design and development is a crucial step in the development of a cost-effective fermentation process.

Composition of Beet Molasses

Water	19.2 %
Sucrose	48.9 %
Glucose + fructose	0.5 %
Raffinose	1.3 %
Organic nonsugars	18.0 %
Ash	12.1 %
Components of ash	
K_2O	6.4 %
CaO	0.21 %
MgO	0.12 %
P_2O_5	0.03 %
Na_2O	1.6 %
Fe_2O_3	0.03 %
Sulphates of SO_3	0.74 %
Cl	0.8 %
Vitamins (mg/100 ml)	
Thiamine (B1)	0.01
Riboflavin (B2)	1.1
Nicotinic acid	8.0
Ca pantothenate	0.7
Folic acid	0.025

Adapted from Vogel and Todaro (1997)

Carbon is one of the vital nutrients since the biomass is typically 50 % carbon on dry weight basis for all organisms. Carbohydrates are excellent sources of metabolic energy apart from carbon, oxygen and hydrogen and generally are present in a concentration range of 0.2–25 % w/v in different media formulations. The complexity of the carbohydrate molecule affects its bioavailability to the microorganisms. These are generally ranked as hexoses > disaccharides > pentoses > polysaccharides. Commercially, sucrose is used for fermentation which is available in the form of molasses. Sugar beet molasses and blackstrap molasses are produced during table and raw sugar production, respectively. They are the most widely used carbon sources in the fermentation industry.

Nitrogen following carbon is the next constituent present in the fermentation media. Nitrogen is present in the cell in the form of amino groups in proteins or in nucleic acids. Microorganisms like algae and fungi are able to assimilate ammonium nitrate and sodium nitrate; however, bacteria and yeasts find difficulty in utilising nitrogen in this form. In commercial fermentation medium, complex organic nitrogen sources like corn steep liquor, dried distillers soluble, yeast, corn germ and digests of casein, yeast and cotton seed are generally used. Corn steep liquor (CSL) is generally used as the preferred source of nitrogen followed by yeast extract. Corn steep liquor is the by-product during starch extraction from the maize and was used in the commercial production of penicillin for the first time. The exact composition of CSL varies depending on the quality of maize and the processing conditions. On average basis, CSL contains about 4 % (w/v) nitrogen including a wide range of amino acids along with minerals and vitamins. Lactose may be present as a residual sugar with a concentration between 9 % and 20 % w/v depending upon the bacterial activity prior to cold storage. Another substrate used for nitrogen supplementation during media development is soya bean meal. Soya bean meal is finely ground defatted soya bean seeds. Soya bean meal consists of 8 % w/w nitrogen. This substrate has been used for the commercial production of streptomycin. Archer Daniels Midland (ADM) Co. has developed some universal fermentation media like PHARMAMEDIA® and PROFLO® which also improve yield and productivity of some fermentation processes. This medium has been designed using yellow flour from the embryo of the cotton seed. PHARMAMEDIA® consists of 55 % moisture-free protein, 24 % carbohydrate, 5 % oil and 5 % ash. It has been used widely for the production of antibiotics like tetracycline and penicillin apart from other fermentation products.

Water is the base liquid in which the medium is formulated and hence required in large quantities for submerged fermentation processes. Prior to use, suspended solids, colloids, microorganisms and hardness are removed. It has been observed that certain products when supplemented to the fermentation medium speed up the process of product formation as well as improve yields of the end product. These are known as precursors or inducers. Phenylacetic acid was first detected in corn steep liquor which enhanced the production of penicillin during fermentation. Further studies proved that phenylacetic acid when used individually improved the product formation process and yield. Similarly, cobalt has been used as a precursor for the production of vitamin B12. Antifoams are also added for prevention of foam formation during the fermentation process. In case the formation of foam is not stopped, it may lead to blocking of filters and may contaminate the medium resulting in loss of aseptic conditions. Natural products as well as synthetic chemicals are used as antifoams and are incorporated into the fermentation medium. Natural antifoaming agents are soya, sunflower and rapeseed oils, while silicone oil, alkylated glycols and polyalcohols serve as chemical antifoam agents.

3.11 Downstream Processing

Downstream processing is an essential part of fermentation technology wherein the desired product is isolated, purified and utilised for different end uses. It accounts for approximately 60 % of the total production cost excluding the cost of the raw materials. The end product could be cells themselves, i.e. the biomass or the spent fermentation medium in case the product is released extracellularly. Thus, envisaging this complexity, downstream processing involves a variety of technologies and methodologies. The bioseparations of end products predominantly are dependent on molecular mass, charge distribution, hydrophobicity and distribution coefficient for their isolation and purification. The major processes involved in downstream processing are provided in Fig. 3.7.

Foam removal is done through a separate outlet in the fermenter during harvesting and is then mechanically broken to recover trapped cells. Solid–liquid separations are essentially carried out by filtration in which the slurry is passed through the filter medium on which the solids are deposited. The filter mesh size is dependent upon the size of the microorganisms used for carrying out the fermentation. Presently, a variety of advanced filtration techniques have been developed, for example, microfiltration, ultrafiltration and reverse osmosis. For cell recovery, the filters used are filter press and rotator drum filters. An alternative method of solid–liquid separation is centrifugation. Tubular centrifuges comprise of a hollow cylindrical rotating element in a stationary casing. The suspension is fed through the bottom, and the clarified liquid is removed from the top leaving solid deposits in the bowl's wall. Disc centrifuges are also used for bioseparations, and their main advantage is continuous operation.

In the case of intracellular end products, cells are ruptured to release the end products. However, cellular disruption is a difficult process because of the strength of the cell walls and high osmotic pressure. Physical, chemical and biological processes are generally used for cellular disruption at different levels. The physical methods of disruption are milling, homogenisation or ultrasonication. An ultrasonicator generates sound waves of about 16 KHz which causes pressure fluctuations to form oscillating bubbles. Cells are also disrupted using chemicals like detergents, alkalis, organic solvents or osmotic shock. The intention of chemical disruption is that it should be easily separable and must be compatible with the products. Surfactants also help in cellular disruption by solubilising the lipid content in the cell wall. Sodium dodecyl sulphate (SDS) and Triton X-100 are generally used in laboratories. Alkali treatment is an expensive and effective method of cell disruption but can be harsh for protein-based products. Organic solvents such as toluene can rupture the cell wall by penetrating into the cell wall lipids thereby causing the cell wall to swell and burst. Biological methods of cell rupturing comprise of enzymatic treatment to degrade the cell wall. However, this method is unsuitable for large-scale industrial processes. The dilute aqueous solution (extracellular spent broth or cell

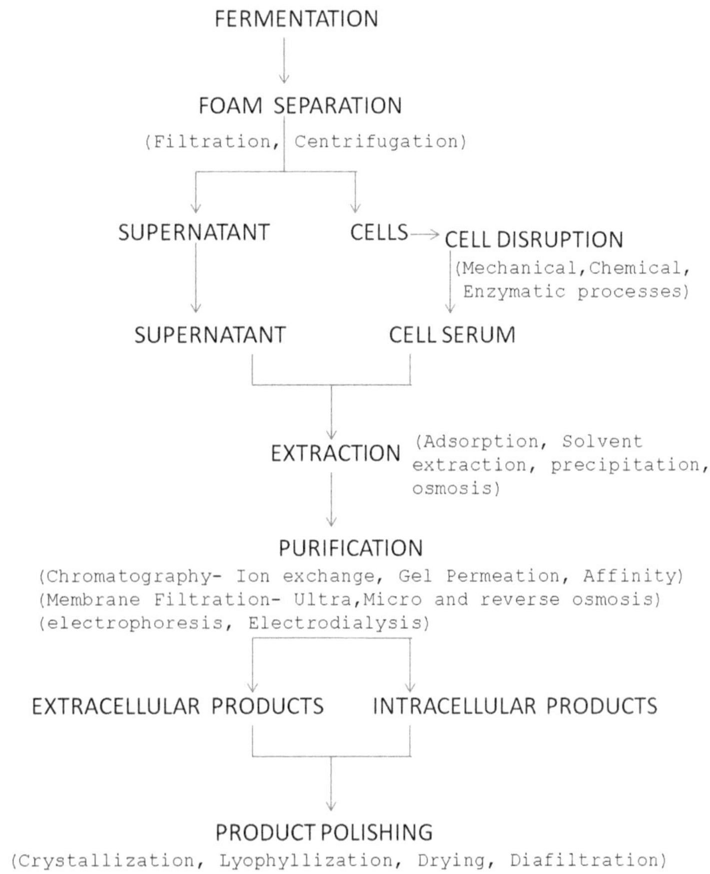

Fig 3.7 Major steps in downstream processing for product recovery

extracts) is subjected to extraction. Extraction is a process of separating the solutes (constituents) in liquid by contact with another insoluble liquid. Extract is a term given to the solvent-rich phase and the residual liquid is referred to as the raffinate. Commercially, solvent extractors are used for the process of extraction. Extraction can be carried out as a single-stage operation either in batch or continuous mode. However, to enhance the recoverability of the product, multistage cross-current or multistage countercurrent extractors are used.

Subsequent to extraction, the specific products could be absorbed by some solids due to physical and chemical interactions. Adsorption can be classified into three broad types, viz., conventional, ion exchange and affinity adsorption. The conventional adsorption process is a reversible process due to intermolecular forces of attraction between the molecules of the substance adsorbed and of the solid. Activated carbon is generally used in the conventional adsorption process for the isolation of valuable products from the fermentation broth by adsorption and then recovery by elution. In ion exchange adsorption, the resin comprises of a polymeric network, ionic network and counterions. The positively charged ions interact with the cation-exchange resin thereby replacing the counterion and getting separated from the solution. These are then eluted out from the resin. In affinity adsorption, the interaction is between the solute and the ligand which is attached to the surface of the carrier particle by covalent or ionic bonds. Examples of these interactions include receptors–hormones, antigens–antibodies, and enzyme inhibitors and enzymes. The interaction is offering a high selectivity during bioseparations; however, the high cost of the resin is a major disadvantage in its use in industrial separation processes.

Once the product is isolated or recovered from the fermentation broth, it may need further purification. This step of purification is accomplished by different methods such as precipitation, chromatography and electrophoresis and ultrafiltration. The precipitation process is predominantly used in the isolation of proteins and antibiotics and is generally achieved by the addition of salts, heat or organic solvents. This process is effective and cost-effective. For the isolation of the majority of proteins, ammonium sulphate is commonly used, but the main disadvantage is its separation from the precipitate. Sodium sulphate is also used, but adequate solubility is achieved at a temperature between 35 and 40 °C. The unwanted or interfering protein in the process of purification can be achieved by heating the recovered product. Organic solvents at temperature below −5 °C decrease the dielectric constant of the solution thereby precipitating proteins.

Chromatographic processes involve separation/purification of solutes in two phases, solid and mobile. The solutes are generally present in the mobile phase and interact with in the stationary phase by different phenomena like adsorption, ion exchange, affinity and gel filtration. The basic principles of ion exchange, adsorption and affinity chromatography have already been discussed in this chapter. Gel filtration chromatography uses cross-linked dextrans, polyacrylamide and agarose at the stationary phase. These gels form the different pore sizes. The smaller molecules penetrate the pore structure to a greater extent and therefore have a longer retention time than the larger molecules. Electrophoresis is also a method largely used for protein separations and is based on their specific migration rates in an electrical field. Electrodialysis is a separation process of ions that occurs due to the imposed potential difference across the ion-selective cationic and anionic exchange membranes. Pervaporation is also a membrane separation technique accompanied by a change of phase of the species transported across the membrane, usually from liquid to vapour.

Crystallisation is a process where the solid particles of specified size and shape are being recovered from the homogeneous phase, and it is generally regarded as the final purification step since crystals so obtained are usually of exceptional purity. The process of crystallisation can be carried out by (a) cooling the solution leading to evaporation (cold evaporation), (b) evaporation of the solvent with little or no cooling and (c) evaporation in an adiabatic or vacuum crystalliser. Subsequently, the products are packed and stored as per their storage conditions.

3.12 Summary

Today, the fermentation industry is in a state of flux due to rapid changes in the product spectrum and the scale of processing. This is attributed to a variety of factors like relocation of large-scale bioprocesses in regions with low manpower costs and development of advance fermentation expression systems with the deciphering of the human genome thereby leading to the development of novel therapeutic proteins, antibodies and vaccines. To drive these new as well as old processes, fermentation knowledge and skills are essential requirements. Hence, the present chapter provides you with an overview of the different aspects of fermentation technology which is currently being exploited in the industry.

Selected Reading

Barbosa M, Soccol CR, Marin B, Todeschini ML, Tonial TM, Flores V (1997) Prospect to production of *Pleurotus sajor-caju* from cassava fibrous waste. In: Roussos S, et al. (ed) Advances in solid state fermentation: edible mushrooms/fungi. Kluwer Academic Publishers, London, pp 515–527

Bryson V, Szybalski W (1952) Microbial selection. Science 18(115):45–51

Bu'Lock JD, Hamilton D, Hulme MA, Powell AJ, Shepherd D, Smalley HM, Smith GN (1965) Metabolic development and secondary biosynthesis in *Penicillium urticae*. Can J Microbiol 11:765–768

Candra TE, Tomasini A, Fernandez FJ, Barrios-Gonzalez J (2004) Produccion de Cefalosporina C por *A. chrysogenum* C10 fermentacion en medico solido de 2 fares. 34 Congreso Nacional de Microbiologica, Cancun, Mexico, p 43

Chisti Y, Moo Young M (1999) Fermentation technology, bioprocessing, scaleup and manufacture. In: Moses V, Cape RE, Springham DG (eds) Biotechnology – the science and business, 2nd edn. Harwood Academic Publishers, New York, pp 177–222

Durand A (2003) Bioreactor design in solid substrate fermentation. Biochem Engg J 13:113–125

Machado CMM, Soccol CR, Oliveira BH, Pandey A (2000) Coffee husk as substrate for the production of gibberellic acid by fermentation. In: Sera T, et al. (ed) Coffee biotechnology and quality. Kluwer Academic Publisher, Dordrecht, pp 401–408

Pandey A, Selvakumar P, Soccol CR, Nigam P (1999) Solid state fermentation for the production of industrial enzymes. Curr Sci 79:149–162

Pandey A, Soccol CR, Mitchell D (2000) New developments in solid state fermentations I: bioprocesses and products. Process Biochem 31:669–678

Prabhakar A, Krishnaiah K, Janaun J, Bono A (2005) Overview of engineering aspects of solid substrate fermentation. Malays J Microbiol 1(2):10–16

Raimbault M (1998) General and microbiological aspects of solid substrate fermentation. Electron J Biotechnol 1(3):174–188

Sadhukhan AK, Ramanamurthy MV, Kumar RA, Mohan EVS, Vandana G, Venkateshwara Rao K (1999) Optimization of mycophenolic acid production in solid substrate fermentation using response surface methodology. Ind Microbiol Biotechnol 22:33–38

Soccol CR, Marin B, Raimbault M, Lebeault JM (1994) Solid state fermentation for production of L (+)-lactic acid by *Rhizopus*. Cond Brasileiro Eng Quim 2:1189–1194

Soccol CR, Ayala LA, Soccol VT, Krieger N, Santos HR (1997) Spore production by entomopathogenic fungus *Beauveria bassiana* from declassified potatoes by solid-state fermentation. Rev Microbiol 28(Suppl 1):34–42

Stanbury PF, Whitakar A, Hall SJ (eds) (1999) Principles of fermentation technology. Butterworth – Heinemann Publishers, Oxford

Vandenberghe LPS, Soccol CR, Pandey A, Lebeault JM (2000) Solid substrate fermentation for synthesis of citric acid by *Aspergillus niger*. Bioresour Technol 74:175–178

Vogel HC, Todaro CL (eds) (1997) Fermentation and biochemical engineering handbook, 2nd edn, Principles of process design and equipment. Noyes Publication, Westwood

4 Agricultural Applications of Microbes

Biofertilisers and Biopesticides

4.1 Introduction

The interaction between plants and microorganisms is complex of which some are beneficial while others are detrimental. However, a beneficial interaction of microorganisms with plants remains largely underexplored. The present chapter discusses the exploitation of microorganisms to enhance the global agricultural productivity with least adverse effects of environment and human health.

To meet the ever increasing food demands, there is a need to develop ecologically compatible environment friendly techniques in agriculture for providing adequate nourishment to humans. To meet the demand for intensification of agricultural production, reducing crop losses is extremely important. This can be achieved by suppressing pests to subeconomic levels using ecofriendly methods and enhancing the nutrient bioavailability to crops to improve yields.

Effective or beneficial microorganisms, which comprise of diazotrophic bacteria, biological control agents (BCAs), plant growth-promoting bacteria (PGPR) and PGPF (plant growth-promoting fungi), can play a key role in overcoming this major challenge of global food demands as they fulfil important ecosystem functions for plants and soils. These microorganisms may have a direct or indirect positive effect on the plants. Endophytic microorganisms dwell intracellularly in the plants for majority of their lifetime (Bacon and White 2000) and have been used for improving plant agronomic characteristics such as nitrogen efficiency, increased drought tolerance and control of various plant diseases.

4.2 Biofertilisers

Nitrogen and phosphorus are the most limiting nutrients for plant growth and development. Commercial man-made fertilisers are added to overcome the bioavailability of nitrogen and phosphorus to increase the crop production. However, intensive use of these synthetic fertilisers have led to the deterioration of soil quality, ground and surface water quality apart from reduction in biodiversity and suppression of ecosystem function. Hence, microorganisms which are beneficial in making nitrogen and phosphorus available effectively to the plant when they are inoculated into cropping systems are referred to as microbial bio-inoculants or biofertilisers (Table 4.1).

4.2.1 Nitrogen-Fixing Microorganisms as Biofertilisers

Nitrogen (N_2) is an essential element which is found in amino acids and proteins and also contributes to the formation of nitrogenous bases in genetic material. The atmosphere is composed of approximately 79 % nitrogen in gaseous state

S. Saxena, *Applied Microbiology*,
DOI 10.1007/978-81-322-2259-0_4,

Table 4.1 Nitrogen-fixing microorganisms

Free living/associative	Symbiotic
Aerobic	Rhizospheric non-nodule forming
Azotobacter	*Azotobacter*
Beijerinckia	*Azospirillum*
Derxia	
Rhizobium	Leguminous
Azospirillum	*Rhizobium*
Herbaspirillum	*Bradyrhizobium*
Facultative anaerobic	*Azorhizobium*
Bacillus	
Klebsiella	Nonleguminous
Anaerobic	*Frankia*
Clostridium	*Nostoc*
Desulfvibrio	*Anabaena*
Arachaebacteria	
Cyanobacteria	
Anabaena	
Calothrix	
Nostoc	
Trichodesmium	

which is inert or nonreactive. Biological nitrogen fixation (BNF) is a process wherein the microorganisms mediate the transformation of nitrogen from the inert gaseous form into ammonia which can be assimilated by some plants directly or is further reduced into nitrate, making it available to plant. BNF constitutes to approximately 65 % nitrogen consumption in agriculture by nitrogen-fixing bacteria (Matiru and Dakora 2004). The nitrogen-fixing microorganisms generally are classified as symbiotic when they are associated with plants and residing in the legumes, while others which exist in the rhizospheric region are known as free living. Yet another class of nitrogen-fixing microorganisms exist that associate with the roots or leaves of the plants and are known as associative nitrogen fixers.

4.2.1.1 Symbiotic Nitrogen-Fixing Microorganisms

Rhizobium, *Mesorhizobium*, *Bradyrhizobium*, *Azorhizobium*, *Allorhizobium and Sinorhizobium* (Rhizobia) form intimate symbiotic relationships with leguminous plants by chemotactically responding to the flavonoid molecules released as signals by the legume host plant. Nodules are formed as a result of interactions between these rhizobial species and leguminous plants. Nitrogenase enzyme present in the bacterium with assistance of nodulins (legume plant proteins) fixes atmospheric nitrogen which is then transferred to the plant in true spirit of symbiosis. This phenomenon has been studied extensively and exploited as a means for enhancing crop yields (Boholool 1990; Sharma et al. 1993). The fixed N_2 is released when the plants die, making it available to other plants, and this helps in fertilising the soil. Studies have indicated that the nitrogen fixation by the rhizobacteria is impressive and ranges between 200 and 300 kg N/ha/year for legume crops and pasture species (Peoples et al. 1995), highlighting the role of rhizobium–legume symbiosis as a major provider of biological nitrogen fixation (Zahran 1999). Recently endophytic association of rhizobia with plants has also been discovered with the identification of *Rhizobium leguminosarum* bv. *trifolii* (Yanni et al. 1997). The hypothesis of endophytic existence of *Rhizobium* is attributed to rotational cropping of berseem clover in the Nile delta for over seven centuries, promoting closer rhizobial affinity for cereals as the host plants. Rhizobia have also been isolated as endophytes from the roots of the other nonleguminous species such as cotton, sweet corn, wheat, maize and canola which are cropped and rotationally cultivated with leguminous plants. Photosynthetic *Bradyrhizobium* were found as natural endophytes of the African wild rice *Oryza breviligulata* (Chaintreuil et al. 2000).

Cyanobacteria also form symbiotic associations which are responsible for nitrogen fixation in cropping systems. Wet rice fields provide ideal conditions for the growth of cyanobacteria wherein they accumulate 20–28 kg nitrogen/ha/crop and can drastically reduce the use of urea as a chemical fertiliser by 25–35 % (Hashem 2001). *Nostoc*, *Anabaena variabilis*, *Aulosira fertilisima*, *Calothrix* sp., *Tolypothrix* sp. and *Scytonema* sp. have been identified from various agro-ecological regions and exploited for rice production (Prasad and Prasad 2001). *Anabaena azollae* forms symbiotic association with water fern Azolla and has been used as a promising biofertiliser by farmers in India.

Frankia is an actinomycete which was first isolated from the nodules of *Comptonia peregrina* in 1978 and symbiotically fixes nitrogen in nonleguminous plants though formation of nodules. It forms symbiotic association with several angiosperms like *Casuarina*, *Hippophae*, *Coriaria* and *Discaria*. This symbiotic relationship is referred as actinorhizal symbiosis which is responsible for enhancing the fertility of temperate forests similar to woody legumes in tropical forests. Commercially *Frankia* has not been used effectively in cropping systems as compared to Rhizobia.

4.2.1.2 Non-symbiotic/Associative Nitrogen-Fixing Microorganisms

Free living nitrogen-fixing microorganisms include *Azotobacter* and *Beijerinckia* which are aerobic heterotrophs, while *Desulfovibrio* and *Clostridium* species are anaerobes. Most common species of Azotobacter are *Azotobacter chroococcum* and *Azotobacter vienlandii. A. chroococcum* under in vitro conditions can fix approximately 10 mg N/g of carbon supplied. *Azotobacter* also produces indoleacetic acid (IAA) and GA (Gibberellic acid), i.e. plant growth hormones, and at the same time, it is recommended as microbial inoculants in agricultural practices. It is found to fix nitrogen in a range of 60–70 kg N/ha in maize, rice and wheat crops. In green house trails, inoculation with Azotobacter replaced around 50 % urea nitrogen under aseptic conditions (Soliman et al. 1995).

Beijerinck discovered *Azospirillum lipoferum* as a soil-inhabiting bacteria which exhibited associative symbiosis during nitrogen fixation apart from the formation of plant growth-promoting substances like gibberellins and IAA. Five species of Azospirillum have been described till date, viz., *Azospirillum brasilense*, *A. lipoferum*, *A. amazonense*, *A. halopraeferens* and *A. irakense*. Azospirillum has been effectively used in rice crops. *A. brasilense* has been found to be associated to the roots of the rice plants with a population of approximately 8×10^7 cells/g dry weight of the soil. *A. lipoferum* and *A. brasilense* have been isolated from the roots and stems of rice plants, while *A. amazonese* has been isolated from the roots only (James et al. 2000). The bacterium was mass produced and inoculated into the soil wherein it was mass produced and inoculated which enhanced the yield from 1.6 to 10.5 g/plant. Apart from its ability to fix nitrogen, *Azospirillum* species also helped in providing the mineral nutrients from the soil, sequestered Fe to survive under harsh environmental conditions and favour beneficial plant–mycorrhizal association (Bashan et al. 2004).

Clostridia are obligately anaerobic heterotrophs which are capable of fixing nitrogen under anaerobic conditions. Clostridia are often found in rice fields, and its activity increases in response to the application of straw in rice fields (Mishustin et al. 1983). *Herbaspirillum seropediacae* was first isolated from rice, maize and sorghum where it existed as an endophyte. The organism is responsible for fixing 31–54 % nitrogen in the rice plant (Baldani et al. 2000). *H. seropedicae* is found in the roots and stem of sugarcane plant. *Herbaspirilla* directly do not contribute to the nitrogen fixation replacing urea; however, they enhance the nitrogen content in the leaves of sugarcane and enhance the cane yield significantly (Muthukumarasamy et al. 1999).

4.2.2 Phosphate Solubilising Microorganisms as Biofertilisers

Unlike nitrogen, phosphorus is a major growth-limiting nutrient which lacks atmospheric source which can be made biologically available. All aspects of plant growth, viz., development of roots, stem and stalk, seed formation, nitrogen fixation, crop quality and resistance to plant diseases, are associated with phosphorus nutrition. Biologically available nitrogen is very low in concentration despite its presence in the soil as orthophosphate, phosphine and phosphonate. Despite large application of phosphate as fertiliser, a very low amount reaches in bioavailable form, i.e. approx. 1 mg/kg of soil. Microbes therefore play a very important role in bioconversion of this unavailable to available form by excreting organic acids that dissolve phosphatic minerals and chelate cationic partners of the phosphate ions, thereby releasing P into solution (He et al. 2002). Phosphate solubilising bacteria (PSB) are being used as biofertilisers since 1950s. The most predominant PSBs belong to the genera *Pseudomonas* and *Bacillus*. Other bacterial genera which also carry out phosphate solubilisation are *Micrococcus*, *Sarcina*, *Escherichia*, *Azospirillum*, *Rhizobium*, *Burkholderia*, *Arthrobacter*, *Alcaligenes*, *Serratia*, *Enterobacter*, *Acinetobacter*, *Flavobacterium and Erwinia*. Phosphate solubilisers produce different organic acids like citric gluconic, lactic, succinic and propionic.

Arthrobacter ureafaciens, *Phyllobacterium myrsinacearum*, *Rhodococcus erythropolis* and *Delftia* species have been for the first time reported as phosphate solubilising bacteria based on their capacity to solubilise tricalcium phosphate under in vitro conditions by secretion of organic acids (Chen et al. 2006). Recently a *Kurthia* species which is a phosphate solubilising novel species has been isolated from rhizospheric soils of tea plantations. Phosphobacterin is a culture of phosphate solubilising bacteria *Bacillus megaterium* var. *phosphaticum* which was absorbed on kaolin. It was first used in Russia for application in agriculture. Similarly Phylazonit M is a product developed from *Bacillus megaterium* with *Azotobacter chroococcum* for enhancing the nitrogen and phosphorus supply to plants used in Hungary. Similarly Indian Agricultural Research Institute (IARI, India) developed microphos which is a combination of *Pseudomonas striata*, *Bacillus polymyxa* and *Aspergillus awamori* (Gaur 1990).

Fungi also are powerful phosphate solubilisers belonging to genera *Penicillium* and *Aspergillus*. Jumpstart® is a commercial product of formulated *Penicillium bilaiae* which was released to the market as a wettable powder in 1999 (Burton and Knight 2005). Similarly *Penicillium radicum* has been isolated from the rhizosphere of wheat roots, and *Penicillium italicum* exhibits potential promise for their exploitation as plant growth promoters. Another product Ketomium® is formulated from *Chaetomium globosum*, and *Chaetomium cupreum* exhibits dual function, i.e. apart from being used as a mycoherbicide as well as plant growth stimulant since it promotes higher growth and high yield in corn, tomato, pepper and in citrus crops (Soytong et al. 2001). *Arthobotrys oligospora* is a nematofungus which also possess the potential to solubilise phosphate (Duponnois et al. 2006).

4.2.3 PGPB (Plant Growth Promoting Bacteria): Plant Growth Promoters

Many microorganisms in plant rhizosphere produce plant growth-promoting compounds like auxins, gibberellins, cytokinins, ethylene and abscisic acid (Table 4.2), apart from providing protection against pathogenic microbes from getting colonised. These are referred to as plant growth producing bacteria/rhizobacteria (PGPB/PGPR) (Kloepper et al. 1989). Studies have indicated that approximately 80 % of the bacteria isolated from the rhizosphere are able to produce auxin, indole-3-acetic acid (IAA) which is responsible for rapid and long-term physiological responses in plants (Patten and Glick 1996).

Pseudomonas putida and *Pseudomonas fluorescens* are most important PGPR which produce

Table 4.2 Some phytohormones producing bacteria

Auxins	Gibberellic acid	Cytokinins
Rhizobium leguminosarum	*Azospirillum* sp.	*Paenibacillus polymyxa*
Azotobacter sp.	*Arthrobacter* sp.	*Pantoea agglomerans*
Azospirillum brasilense	*Rhizobium meliloti*	*Rhodospirillum rubrum*
Pseudomonas fluorescens	*Rhizobium phaseoli*	*Bacillus subtilis*
Bacillus cereus	*Acetobacter diazotrophicus*	*Pseudomonas fluorescens*
Herbaspirillum sp.	*Herbaspirillum seropedicea*	*Arthrobacter* sp.
Acetobacter sp.		*Azotobacter* sp.

Auxins and promote the yield in plants. The other auxins like the indole-3-butyric acid (IBA) and indole-3-ethanol (TOL) are produced by bacteria belonging to the genera *Paenibacillus* and *Azospirilla* and indirectly contribute to plant growth (Lebuhn et al. 1997). The synthesis of gibberellins was first reported by *Azospirillum brasilense* and later in rhizobium. Currently there exists is ample literature which highlights that gibberellins are being produced by a variety of PGPRs like *Acetobacter diazotrophicus*, *Azospirillum brasilense*, *Azospirillum lipoferum*, *Bacillus pumilus*, *Herbaspirillum seropedicea* and *Rhizobium phaseoli*, producing GA1, GA3, GA4 and GA20 (MacMillan 2002).

Similarly cytokinins have been produced by some strains of *Azotobacter* spp., *Rhizobium* spp., *Paenibacillus polymyxa*, *Pantoea agglomerans*, *Rhodospirillum rubrum*, *Bacillus subtilis* and *Pseudomonas fluorescens*. Plant growth-promoting bacteria also fix nitrogen and solubilise phosphorus and therefore have been exploited commercially. They improve the plant yield by indirect mechanisms which generally include production of antibiotics and lytic enzymes and provide resistance to biotic as well as abiotic stresses. PGPRs especially Pseudomonads and Bacilli produce a variety of antibiotics to suppress pathogenic microorganisms in the rhizospheric region and improve plant growth. Some commonly produced antibiotics of PGPR are 2, 4-diacetylphloroglucinol, phenazine-1-carboxylic acid, viscosinamide, butyrolactones and cepaciamide A. Pyoluteorin, phenolic polyketide was first isolated from *Pseudomonas aeruginosa* followed by *Pseudomonas fluorescens* P45 (Takeda 1958) and possess bactericidal, herbicidal and fungicidal properties. Pyrrolnitrin is produced by several fluorescent and nonfluorescent pseudomonads and is a chlorinated phenylpyrrole antibiotic first isolated from *Burkholderia pyrrocinia* (Arima et al. 1964). These are effective in control of the postharvest disease in apples, pears and cut flowers caused by *Botrytis cinerea* (Janisiewicz and Roitmann 1998). Bacillomycin D is an antifungal lipopeptide produced by *Bacillus subtilis*.

Chitinolytic enzymes produced by *Pseudo-monas stutzeri* act as biocontrol agent. These enzymes digest and lyse mycelia of *Fusarium solani*, thereby preventing the fungus from causing crop loss owing to root rot (Lim et al. 1991). β-1, 3-glucanase producing strain of *Pseudomonas cepacia* was able to damage the mycelium of *Rhizoctonia solani*, *Sclerotium rolfsii* and *Pythium ultimum* (Friedlander et al. 1993), thereby serving as a biocontrol agent.

Thus, PGPRs and PGPBs have been used either singly or as consortia based on their functions in agricultural applications as microbial inoculants for improving plant growth and yield.

4.3 Biopesticides

Sustainable agriculture not only aims at increasing the yield of food and fibre crops but also reducing the incidence of pests and diseases to subeconomic levels to meet the growing food demand of the population which is not going to stabilise by 2035 and would be around 8.6 billion in 2035. At present all agriculturally productive land is producing food; however, there is a demand for further escalation of the agricultural production to meet the global food demand. Thus, reducing crop losses in agriculture both at preharvest and postharvest levels would significantly help in meeting the growing food demand. These crop losses are attributed to pests (weeds, plant pathogens, insects and rodents). A variety of chemicals were brought into use in early 1940s to suppress these pests to subeconomic levels. However, to meet the ever increasing demands for food, these chemicals were indiscriminately used leading to hazardous side effects to millions of people due to their persistence in the food as well as were responsible for deterioration of the environment. The residual toxicity of synthetic agrochemicals had effect on nontarget organisms. Further overuse of these agrochemicals has led to the development of resistance in the pests. It has been reported that over 447 species of mites, 200 species of plant pathogens and over 49 species of weeds have become resistant to one or more than one agrochemical. Thus, there is a societal concern regarding further use of these chemical pesticides.

There is a need to develop alternate strategies for pest management by development of new environmentally and toxicologically benign products to replace these synthetic chemicals. Currently research groups while designing pesticides have a complex set of questions: 'What form should the crop protection agent take so that it causes no harm to the end user, can be distributed in water, can be sprayed without difficulty, takes effect against the pest in crops in lowest possible concentration and in the process cause as little burden to the environment as possible?'

Recently there has been upsurge in exploration and exploitation of naturally occurring microorganisms to control crop pests, weeds and diseases. The concept behind this strategy is biological control which can be defined as 'direct and purposeful manipulation of natural enemies to increase pest competition (in whole or in part) or resource requirements by these organisms for the reduction of negative pest effects or pest species density to levels at or below economic thresholds'. There are two broad groups of biological control: classical and innundative.

4.3.1 Bio-weedicides

4.3.1.1 Classical Biological Control

This approach establishes new natural enemies in a region for the eradication of the pest. This is generally adopted for controlling exotic pest which have become problematic due to absence of their natural enemies. Hence, the biological control agents (BCAs) are explored from the original geographic range of the pest. The promising BCAs are screened for their potential to be introduced to the new region with the expectation that the organism would provide pest suppression without complete eradication. The outstanding example of this approach has been the development of the fungus *Puccinnia chondrillina* for the control of *Chondrilla juncea* (skeleton weed). *Puccinia* is a rust fungus found in Mediterranean and tested on several crop plants as well as members of Asteraceae, which were closely related to the skeleton weed. The fungus was closely related to skeleton weed, and finally the first strain of this fungus was released in Australia, capable of suppressing the skeleton weed (Hasan 1988). *Puccinia chondrillina* also proved to be effective in the control of skeleton weed when released in western USA (Suopkoff et al. 1988). The other examples of classical biological control agents are smut fungus *Entyloma ageratinae* imported from Jamaica to control *Hamakua pamakani* (*Ageratina riparia*, Asteraceae family) in Hawaiian forests and rangelands, and *Uromycladium tepperianum* has been used to control the invasive tree *Acacia saligna* in South Africa (Morris 1997). However, the major concern is the cost of importing the microorganism and its safety aspects when released in a new geographical region as it may also effect other nontarget plants or beneficial plants leading to huge economic losses.

4.3.1.2 Innundative or Bioherbicidal Approach

This approach involves exploration and use of natural enemies from the region of eradication of the pest by repeated release so as to temporarily boost their abundance and hence also referred as the augmentative approach. This approach gained importance with commercial development of DeVine® wherein chlamydospores of the pathogen *Phytophthora palmivora* were applied as a liquid suspension to control the weed *Morrenia odorata* in the citrus orchards of Florida (Kenney 1986). The formulation suppressed 96 % weed population within 10 weeks. Since then many bacteria and fungi have been formulated into bioherbicides and have been registered for commercial applications (Table 4.3).

Chondrostereum purpureum (Pers. Ex fr) is a wood-rotting basidiomycete which has been formulated under the trade name Biochon in Netherlands for control of invasive North American *Prunus* and *Populus* spp. in natural and commercial forests. Native isolates of *C. purpureum* from Canada have been developed for stump treatment of hardy wood plants like red alder, red maple, aspen and birch and were registered as CHONTROL™ and MYCOTECH™ PASTE. This example exhibits the exploitation of indigenous microflora for exploitation in different geographical regions. A variety of bacterial as well as fungal isolates are under evaluation for

Table 4.3 Commercially registered bioherbicides

Biological control agent	Commercial name	Target weed	Company/country where used
Phytophthora palmivora	DeVine®	*Morrenia odorata*	USA (1981)
Colletotrichum gloeosporioides f. sp. *aeschynomene*	COLLEGO®	Northern Jointvetch	Encore Technologies USA (1982)
Alternaria cassia	CASST™	Sickle pod and coffee senna (*Cassia* sp.)	USA (1983)
Cercospora rodmanii	ABG-5003	Water hyacinth (*Eicchornia crassipes*)	USA (1991)
Colletotrichum gloeosporioides f. sp. *malvae*	BIOMAL®	*Malva pusilla*	USA (1992)
Cylindrobasidium leave	STUMPTOUT™	Acacia species	South Africa (1997)
Chondrostereum purpureum	BIOCHON™	Woody weeds like *Prunus serotina*	Netherlands (1997)
Xanthomonas campestris pv poae	COMPERICO™	*Poa annua* (Turf grass in golf courses)	Japan (1997)
Colletotrichum acutatum	HAKATAK	*Hakea gummosis* and *H. sericea*	South Africa (1999)
Puccinia thlaspeos	WOAD WARRIOR	Dyer's woad (*Isastis tinctoria*) in farms, rangeland, waste areas and roadsides	USA (2002)
Chondrostereum purpureum	CHONTROL™ ECOCLEAR™	Alders, aspen and other hard-woods	Canada (2004)
Chondrostereum purpureum	MYCOTECH™ PASTE	Deciduous tree species in rights of way and forests	Canada (2004)
Alternaria destruens	SMOLDER™	Cuscuta	USA (2005)
Sclerotinia minor	SARRITOR®	Dandelion (*Taraxacum officinale*) in lawns/turf	Canada (2007)

their possible commercialisation as bioherbicide. *Phoma macrostoma* is being developed for commercialisation by the Scott Miracle Gro Company, Canada, for the biological control of Canada thistle. The fungus has been provisionally registered with the pesticide regulatory authority (PMRA) in June 2011 in Canada. A new facet which has fascinated the plant pathologists, microbiologists and crop protection specialists is exploitation of deleterious rhizobacteria (DRB). DRB colonise in plant roots; they are nonparasitic and generally possess the growth suppressing activity. The association of DRB with weeds was first discovered in downy brome (*Bromus tectorum*). More than 2,000 isolates of Rhizobacteria obtained from the prairie soils in Canada are being investigated for their potential as a bioherbicide (Boyetchko 1999). Like fungi, DRBs are also applied innundatively in the soil prior to the emergence of weed. The method does not eradicate but reduces their competitive ability. *Pseudomonas trivalis* X33D is a promising biocontrol agent against great brome (*Bromus diandrus*) in durum wheat. This strain when formulated reduced the growth of brome and increased the growth of wheat (Mejri et al. 2013)

4.3.2 Bioinsecticides

To control insects which cause crop losses, entomopathogenic fungi has been employed as biological control agents using the innundative approach. Some entomopathogenic fungi formulated into mycoinsecticides are *Beauveria bassiana*, *Verticillium lecanii*, *Nomurea rileyi* and *Lagenidium giganteum* (Table 4.4). *Beauveria bassiana* commonly inhabits the soil and has a broad host range including beetle and fire ants. *Nomurea rileyi* is being developed for the control

Table 4.4 Fungal biocontrol agents used as a registered bioinsecticides

Biological control agent	Commercial name	Target pest	Company/country where used
Sporothrix insectorum	Sporothrix ES	Hemiptera	Biocarto Ind. Com Prod. Agrop. Ltda., Brazil
Metarhizium anisopliae var. *acridum*	Green Muscle OF	Orthoptera	Biological Control Prod. SA(Pty) Ltd., South Africa
	Green Guard ULV		
	Green Guard SC		
Metarhizium anisopliae	Granment-P	Coleoptera	Kwizda Agro GmbH, Austria
	BIO 10210	Coleoptera (Curculionidae)	Intrachem Bio SA, Italy
	Metarhizium Schweizer	Coleoptera (Scarabaeidae)	Eric Schweizer Samen AG, Switzerland
	Bio-Magic	Coleoptera, Hemiptera and other plant hopper	T. Stanes & Company Limited, India
	Bio-Blast Biological Termiticide	Isoptera (Kalotermitidae, Rhinotermitidae, Termopsidae)	EcoScience Corporation, USA
	Taenure Granular Bioinsecticide	Coleoptera (Curculionidae, Scarabaeidae), Diptera (Ephydridae, Mycetophilidae, Sciaridae, Tipulidae)	Novozymes Biologicals Inc., USA (previously: Earth BioSciences Inc., CT, USA; Taensa Co., USA)
	Tick-EX EC	Acari (Ixodidae) + Coleoptera (Scarabaeidae)	Novozymes Biologicals Inc., USA (previously: Earth BioSciences; Taensa Co., USA)
	Metadieca	Hemiptera (Cercopidae)	Liga Agricola Industrial de LaCan˜a de Azucar (LAICA), Costa Rica
Lecanicillium species (*V. lecanii*)	Trichovent		Trichodex S A, Spain
	MicroGermin Plus	Hemiptera (Aleyrodidae, Aphididae), Thysanoptera (Thripidae) + Acari (Tetranychidae)	Omya (Schweiz) AG, Switzerland
	Bio-Catch	Hemiptera (Aleyrodidae, Aphididae, Pseudococcidae)	T. Stanes & Company Limited, India
	Biovert Rich	'Insects' + Nematoda	Plantrich Chemicals & Biofertilizers Ltd, India
	Verti-Sin	Hemiptera (Aphididae), Thysanoptera (Thripidae)	Agrobiologicos del Noroeste S.A. de C.V. (Agrobionsa), Mexico
Lecanicillium muscarium	Mycotal	Hemiptera (Aleyrodidae), Thysanoptera (Thripidae)	Koppert Biological Systems, Netherlands (previously: Tate and Lyle, UK)
Lecanicillium longisporum	Vertalec	Hemiptera (Aphididae)	
	Vertirril WP 1300	Hemiptera (Aleyrodidae, Ortheziidae)	Itaforte Industrial de BioProdutos Agro-Florestais Ltda., Brazil
Lagenidium giganteum	Laginex AS	Diptera (Culicidae)	AgraQuest, USA (now Bayer CropScience)
Isaria sp. (*formerly Paecilomyces* sp.)	PaciHit Rich	Hemiptera (Aleyrodidae), Thysanoptera (Thripidae) + Nematoda	Plantrich Chemicals & Biofertilizers Ltd, India

Isaria fumosorosea	PreFeRal	Hemiptera (Aleyrodidae)	Biobest n.v., Belgium
	Priority	Acari (Eriophyidae, Tetranychidae)	T.Stanes & Company Limited, India
	PFR-97 20 % WDG	Hemiptera (Aleyrodidae, Aphididae), Thysanoptera (Thripidae) + Acari (Tetranychidae)	Certis, Inc., USA (previous owner: Thermo Trilogy Corp., USA)
	Successor	Hemiptera (Aleyrodidae, Aphididae), Thysanoptera (Thripidae) + Acari (Tetranychidae)	Live Systems Technology S.A., Colombia
Conidiobolus thromboides	Vektor 25 SL	Hemiptera (Aleyrodidae, Ortheziidae)	Laverlam S.A., Colombia
Beauveria brongniartii	Melocont-Pilzgerste	Coleoptera (Scarabaeidae)	Kwizda Agro GmbH, Austria/Agrifutur s.r.l., Italy
	Myzel	Coleoptera (Scarabaeidae)	LBBZ Arenenberg, Switzerland
	Beauveria Schweizer	Coleoptera (Scarabaeidae)	Eric Schweizer Samen AG, Switzerland
	Betel	Coleoptera (Scarabaeidae)	Betel Reunion S.A., Reunion Island (subsidiary of Natural Plant Protection, France)
	BiolisaKamikiri	Coleoptera (Cerambycidae)	Nitto Denko, Japan
Beauveria bassiana	Trichobass-L/P	Coleoptera (Curculionidae, Scarabaeidae), Lepidoptera (Castniidae, Pieridae), Hemiptera (Aleyrodidae), Thysanoptera (Thripidae) + Acari (Tetranychidae)	Trichodex S.A., Spain
	Bb Plus	Hemiptera (Aphididae) + Acari (Tetranychidae)	Biological Control Products SA (Pty) Ltd, South Africa
	Bb Weevil	Coleoptera (Curculionidae)	
	BioGuard Rich	Coleoptera (Curculionidae, Scarabaeidae), Hemiptera (Aleyrodidae, Aphididae), Lepidoptera (Crambidae), Thysanoptera (Thripidae)	Plantrich Chemicals & Biofertilizers Ltd, India
	Bio-Power	Coleoptera (Curculionidae, Scarabaeidae), Hemiptera: Auchenorrhyncha (Cicadellidae, Delphacidae), Lepidoptera (Plutellidae)	T.Stanes & Company Limited, India
	BotaniGard ES	Coleoptera (Curculionidae, Scarabaeidae), Hemiptera (Miridae, Aleyrodidae, Aphididae, Pseudococcidae, Psyllidae), Thysanoptera (Thripidae)	Laverlam International Corporation, USA (previously: Emerald BioAgriculture Corp., USA; Mycotech Corp., USA)
	Mycotrol ES	Coleoptera (Chrysomelidae, Curculionidae, Scarabaeidae), Hemiptera (Miridae, Cicadellidae, Fulgoridae, Aleyrodidae, Aphididae, Pseudococcidae, Psyllidae), Lepidoptera (Crambidae), Orthoptera (Acrididae, Tettigoniidae), Thysanoptera (Thripidae)	Laverlam International Corporation, USA (previously: Emerald BioAgriculture Corp. USA; Mycotech Corp., USA)
	Brocaril 50 WP	Coleoptera (Curculionidae)	Laverlam S.A., Colombia
	Nativo 2 SC	Coleoptera (Curculionidae)	Bayer Cropscience S.A., Colombia

of caterpillars in soyabean plantations, and Laginex is the antimosquito formulation of the fungus *Lagenidium giganteum* being manufactured and marketed by AgraQuest USA (now acquired by Bayer CropScience, USA).

Apart from fungi, bacteria also are found to possess insecticidal properties which could be used for the development of strategies to introduce insect resistance to plants. *Bacillus thuringiensis* (abbreviated as Bt) is the best known and most widely used insecticides formulated from *Bacillus thuringiensis* var. *kurstaki* isolates that are pathogenic and toxic only to larvae of the butterflies and moths. Most of the *Bt* products that have been registered with the United States Environmental Protection Agency are Dipel®, Javelin®, Thuricide®, Worm Attack®, Caterpillar Killer®, Bactospeine® and SOK-Bt®. These products are used to control many common leaf-feeding caterpillars, including caterpillar pests on vegetables, bagworms and tent caterpillars on trees and shrubs, larvae of the gypsy moth and other forest caterpillars. One product with a very specific target is Certan®, formulated from *Bacillus thuringiensis* var. *aizawai* and used exclusively for the control of wax moth larvae in honeybee hives. Vectobac®, Teknar®, Bactimos®, Skeetal® and Mosquito Attack® are products utilising *Bacillus thuringiensis* var. *israelensis* (*Bti*), a subspecies that kills the larvae of certain Diptera (the insect order containing the flies and mosquitoes). The main targets for *Bti* are the larval stages of mosquitoes, black flies and fungus gnats; it does not control larval stages of 'higher' flies such as the housefly, stable fly or blowflies. Mosquitoes that are most susceptible to *Bti* include species in the genera *Aedes* and *Psorophora. Anopheles* and *Culex* species are controlled only when higher than normal rates of *Bti* are applied. Another bacterium which is especially active against the larval stages of mosquitoes belonging to the genera Culex, Psorophora and Culiseta is *Bacillus sphaericus*. Further *Bacillus thuringiensis* var. *san diego* has been formulated and registered under the trade name M-One® to control the larvae of Colorado potato beetle. *Bacillus lentimorbus* and *Bacillus papillae* are commercialised under the trade name Doom®, Japidemic® and GrubAttack®. The major hindrance in their commercial production is that they cannot be grown in fermentation tank like other microorganisms but have to be cultivated on laboratory reared insect larvae. *Serratia entomophila* (Enterobacteriaceae) has been applied as a biocontrol of the grass grub and is registered under the trade name Invade®. GRANDEVO® is a next-generation bioinsecticide for controlling broad spectrum of chewing and sucking insects and mites developed by Marrone BioInnovations, USA. This formulation is made using a new bacterium *Chromobacterium subtsugae*.

4.3.3 Biofungicides

Biofungicides comprise that group of bacteria and specialised fungi which are exploited to control diseases caused by plant pathogenic fungi by innundative approach. These generally inhabit soil. A variety of biofungicides have been developed for preharvest and postharvest diseases caused by fungi in crops. Some of the prominent biofungicides SoilGard™, AQ10™, Campanion™ and Mycostop™ have been developed and commercialised as alternative to chemical fungicides (Table 4.5). The four basic mechanisms by which biofungicide controls the pathogenic microorganisms are direct competition, antibiosis, predation or parasitism or inducing resistance to the host plants. The effects of synthetic agrochemicals on biological control agents are extremely important for them to be successfully incorporated into agricultural practices. It is also possible that the efficacy of microbial control agents could be enhanced when used in combination with chemicals (Smith 1991).

4.4 Precincts of Biological Control

Despite development and commercial deployment of a variety of biopesticides, there exist several intrinsic limitations which restrict many potential candidates for commercial development. In case of classical biological control, the cost of importing a pathogen apart from potential risk to other crop plants which require stringent safety

Table 4.5 Microorganisms used as registered biofungicides

Biological control agent	Commercial name	Target pest	Company/country where used
Pseudomonas chlororaphis strain 63-28	ATEze	Suppression of *Rhizoctonia solani* and *Pythium* spp.	Agrium US Inc., www.agrium.com
Bacillus subtilis QST 713	CEASE	Bacterial diseases, powdery mildew, *Botrytis*, anthracnose, *Alternaria* and *Entomosporium*	Bioworks, www.bioworksinc.com
Bacillus subtilis (strain GB03)	Companion	*Rhizoctonia*, *Pythium*, *Fusarium* and *Phytophthora*	Growth Products, www.growthproducts.com
Bacillus subtilis	Kodiak, Kodiak HB, Kodiak AT	*Rhizoctonia solani*, *Fusarium* spp., *Alternaria* spp. and *Aspergillus* spp.	Gustafson Inc., www.bayercropscience.com/gustafson
Streptomyces griseoviridis strain K61	Mycostop	*Fusarium* spp., *Alternaria brassicola*, *Phomopsis* spp., *Botrytis* spp., *Pythium* spp. and *Phytophthora* spp.	AgBio Development
Trichoderma harzianum Rifai strain KRL-AG2 (T-22)	PlantShield	*Pythium* spp., *Rhizoctonia solani*, *Cylindrocladium*, *Thielaviopsis*, *Fusarium* spp. and *Botrytis*	Bioworks, www.bioworksinc.com
Gliocladium catenulatum	PreStop, Primastop	*Pythium* spp., *Rhizoctonia solani*, *Botrytis* spp., *Didymella* spp.	AgBio Development. www.agbio-inc.com
Bacillus subtilis QST 713	Rhapsody	Bacterial diseases, powdery mildew, *Botrytis*, anthracnose, *Alternaria* and *Entomosporium*	AgraQuest, www.agraquest.com
Trichoderma harzianum Rifai strain KRL-AG2 (T-22)	RootShield	*Pythium*, *Rhizoctonia* and *Fusarium*	Bioworks, www.bioworksinc.com
Trichoderma virens	Soilgard 12G	*Pythium* and *Rhizoctonia*	Certis USA, www.certisusa.com
Bacillus subtilis 44 var. *amyloliquefaciens strain FZB24*	Taegro	Damping off and root rot pathogens, *Rhizoctonia solani* and *Fusarium* spp.	Taensa Inc., www.taensa.com
Streptomyces lydicus WYEC108	ACTINOVATE	*Pythium ultimum*, *Fusarium oxysporum f.* sp. *lycopersici*, *Alternaria porri* and *Botrytis aclada.*	Futureco Bioscience, Barcelona, Spain www.futureecosbioscience.com
Trichoderma lignorum strain TL-0601	Mycotric	Rhizoctonia, Verticillium, Sclerotinia, Sclerotium, Phoma, Fusarium	Futureco Bioscience, Barcelona, Spain
Bacillus subtilis	Serenade	Powdery mildew, other foliar diseases	AgraQuest
Bacillus pumilus	Sonata	Downy and powdery mildew, rust	AgraQuest
Ampelomyces quisqualis isolate Q-10	AQ10	Powdery mildew	Belchim Crop Protection Limited, UK

evaluation prior to release by the quarantine agencies is a major concern. It demands setting up of a sophisticated quarantine laboratory which further makes the approach more cost intensive. In augmentative or innundative approach, the limited host specificity reduces its commercial viability apart from loss of virulence, short shelf life and suitable formulation and application technology. Further application rates of BCAs vary depending upon the environmental conditions. Commercial interest in biological control products for pest control and management is slowly increasing with global players like BayerCropScience, acquiring small companies producing specific high-end application products in their portfolio with increasing societal concern on the use of synthetic agrochemicals and more reliance on organically grown food. Thus, environmental window for the success of this approach is slowly broadening for sustainable agriculture and forestry.

4.5 Biorational Pesticides of Microbial Origin

Biorational pesticides is a term which essentially refers to chemicals of natural origin which function as pesticides but have no limited or no adverse effects on the environment, nontarget organisms including humans. The origin of biorational pesticides is due to the phenomenon of allelopathy which essentially refers to biochemical interactions occurring among plants those mediated by microorganisms. The term allelopathy also comprises of fungistasis, antibiosis between microorganisms, development of disease symptoms, promotion of infection and host resistance to pathogens. Thus, allelochemicals could be defined as organic compounds produced by microbes or plants that stimulate or inhibit the neighbouring plant or microorganism. Thus, natural products offer a variety of structural diversity for designing agrochemicals, which would be less persistent in the environment, more selective and environmentally safe.

Microbial secondary metabolites possess potential to provide agricultural researchers with novel structures, which could be directly used, or as leads in developing biorational and ecofriendly structures. They have been referred to as toxins, broadly differentiated as phytotoxin, zootoxins and antibiotics. Phytotoxins are toxic to plants; zootoxins have toxicity towards animals and insects, while the antibiotics are used against plant pathogens.

4.5.1 Bacterial Secondary Metabolites as Agrochemicals

A variety of bacterial secondary metabolites have been isolated and tested for their potential to be used as agrochemicals. The most effective bacterial toxins which have been exploited to date are the endotoxins produced by the *Bacillus thuringiensis* and related species which has already been exploited as a bioinsecticide. These toxins are collectively referred as Bt-toxins and are proteins which are poisonous to insects which creates ulcers in the stomach's lining as a result of which the insect stops eating and eventually dies. These crystal proteins have specific activities against Lepidoptera (moths and butterflies), Diptera (flies and mosquitoes), Coleoptera (beetles), Hymenoptera (wasps, bees, ants and sawflies) and nematodes. Diabroticin A is a polar insecticide produced by *Bacillus cereus* and *Bacillus subtilis* which is active against southern corn rootworm *Diabrotica undecimpunctata* through diet. Recently a symbiotic bacterium *Photorhabdus asymbiotica residing in nematodes* of the genus *Heterorhabditis* produce many toxins and other potential virulence factors. *Photorhabdus* insect-related protein (Pir A/B) from *Photorhabdus asymbiotica* shows larvicidal activity against both *Aedes aegypti* and *Aedes albopictus* larvae which cause dengue fever.

4.5.2 Agroactive Compounds from Actinomycetes

Actinomycetes are prolific producers of bioactive secondary metabolites, majority of which are being exploited as pharmaceutically active agents as well as leads for the development of biorational pesticides. Over 500 interesting and intriguing chemicals have already been reported from them and have been exploited for development of antibacterial, antifungal and anticancer agents.

Agricultural application of secondary metabolites of Actinomycetes came into limelight with the discovery of Bialaphos from *Streptomyces hygroscopicus* and *S. viridochromogenes* in the late 1970s. Bialaphos is converted into phosphinothricin inside the plant due to cleavage of two alanyl groups and irreversibly binds to enzyme glutamine synthetase (GS) which plays a key role in ammonia assimilation leading to death. Another secondary metabolite produced from *Streptomyces avermitilis* was avermectin. It is used for cockroaches, mites and leafminers and sold under the commercial name Avid/Agrimek by Syngenta (previously Novartis). Another second-generation avermectin class of compounds launched by Syngenta is Emamectin. Emamectin is a 4-deoxy-4-methyl amino derivative of abamectin, a macrocyclic lactone fermentatively produced by *Streptomyces avermitilis*. Emamectin is used for controlling lepidopterous pests. The product is significantly popular among farmers due to low application rate of the active ingredient, approx. 6 g/acre and broad spectrum applicability as an insecticide. Other compounds from *Streptomyces* species have also been exploited as herbicides and fungicides. Methoxyphenone (3, 3-dimethyl-4-methoxybenzophenone) was the first biorational herbicide developed on the template of Anisomycin which was isolated from *Streptomyces* species (Fig. 4.1).

Anisomycin and Methoxyphenone exhibit excellent activity against barnyard grass and crabgrass. Other promising phytotoxins produced by *Streptomyces* species are herboxidiene, isoxazole-4-carboxylic acid, nigericin and vulgamycin. Blasticidin-S has been isolated from *S. griseochromogenes* and is used as biofungicide for the control of rice blast caused by *Pyricularia oryzae* (Misato et al. 1959). It is used as a foliar application as benzylamino-benzenesulfonic acid salt. Kasugamycin is another secondary metabolite produced by *S. kusgaensis* hydrochloride hydrate by Hakko Chemical Industry Co. ltd (Umezawa et al. 1965). It is used to control rice blast and leaf spot in sugar beet and celery.

Spinosad is produced by the actinomycetes *Saccharopolyspora spinosa* introduced by DowAgrosciences (Krist et al. 1992). It is used for the control of a wide range of caterpillars, leafminers and thrips. The trade name of this biorational fungicide is Conserve®, Entrust® and SpinTor®. *S. hygroscopicus* also produces milbemycin which is insecticidal as well as acaricidal (Takiguchi et al. 1980). The pesticidal activities of actinomycetes group in laboratory and green house evaluations hold a promising future for using them as biorational pesticides in integrated pest management.

4.5.3 Fungal Secondary Metabolites as Agrochemical

Fungi by virtue of their secondary metabolites are capable of inducing disease symptoms in their respective hosts and are frequently referred as phytotoxins or zootoxins.

4.5.3.1 Phytotoxins as Mycoherbicide

Over the last decade, phytotoxins have also been studied for their possible role in biorational herbicide development by a few research groups. Phytotoxins have been broadly classified as host-specific and non-host-specific toxins. Phytotoxins offer unique chemistries and mode of action which have not been exploited commercially for herbicide development. Maculosin, a cyclic dipeptide, is a host-specific phytotoxin produced by *Alternaria alternata* on spotted knapweed (*Centaurea maculosa*) (Stierle et al. 1989); AAL-toxin from *Alternaria alternata* f. sp. *lycopersici* and AM toxin produced by *Alternaria alternata* f. sp. *Mali* are other host-specific phytotoxins. Bipolaroxin from *Bipolaris cynodontis* (Marignoni) shoemaker, a fungal pathogen of bermuda grass (*Cynodon dactylon*), has been found to be host selective in low concentrations (Sugawara et al. 1985). Concentrations 20 times higher than required to affect Bermuda grass cause phytotoxicity to wild oats, sugarcane and maize. Host-specific phytotoxins like maculosin

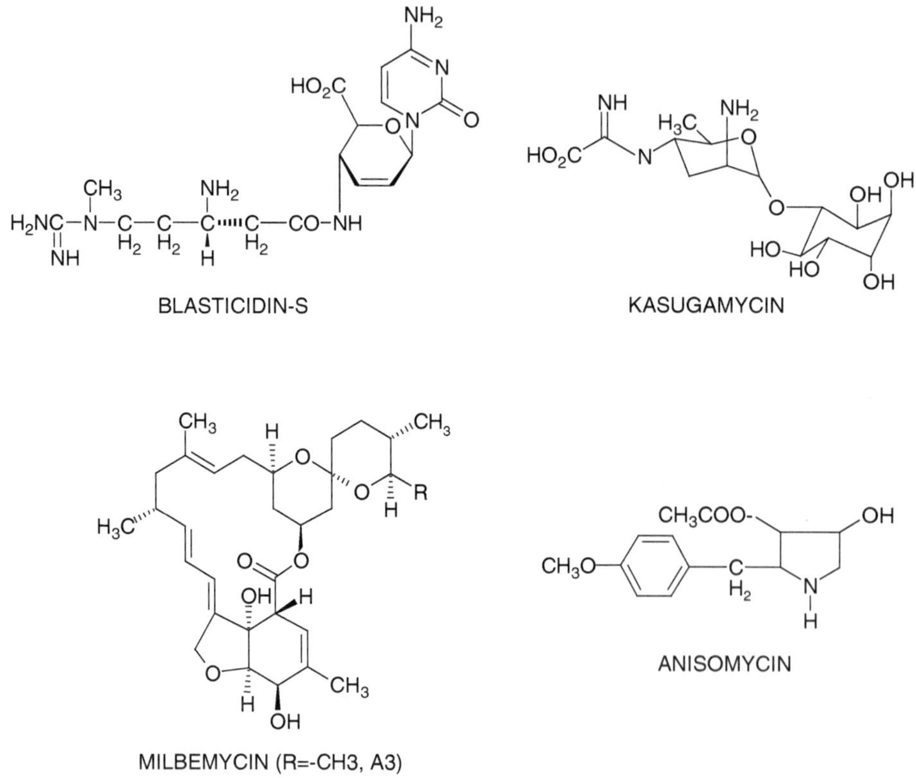

Fig 4.1 Biopesticidal compounds produced by actinomycetes

and phomalairdenone are prohibitively expensive to be developed and commercialised for a single plant species. Hence non-host-specific phytotoxins which have a broad spectrum of activity against more than two weeds are generally preferred.

Cornexistin from *Paecilomyces variotii* and Irpexil from *Irpex pachydon* have been discovered and patented for their use as biorational herbicides. Cornexistin is active against both monocots and dicots with selective protection to corn (Nakajima et al. 1991). *Helminthosporium* sp., a pathogen on Johnsongrass, produces prehelminthosporal and dihydroprehelminthosporal which have been tested against Sorghum (*Sorghum bicolor*) and Johnsongrass (*Sorghum halpense*). During in vitro assays, it was observed that prehelminthosporal was more active than dihydroprehelminthosporal. Prehelminthosporal doses up to 450 mg/kg in 1 day chicks did not produce any visible affects and therefore is a candidate for agricultural development. Colletotrichin is also a potential phytotoxin which is produced by several *Colletotrichum* species and causes damage of plasma membrane of the host plant followed by massive cellular leakage. Similarly zinniol is also a phytotoxic product of several *Alternaria* species and of *Phoma macdonaldii* which cause necrosis and death of tissues by modulating calcium levels in the plants. Tentoxin is a cyclic tetrapeptide which is elaborated by *Alternaria alternata* and is responsible for causing phytotoxic damage to monocots as well as dicots. It inhibits the CF1 ATPase activity (Fig. 4.2).

Phytotoxin 1233A is produced by *Scopulariopsis candidus*, *Cephalosporium* sp. and *Fusarium* species which act on HMG-CoA (3-hydroxy-3methylglutaryl coenzyme A) synthetase, thereby acting as a broad spectrum herbicide. Putaminoxin and pinolidoxin are two phytotoxic nonenolides produced by phytopathogenic *Phoma* and *Ascochyta* species exhibiting

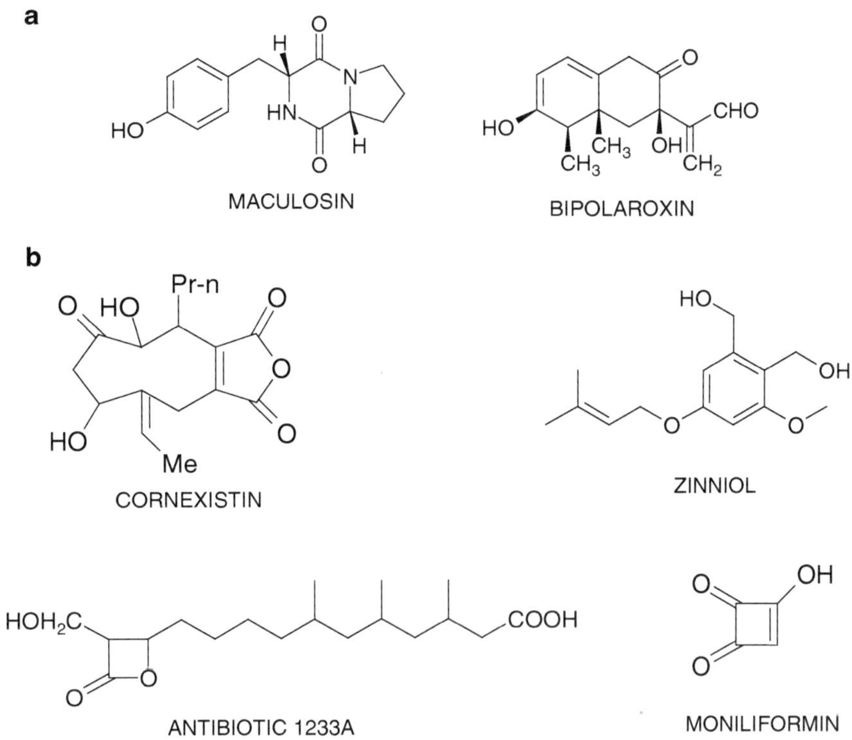

Fig. 4.2 Fungal phytotoxins as herbicides (**a**) host-specific phytotoxins (**b**) non-host-specific phytotoxins

potent herbicidal activity (Evidente et al. 1998). Ascochytine, pyrenolide A and hyalopyrone are phytotoxins produced by *Ascochyta hyalospora*, the causal agent of leaf spot on Lambsquarters (Venkatasubbaiah and Chilton 1992). These phytotoxins exhibited broad spectrum phytotoxic activity against *Sida spinosa*, *Ipomea* sp., *Sorghum halpense* and *Chenopodium album*.

Thus, there exists a huge potential to exploit fungal phytotoxins in designing biorational herbicides which have least environmental burden and are safe for the end user.

4.5.3.2 Toxins of Entomopathogenic Fungi

A very limited spectrum of entomopathogenic fungi have been studied for their use as bioinsecticides. These also produce an array of secondary metabolites which could be explored for their potential as biorational pesticides. The most common entomopathogenic biocontrol agents are Metarhizium sp., *Beauveria* sp., *Lecanicillium* sp., *Paecilomyces fumosoroseus and Tolypocladium* spp. The first entomopathogenic metabolites were produced by M. *anisopliae* known as destruxin A and B. Subsequently varieties of destruxin isomers were isolated and studied. At present, 28 structurally different destruxins have been explored. Destruxins A1, A4 and A5 and homodestruxin B production have been reported from entomopathogenic fungus *Aschersonia* spp. (Krasnoff and Gibson 1996).

Oosporin is a dibenzoquinone red-coloured toxin produced by *Beauveria bassiana* (Eyal et al. 1994). It has also been reported from three commercial strains of *Beauveria brongniartii* on submerged cultures and on sterilised barley kernels. *Tolypocladium* species produce efrapeptins which exhibit miticidal and insecticidal activities against arthropod pests like spider mites, potato beetles and diamondback moth. Beauvericin is a hexadepsipeptide produced by entomopathogenic fungi *Beauveria bassiana*, *Paecilomyces* spp. It exists in two forms – beauvericin A and B. Beauvericin has been found highly toxic to murine as well as human cell lines apart from insecticidal activity.

However, the entomopathogenic secondary metabolites produced from fungi have not been successful in the development of a commercial biorational agrochemical till date.

4.5.3.3 Fungal Secondary Metabolites as Biofungicides

The only compounds of fungal origin which has been exploited in the development of commercial fungicides are strobilurins. Strobilurins are a chemical class produced by the fungi *Oudemansiella mucida* and *Strobilurus tenacellus* strain no. 21602. The antifungal potential of strobilurins was recognised in early 1980s. Strobilurins are photochemically unstable, rendering them unfit for their commercial use. Structural modifications of strobilurins resulted in the development of three analogues, viz., azoxystrobin, kresoxim methyl and metominostrobin (Fig. 4.3).

These compounds possess remarkable broad spectrum activity against many foliar pathogens from the ascomycetes, basidiomycetes and oomycetes in cereals, rice, grapevine, vegetables and turf grass. BASF has developed commercial biofungicides from strobilurin analogues like Cabrio® Plus which comprises of pyraclostrobin as the first ingredient which inhibits respiration in the mitochondria and Cabrio EG® which is also a strobilurin which prevents energy generation within the mitochondrion.

Microbial secondary metabolites thus offer the best readily accessible and rewarding source of novel chemical class or chemical structures which could be directly or indirectly used as alternative to conventional agrochemicals which are still being used. The microorganisms need not exhibit a similar metabolite profile if they are being used as biopesticide on the same host. There could be marked differences in quantity and quality of the types of toxins produced which is attributed to variations at inter- and intraspecific levels.

4.6 Summary

Microbes have tremendous applications in agricultural field for improving yield, plant characteristics, resistance to pests, drought and in fighting diseases. They have the potential to serve as biofertilisers in crops for nitrogen fixation and phosphate solubilisation, thereby enhancing their bioavailability to plants and lowering the need of chemical fertilisers apart from being ecofriendly.

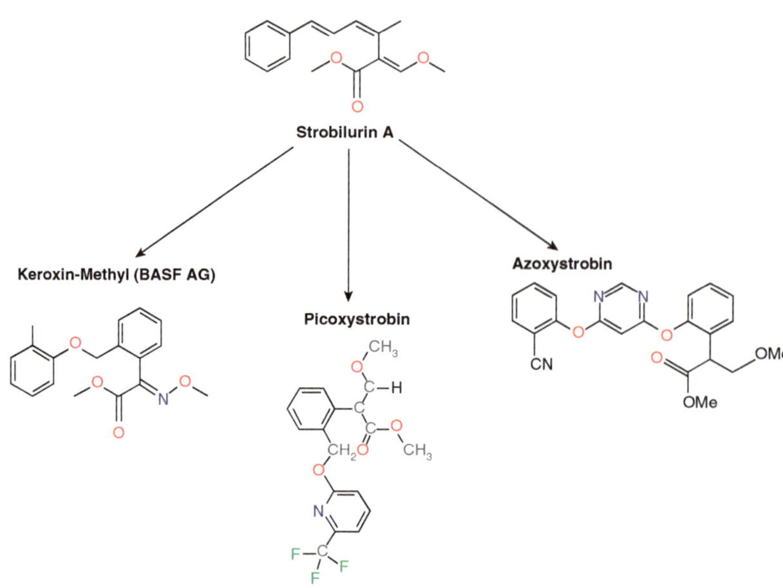

Fig. 4.3 Strobilurin A and its derivatives as biofungicides

Some microbes or their secondary metabolites have been used as biopesticides for the control of pests. They have been successfully used as herbicides, as well as insecticides. Thus, the role of effective and beneficial microorganisms is tremendous in boosting the agricultural as well as forestry practices. The search of these microbes is never ending to improve food production, safety and preservation in order to meet the ever increasing global food demand.

Selected Reading

Arima K, Imanaka I, Kousaka M, Fukuta A, Tamura G (1964) Pyrrolnitrin, a new antibiotic substance produced by *Pseudomonas*. Agric Biol Chem 28:575–576

Bacon CW, White JF (2000) Microbial endophytes. Marcel Dekker, New York

Baldani VLD, Baldani JI, Dobereiner J (2000) Inoculation of rice plants with the endophytic diazotrophs *Herbaspirillum seropedicae*. Biol Fertil Soils 30:485–491

Bashan Y, Holguin G, de-Bashan LE (2004) *Azospirillum-plant* relationships: physiological, molecular, agricultural, and environmental advances (1997–2003). Can J Microbiol 50:521–577

Boholool B (1990) Introduction to nitrogen fixation in agriculture and industry: contribution of BNF to sustainability of agriculture. In: Chalk PM, Gresshoff MM, Roth LE, Stacey G, Newton WE (eds) Nitrogen fixation: achievements and objectives. Chapman and Hall, New York, pp 613–616

Boyetchko SM (1999) Innovative applications of microbial agents for biological weed control. In: Mukerji KG, Chamola BP, Upadhyaya RK (eds) Biotechnological approaches in biocontrol and plant pathogens. Plenum, London, pp 73–97

Burton EM, Knight SD (2005) Survival of *Penicillium bilaiae* inoculated on canola seed treated with Vitavax RS and extender. Biol Fertil Soils 42:54–59

Chaintreuil C, Giraud E, Prin Y, Lorquin J, Ba A, Gillis M, deLajudie P, Dreyfus B (2000) Photosynthetic brady*rhizobia* are natural endophytes of the African wild rice *Oryza breviligulata*. Appl Environ Microbiol 66:5437–5447

Chen YP, Rekha PD, Arun AB, Shen FT, Lai WA, Young CC (2006) Phosphate solubilizing bacteria from subtropical soil and their tricalcium phosphate solubilizing abilities. Appl Soil Ecol 34:33–41

Duponnois R, Kisa M, Plenchette C (2006) Phosphate solubilizing potential of the nematofungus *Arthrobotrys oligospora*. J Plant Nutr Soil Sci 169:280–282

Evidente R, Capasso A, Andolfi VM, Zonno MC (1998) Structure- activity relationship studies of putaminoxins and pinolidoxins: phytotoxic nonelides produced by phytopathogenic *Phoma* and *Ascochyta* species. Nat Toxins 6:183–188

Eyal J, Mabud A, Fischbein KL, Walter JF, Osborne LS, Landa Z (1994) Assessment of *Beauveria bassiana* Nov., EO-1strain, which produces a red pigment for microbial control. Appl Biochem Microbiol 44:65–80

Friedlander M, Inbar J, Chet I (1993) Biological control of soil borne plant pathogens by a β-1,3 glucanase-producing *Pseudomonas cepacia*. Soil Biol Biochem 25:1121–1221

Gaur AC (1990) Phosphate solubilising microorganisms as biofertilizer. Oxford Publishing Co, New Delhi, pp 26–29

Hasan S (1988) Biocontrol of weeds with microbes. In: Mukerji KG, Garg KL (eds) Biocontrol of plant diseases, vol 1. CRC Press, Boca Raton, pp 129–151

Hashem MA (2001) Problems and prospects of cyanobacterial biofertilizer for rice cultivation. Aust J Plant Physiol 28:881–888

He ZL, Bian W, Zhu J (2002) Screening and identification of microorganisms capable of utilizing phosphate adsorbed by goethite. Commun Soil Sci Plant Anal 33:647–663

James EK, Gyaneshwar P, Barraquio WL, Manthan N, Ladha JK (2000) Endophytic diazotrophs associated with rice. In: Ladha JK, Reddy PM (eds) The quest for nitrogen fixation in rice. International Rice Research Institute, Los Banos, pp 119–140

Janisiewicz WJ, Roitmann J (1998) Biological control of blue mold and grey mold on apple and pear with *Pseudomonas cepacia*. Phytopathology 78:1697–1700

Kenney DS (1986) DeVine: the way it was developed – an industrialist view. Weed Sci 34(Suppl 1):15–16

Krist HA, Michel KH, Mynderse JS, Chio EH, Yao RC, Nakasukasa WM et al (1992) In: Baker DR, Fenyes JG, Steffens JJ (eds) Discovery, isolation and structure elucidation of a family of structurally unique, fermentation derived tetracyclic macrolides in synthesis and chemistry of agrochemicals III, ch.20. American Chemical Society, Washington, DC, pp 214–225

Kloepper JW, Lifshitz R, Zablotowicz RM (1989) Free-living bacterial inocula for enhancing crop productivity. Trends Biotechnol 7:39–43

Krasnoff B, Gibson DM (1996) New destruxins from entomopathogenic fungus *Aschersonia* sp. J Nat Prod 59:485–489

Lebuhn M, Heulin T, Hartmann A (1997) Production of auxin and other indolic and phenolic compounds by *Paenibacillus polymyxa* strains isolated from different proximity to plant roots. FEMS Microbiol Ecol 22:325–334

Lim HS, Kim YS, Kim SD (1991) *Pseudomonas stutzeri* YPL-1 genetic transformation and antifungal mechanism against *Fusarium solani*, an agent of plant root rot. Appl Environ Microbiol 57:510–516

MacMillan J (2002) Occurrence of gibberellins in vascular plants, fungi, and bacteria. J Plant Growth Regul 20:387–442

Matiru VN, Dakora FD (2004) Potential use of rhizobial bacteria as promoters of plant growth for increased yield in landraces of African cereal crops. Afr J Biotechnol 3(1):1–7

Mejri D, Gamalero E, Souissi T (2013) Formulation development of deleterious rhizobacterium Pseudomonas trivalis X33d for biocontrol of brome (Bromus diandrus) in durum wheat. J Appl Microbiol 114(1):219–228

Misato T, Ishii I, Asakawa M, Okimoto Y, Fukunaga K (1959) Antibiotics as protectant fungicides against rice blast. II. The therapeutic action of blasticidin. S Ann Phytopathol Soc Jpn 24:302–303

Mishustin EN, Emtsev VT, Lockmacheva RA (1983) Anaerobic nitrogen-fixing microorganisms of the Clostridium genus and their activity in soil. Biol Bull 9:548–558

Morris MJ (1997) Impact of gall forming rust fungus *Uromycladium tepperianum* on the invasive tree *Acacia saligna* in South Africa. Biol Control 10:75–82

Muthukumarasamy R, Revathi G, Lakshminarasimhan C (1999) Diazotrophic associations in sugarcane cultivation in South India. Trop Agric 76:171–178

Nakajima M, Itoi K, Takamatsu Y, Sato S, Furukawa Y, Honwa T, Furuya K, Kadotini J, Kosaza M, Haneishi T (1991) Cornexisitin: a new fungal metabolite with herbicidal activity. J Antibiot 44:1065–1072

Patten C, Glick BR (1996) Bacterial biosynthesis of indole-3-acetic acid. Can J Microbiol 42:207–220

Peoples MB, Herridge DF, Ladha JK (1995) Biological dinitrogen fixation: an efficient source of nitrogen for sustainable agriculture production. Plant Soil 174:3–28

Prasad RC, Prasad BN (2001) Cyanobacteria as a source Biofertilizer for sustainable agriculture in Nepal. J Plant Sci Bot Orientalis 1:127–133

Sharma PK, Kundu BS, Dogra RC (1993) Molecular mechanism of host specificity in legume-*Rhizobium* symbiosis. Biotechnol Adv 11:714–779

Smith RJ Jr (1991) Integration of biological control agents with chemical pesticides. In: TeBeest DO (ed) Microbial control of weeds. Chapman & Hall, New York, pp 189–208

Soliman S, Seeda MA, Aly SSM, Gadalla AM (1995) Nitrogen fixation by wheat plants as affected by nitrogen fertilizer levels and non-symbiotic bacteria. Egypt J Soil Sci 35:401–413

Soytong K, Kanokmedhakul S, Kukongviriyapa V, Isobe M (2001) Application of *Chaetomium* species (Ketomium®) as a new broad spectrum biological fungicide for plant disease control: a review article. Fungal Divers 7:1–15

Stierle A, Cardellina H, Strobel G (1989) Maculosin, a host-specific phytotoxin from *Alternaria alternata* on spotted knapweed. Am Chem Soc Symp Ser 439:53

Sugawara F, Strobel G, Fischer LE, van Dyke GD, Clardy J (1985) Bipolaroxin: a new selective phytotoxin produced by *Bipolaris cynodontis*. Proc Natl Acad Sci U S A 77:5904

Suopkoff DM, Joley DB, Marios JJ (1988) Effect of introduced biological control organisms on the density of *Chondrilla juncea* in California. J Appl Ecol 25:1089–1095

Takeda R (1958) Pseudomonas pigments I. Pyoluteorin, a new chlorine containing pigment by *Pseudomonas aeruginosa*. Hakko Kogaku Zasshi 36:281–290

Takiguchi Y, Mishima H, Okuda M, Terao M, Aoki A, Fukuda R (1980) Milbemycins, a new family of macrolide antibiotics: fermentation, isolation and physicochemical properties. J Antibiot 33:1120–1127

Umezawa H, Okami Y, Hashimoto T, Suhara Y, Otake N (1965) A new antibiotic kasugamycin. J Antibiot Ser A 18:101–103

Venkatasubbaiah P, Chilton WS (1992) Phytotoxins of *Ascochyta holospora*, causal agent of lambsquarters leaf spot. J Nat Prod 55:461–467

Yanni YG, Rizk RY, Corich V, Squartini A, Ninke K, Philip- Hollingsworth S, Orgambide G, de Bruijn F, Stoltzfus J, Buckley D, Schmidt TM, Mateos PF, Ladha JK, Dazzo FB (1997) Natural endophytic association between *Rhizobium leguminosarum* bv. *trifolii* and rice roots and assessment of its potential to promote rice growth. Plant Soil 194:99–114

Zahran HH (1999) *Rhizobium*–legume symbiosis and nitrogen fixation under severe conditions and in an arid climate. Microbiol Mol Biol Rev 63:968–989

Environment and Microbes

5

Microbial Bioremediation, Biomining, and Microbially Enhanced Oil Recovery (MEOR)

5.1 Introduction

Fundamentally, microbes play a vital role in global recycling of matter in order to obtain energy and grow, thereby releasing carbon, nitrogen, phosphorus and sulphur. The microbial buffet comprises diverse form of materials like dead plants and animals, wastes from human and animals and food leftovers. These interactions of microbes help in nutrient recycling and are referred as biogeochemical cycles. Microbes have been found to be surviving in extreme environmental conditions which at times are extremely difficult for survival of other life forms including human beings. The metabolic versatility of these microorganisms not only helps in removing harmful wastes from the environment but also propels recovery of important products during geological explorations. In this chapter we explore the role of microbes in degradation of xenobiotics and petrochemicals, in wastewater treatment, in recovery of minerals and crude oil and finally bio-concentration of toxic heavy metals from the environment. The chapter also dwells upon harnessing the microbial potential in recovery of petrochemical resources and important elements from the earth's crust.

5.2 Microbial Bioremediation

Microbial bioremediation refers to the destruction or removal or reduction of the concentration of hazardous elements below toxic levels with the help of microorganisms. Microbial bioremediation may involve a single microorganism or a group of microorganisms (referred as a consortium) for carrying out the detoxification or remediation process. The technologies involved in microbial bioremediation could be broadly classified as *in situ* (on site) and *ex situ* (away from the site, i.e. physical removal of the contaminated substrate and subsequent treatment for decontamination).

5.2.1 *In Situ* Bioremediation by Microbes

It is generally assumed that the contaminated sites are generally dominated by the microorganisms which have adapted to the organic chemical waste and can degrade them partially or completely. During *in situ* bioremediation a significant challenge is the supply of nutrients or chemicals required by the microbes and their appropriate mixing with contaminants which are to be degraded. This can be achieved through two processes – biostimulation and bioaugmentation.

Biostimulation refers to the stimulating activity of the microorganisms by providing nutrients like phosphorus, nitrogen, oxygen and other electron acceptors. Bioaugmentation refers to the incorporation of microbes in the subsurface environment to overcome the deficiency of decontaminating microorganisms to degrade specific decontaminants. During this process the microorganism sceded may be a natural inhabitant of the

S. Saxena, *Applied Microbiology*,
DOI 10.1007/978-81-322-2259-0_5, © Springer India 2015

site which has been multiplied in a bioreactor or specifically cultivated strains which have known degradative capabilities for a specific contaminant.

Oil spills are potentially the most destructive pollution source which threatens the beaches and the coastline. Oil spills have increased due to seaborne oil trade which has increased steadily since the 1970s. Some notable examples are the Arrow oil spill which spread across the 305-km coastline of Chedabucto Bay in Nova Scotia, Canada, in 1970 and the Amoco Cadiz oil spill in Brittany, France, which involved 320 km of the coastline in 1978.

The Exxon Valdez oil spill in March 1989 spilled some 38 tonnes of Alaskan North Slope crude oil, polluting approximately 2,000 km of rocky intertidal shorelines within Prince William Sound (PWS) in Alaska, USA. Two decades after the Exxon Valdez oil spill, patches of subsurface oil persist in the most polluted beaches along PWS. Approximately 250,000 seabirds, 3,000 otters, 300 harbour seals, 250 bald eagles, 22 killer whales and billions of salmon and herring eggs were destroyed due to the oil spill.

It has been convincingly demonstrated that fertilisers can be applied to the oil beaches to overcome nutrient limitations, thereby enhancing the growth and biodegradation of oil by stimulating the growth of microflora. Scientists at the US Environment and Protection Agency have demonstrated that oil degradation by indigenous microflora on beaches of PWS was accelerated by adding fertilisers to the surface of oil-contaminated beaches (Pritchard et al. 1992). Nitrogen was found to be the main factor limiting biodegradation of oil in beaches of PWS (Swanell et al. 1996).

Exxon spread 50,000 lb of fertilisers over 74 miles of the beach to supplement naturally available phosphorus and nitrogen. Exxon used two fertilisers for the cleanup process – Customblen (solid fertiliser, slow release type) and Inipol. Inipol was composed of oleic acid and urea. As oleic acid could stick to the oil surface, nitrogen was delivered close to it; therefore it was supplied on the surface, while Customblen was supplied subsurface to remediate the soil.

In situ bioremediation of groundwater for hydrocarbons has been used for four decades now. There are recent reports highlighting the role of bioaugmentation for carrying out dehalorespiration. *In situ* bioaugmentation has several advantages, viz. reduction in adaptation time, efficiency of taking care of recalcitrant contaminations which are very limited to the number of naturally occurring organisms capable of transforming or degrading, environmental constraints do not allow a critical biomass of microbes to get established and maintain themselves to remediate the contaminant.

In situ bioaugmentation has been successfully demonstrated for complete reductive dechlorination of tetrachloroethane (PCE) and trichloroethane (TCE) to ethane using a microbial consortium KB-1 under anaerobic conditions in a chlorinated ethane aquifer near Kelly Airforce Base, Texas, USA. The known dehalorespiring bacteria are *Dehalospirillum multivorans*, *Dehalococcoides ethenogenes* and *Dehalobacter restrictus. Dehalococcoides ethenogenes* is the only organism known to completely and rapidly dechlorinate chlorinated ethers to ethene by dehalorespiration.

The pilot scale experimentation conducted at a chlorinated-ethane-contaminated aquifer near Kelly Airforce Base, Texas, indicated that the indigenous microbes when biostimulated were capable of reductive chlorination of PCE and TCE to cis-1,2-dichloroethene. However, further dechlorination of cis-1,2-dichloroethene to ethene occurred only when natural dehalorespiring microbial consortium KB-1 was added. Within a few hours of inoculation, KB-1 effectively reduced chlorinated solvents as observed by their half-life. Molecular probe using 16S RNA confirmed the survival and establishment of dehalorespiring organism within the test area.

5.2.2 *Ex Situ* Bioremediation by Microbes

Oil contamination of soil leads to loss in its fertility for over 20 years. Effluents from oil industry, oil sludge and oil spills are a major threat to the environment since the constituents are toxic, mutagenic and carcinogenic. The Energy and Resources Institute (TERI) has developed an indigenous bacterial consortium named 'Oilzapper' isolated from the various contaminated sites of India which could degrade the total petroleum hydrocarbon (TPH) of the oily waste. Oilzapper is being used for remediation of oily sludge, oil-contaminated drill cuttings and oil-contaminated soils.

TERI with IOCL (Indian Oil Corporation Limited) has jointly developed Oilivorous-S and Oilivorous-A for application to the specific quality of oily sludge. Oilivorous-A is effective against oily sludge which is acidic in nature, while Oilivorous-S for sludge with high oil content. Oilzapper has been used for *ex situ* bioremediation of an accidental oil spill near Gujarat due to crude oil trunk line rupture transporting crude oil from the oil producing field to Gujarat Refinery Baroda, India. The immediate steps were stopping crude oil pumping and barricading the oil spill site to prevent the spread. Subsequently, ONGC (Oil and Natural Gas Commission) TERI Biotech Ltd excavated 14,694 m^3 of oil-contaminated soil and transported it to a secured bioremediation pit fitted with HDPE liners. Subsequently, 74.5 tonnes of Oilzapper (crude oil-containing bacterial consortium) was applied as powder for the degradation of TPH in the contaminated soil. Nutrient recipe was sprayed on the oil-soaked soil and tilling done at regular intervals.

At the beginning of the bioremediation, the oil content of the soaked soil was 14.5 % which was reduced to 7.31 % after 2 months. Further reduction was observed in oil content at the end of 3 months when the oil content reached 3.12 %. At the end of 3½ months the oil content of the soil was 1.7 % which was reduced to 0.58 % (5,800 ppm) after the end of 4 months. Similarly, the aromatic fraction of the soil was also degraded within 4 months. The fish toxicity test of the bioremediated soil was tested, and it was found that the fish survived, indicating the effective *ex situ* bioremediation. This study also revealed that bioremediation of oil spill, oil-soaked soil is eco-friendly as well as cost effective as compared to other remediation strategies.

5.3 Biodegradation of Xenobiotic Compounds

Xenobiotic compounds are chemically synthesised, unnatural compounds in occurrence which are foreign to biosphere. They are stable in environment under both aerobic and anaerobic conditions. These broadly belong to halogenated aliphatic compounds, aromatic hydrocarbons, phthalate esters and polycyclic aromatic hydrocarbons. There are xenobiotic compounds which are toxic to living organisms and persist in the environment. Microorganisms are also involved in degradation/remediation of xenobiotic compounds.

Nitro-aromatics are prominent organic compounds which are used in different industrial processes and as industrial feedstocks. However despite their immense value to the industry, properties like stability, persistence and toxicity render them hazardous when released in the environment and could be converted into potentially carcinogenic/mutagenic derivatives and hence need to be bioremediated. *Pseudomonas* sp., *Nocardia* sp. and *Arthrobacter* have evolved degradative pathways to catabolise the nitro-substituted aromatic rings (Table 5.1). In the agricultural field nitropesticides like methylparathion/parathion are used. A bacterial isolate *Arthrobacter protophormiae* has been identified which is capable of utilising nitroaromatic compounds o-nitrobenzoate (ONB), p-nitrophenol (PNP) and 4-nitrocatechol as sole source of carbon, nitrogen and energy. It exhibits promising potential in degrading PNP in soil microcosms and in small-scale field studies. *Phanerochaete chrysosporium* also degrades trinitrotoluene (TNT) into amino dinitrotoluenes via electron transport through a membrane-bound electron transport chain.

Aliphatic functions of hydrocarbons consist of straight chain, branched chain and cyclic chain carbon moieties. The microorganisms which degrade aliphatic hydrocarbons are *Acinetobacter*, *Pseudomonas*, *Burkholderia*, *Flavobacillus*, *Bacillus*, etc. Cyclic hydrocarbons are resistant to microbial degradation; however, *Pseudomonas citronellolis*, *Brevibacterium erythrogenes* and *Saccharomyces cerevisiae* have been found to

Table 5.1 Microorganisms degrading nitro-aromatics

Nitro-aromatic class	Microorganisms
Nitrobenzene	*Pseudomonas pseudoalcaligenes*
	Pseudomonas putida 2NP8
	Pseudomonas mendocina
	Pseudomonas pickettii
Nitrobenzoates (NBA)	*Arthrobacter protophormiae RKJ100*
	Pseudomonas fluorescens KU-7
	Comamonas acidovorans NBA-10
	Pseudomonas sp. *strain J51*
	Comamonas sp. *strain JS40*
Nitrotoluenes	*Pseudomonas* sp. *strain JS42*
	Burkholderia cepacia JS850
	Hydrogenophaga haleronii JS863
Nitrophenols	*Pseudomonas putida B2*
	Ralstonia eutropha JMP134
	Moraxella sp.
	Pseudomonas sp. *YTK17*
1,3,5-Trinitroperhydro-1,3,5-triazine (RDX)	*Rhodococcus* species *YTK32*
	Rhodococcus rhodochrous 11Y
Atrazine	*Rhodococcus* sp.
	Phanerochaete chrysosporium
	Klebsiella pneumonia
Pentaerythritol tetranitrate (PETN)	*Enterobacter cloacae PB2*
	Pseudomonas putida IIB

degrade cycloalkanes. Polycyclic aromatic hydrocarbons (PAH) have also been extensively studied for their biodegradation due to their persistence in the environment and potential deleterious effects on humans.

Commonly reported species which have exhibited potential to degrade PAH's are *Pseudomonas putida*, *Pseudomonas fluorescens*, *Pseudomonas vesicularis*, *Corynebacterium renale* and *Alcaligenes denitrificans*. Majority of them are usually gram negative belonging to the genus *Pseudomonas*. *Arthrobacter sulfureus* isolated from oil fields in Gujarat could utilise phenanthrene as a sole source of carbon and energy. *Pseudomonas putida* can be genetically engineered and manipulated for expressing genes encoding several PAH-degrading enzymes. Naphthalene is the simplest and most soluble PAH, and hence microorganisms degrading naphthalene can be easily isolated (Fig. 5.1). *Corynebacterium renale* uses naphthalene as the main source of carbon and energy. Anthracene has been completely degraded by *Pseudomonas*, *Sphingomonas*, *Nocardia* and *Beijerinckia* with dihydriol as initial oxygenated intermediate. Fluorene has been degraded via 3,4-dihydroxyfluorene and extradiol fission to 1-indanone as terminal metabolite by *Pseudomonas cepacia* strain F297.

Mechanisms of PAH degradation with more than three aromatic rings is still unclear. This is generally attributed to PAH like benz-α-anthracene, benz-α-pyrene and pyrene. Microorganisms belonging to genera *Mycobacterium* and *Alcaligenes* have been found to degrade Fluoranthene. Benz-α-anthracene degradation has been reported by *Beijerinckia* and *Mycobac-terium. Rhodococcus* sp. UW1 is capable of utilising pyrene and chrysene as a sole source of carbon and energy. PAH bioremediation has been found successful in cleanup of municipal and industrial waste waters.

5.4 Bioremediation of Heavy Metals

Discovery and exploitation of metals has played a significant role in the development of human civilisations. Metals not only contribute to our industrial heritage but have also been exploited in different fields like medicine, electronics, catalysis and generation of nuclear power. It is not surprising that the abuse of metals has led to severe environmental problems which need to be addressed. Metals also play an important role in the life processes of microbes as micronutrients, thereby performing a variety of metabolic functions. Metals which play a role as micronutrients are, calcium (Ca), chromium (Cr), magnesium (Mg), sodium (Na) and copper (Cu). However there are other metals which have no biological role and which could be potentially toxic. At present heavy metal pollution is of great concern since they are non-biodegradable and persist both in soil and water. These serve as a sink for heavy metal pollution, and thus environmentalists consider microbes as ecofriendly nano-factories which could be effectively used for metal bioremediation.

Microbial resistance to heavy metal is defined as the ability of the microbe to survive toxic

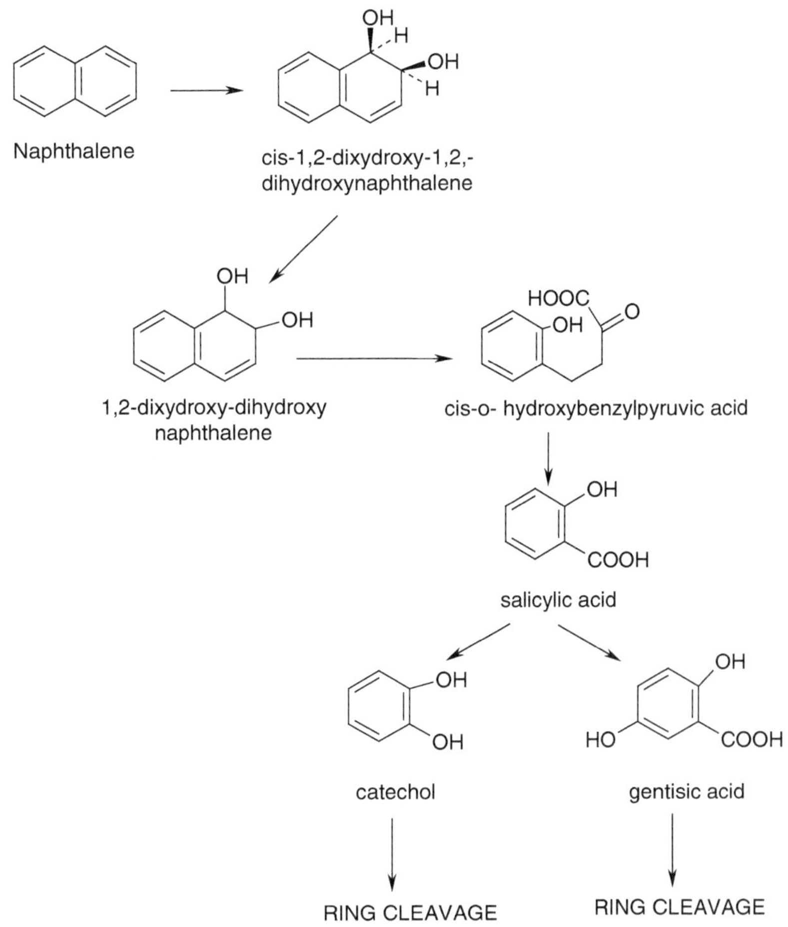

Fig. 5.1 Degradation pathway of naphthalene

concentration of metal exposure as they possess a detoxification mechanism which responds to heavy metal species.

Biosorption phenomenon is a passive sequestration by non-growing biomass. It comprises of cell surface complexation, ion exchange and micro-precipitation methods. Cell surface carries a net negative charge at neutral pH due to the presence of carboxyl, amine, hydroxyl, phosphate and sulphydryl groups and can absorb appreciable quantities of positively charged cationic metals. Fungi and yeasts which exhibit immense potential for metal sorption are *Rhizopus*, *Aspergillus*, *Streptoverticillium* and *Saccharomyces*. Heavy metals like lead, cadmium and copper have been removed by *Aspergillus niger*. Pretreatment of biomass of *Aspergillus versicolor* with dimethyl sulphoxide, hydrogen peroxide and glutaraldehyde enhanced the biosorption of lead. *Hirsutella* and *Aspergillus* species isolated from tanning effluents have higher potential to uptake chromium. Mushrooms also exhibit the phenomenon of biosorption. *Volvariella volvacea* has been reported to uptake cadmium, lead, copper and cobalt in mycelia and sporocarps. *Agaricus macrocarpus* has been efficiently used for the extraction of cadmium, mercury and copper from contaminated substrates.

Among bacteria *Zoogloea ramigera*, *Pseudomonas* sp. and *Streptomyces* sp. have been reported to exhibit biosorption of metals. *Staphylococcus saprophyticus* removes chromium, lead and copper ions from industrial wastes. *Saccharomyces cerevisiae* has been used as a promising biosorbent for heavy metal bioremediation.

Exploitation of an appropriate immobilisation technique is mandatory for the biomass to absorb the metals. Studies using free cells provide valuable information about their bioremediation potential. However due to small particle size, low mechanical strength and being free, the individual cells cannot withstand the hydrostatic pressure required for the flow rates in the industrial settings. Thus, immobilised biomass offers advantages of better usability, high biomass loading and minimal clogging. Immobilised cells of *Chlorella salina* exhibited better binding of cobalt, zinc and manganese than free cells. Polyacrylamide-immobilised *Citrobacter* exhibited a very high uptake potential of uranium, cadmium and lead. Alga SORB is a silica-immobilised algal preparation which is being commercially used for metal uptake and retains over 90 % efficiency beyond 18 months. AMT-BIOCLAIM is also a metal sorption product which exploits biomass of *Bacillus* to manufacture the granulated material for metal and wastewater recovery. Stripping of the metals is carried out with sulphuric acid and sodium hydroxide, and these granules are regenerated for their repeated use.

Radionuclides are being used for the production of energy as well as for the production of nuclear weapons. Their subsequent movement in the environment is of intense public concern and has prompted research on the environmental fate of key radionuclides. The most common radioactive nuclide for energy production is uranium. It occurs naturally in low concentration in rocks, water and soil and in higher concentrations in uraninite and other uranium-bearing minerals or ores from where it is extracted for the purposes of fuel production, weapon manufacturing and nuclear research. The processes for uranium bioremediation basically involve its conversion into a less soluble form or its accumulation in dead biomass. These processes basically immobilise or concentrate uranium and prevent it from leaching and contaminating the groundwater. The first process of uranium bioremediation comprises of dissimilatory metal bioreduction of soluble uranium (VI) to sparingly soluble uranium (IV). The second process comprises of microbially generated by products, and in the third process it is bisorped on the surface. Sulphate-reducing bacteria (SRB) and iron-reducing bacteria (FeRB) carry out the direct enzymatic reduction of U (VI) as they contain metal reductases. δ- Proteobacteria also carries out the enzymatic reduction of uranium (VI). U (VI) has been actively adsorbed by the nonliving brown marine alga *Cystoseira indica*. *Deinococcus radiodurans* is an exceptional organism as it is resistant to ionising radiations. Apart from this *Deinococcus radiodurans* also enzymatically reduces uranium (VI) using the electron shuttle anthraquinone-2,6-disulfonate.

5.5 Biomining

Microbes can be helpful in harnessing the recovery of minerals and metals from the earth's crust. This process is known as biomining. This process is getting popularised in extraction of metals as using mechanical and chemical methods is difficult and expensive apart from being non-environment friendly. Biomining comprises of two microbial processes which have been utilised for the extraction of metals, viz. bioleaching and bio-oxidation. Leaching is generally defined as the process of solubilisation of one or more components of solids by coming in contact with liquid. The ability of microorganisms to solubilise metals from their insoluble state is known as bioleaching. Bio-oxidation on the other hand refers to bacterial oxidation of reduced sulphur species associated with the metal of interest. The first patent was filed on the potential of *Thiobacillus* spp. to oxidise iron pyrites and copper sulphide. Bioleaching is considered as the main process for the large-scale operations in recovering of metals and minerals. Bioleaching has been successfully employed for the extraction of gold, copper, iron and uranium. The metal-leaching microorganisms generally use ferrous ion and reduce sulphur compounds (chemoautolithotrophs) as electron donors and fix carbon dioxide. Majority of the microorganisms are acidophiles since during the formation of metal sulphides they produce sulphuric acid. *Thiobacillus ferrooxidans* and

Thiobacillus thiooxidans uses reduced sulphur compounds for the bioleaching process, while *Leptospirillum ferrooxidans* uses only ferric ions.

5.5.1 Extraction of Copper

Extraction of copper by leaching is practised in several countries like the USA, Australia, Canada, Chile, Mexico, Peru and Russia and accounts for ca. 25 % of the world's copper production. Copper leaching is done through the bacterium *Thiobacillus ferrooxidans* (now known as *Acidithiobacillus ferrooxidans*). Copper ore mined from open pits is segregated as a higher grade ore or lower grade ore. The higher grade material is directly concentrated to produce feed for smelting, while the lower grade material is subjected to biological leaching. The low-grade material is piled over an impermeable surface in a suitable dimension and then leach solution, i.e. mild acidic solution, is sprayed on the dump. This promotes the growth of *Acidithiobacillus ferrooxidans* and other leaching microorganisms. The enhanced bacterial activity and colonisation mainly in the top one meter or so raises the temperature of the dump to 90 °C in the interior and supports a range of anaerobic or microaerophilic thermophiles which also promote the process of metal leaching. Copper upon oxidation is dissolved in the dilute acid and is collected as copper sulphate at the bottom of the dump. The leach solutions enriched which copper also are collected at the base of the dump and then conveyed to central recovery facility. In large-scale operations the concentration of copper in leach solution varies from 0.5 to 2.0 g of copper per litre. The common method of copper recovery is precipitation using large cementation unit, electrowinning or solvent extraction followed by electrowinning.

5.5.2 Extraction of Uranium

Conventionally uranium is extracted by using strong acid and large amounts of energy. However, this process creates a lot of environmental problems and issues. Hence, bioleaching is an efficient method of uranium extraction from ores and largely depends upon the mineralogy of uranium ore, bearing rock type, level of toxic material and leaching variables. Biologically generated ferric sulphate is in trend for heap and dump leaching operations to recover uranium and copper. *Acidithiobacillus ferrooxidans* has been found to be the dominant bacterial species which is associated with bioleaching of uranium by enhancing the oxidation rate of insoluble uranium into soluble uranium sulphate. The bacterium generally generates Fe (III) from pyrite and soluble Fe (II). Fe (III) readily attacks minerals incorporating U (IV), thus converting it to U (VI) which is soluble in dilute sulphuric acid. The biological oxidation process is 10^5–10^6 times faster than the chemical oxidation. Uranium solubilisation by the indirect method is

$$FeS_2 + H_2SO_4 \rightarrow 2FeSO_4 + H_2O + 2S^{o}$$

$$FeSO_4 + H_2SO_4 + O_2 \rightarrow Fe_2(SO_4)_3 + H_2O$$

The Fe (II) is re-oxidised by microbes to Fe (III) which again takes part in the oxidation process. The sulphur so formed is converted into H_2SO_4 and helps in the dissolution of uranium:

$$2S^{o} + 3O_2 + 2H_2O_2 \rightarrow H_2SO_4$$

The insoluble uranium (IV) is oxidised to the water-soluble uranium (VI) sulphate by the reaction given below:

$$UO_2 + Fe_2(SO_4)_3 \rightarrow UO_2SO_4 + 2FeSO_4$$

Fungi have also been used for bioleaching process for recovery of uranium from ores in Egypt. The fungi *Aspergillus terreus* and *Penicillium spinulosum* were found to intensify the ore concentrations on growth media reaching to a maximum concentration of 4 % (w/v). Apart from bioleaching the fungal mycelium also carries out the biosorption of uranium, thereby bioconcentrating it.

5.5.3 Extraction of Gold

Cyanide solution is generally used for the extraction of gold in the form of solution. However, cyanide usage poses environmental concerns; therefore, bio-oxidation by microorganisms is generally preferred for the recovery of gold from ores. In refractory ores the small peptides of the gold are encased in a matrix of arsenopyrite/pyrite, and hence after milling the gold is not recoverable. Hence, the ore is treated with liquid cyanide and a gold-bearing concentrate is prepared by floatation. Previously the concentrate was roasted at 700 °C in the presence of oxygen or digested with acid under pressure in an oxygen-enriched atmosphere. Some of the microorganisms known to oxidise cyanide include species of the genera *Actinomyces*, *Alcaligenes*, *Arthrobacter*, *Bacillus*, *Micrococcus*, *Neisseria*, *Paracoccus*, *Thiobacillus* and *Pseudomonas.*

Biomining bacteria decomposed ores and concentrates at atmospheric pressure and at ambient temperature. The first stirred-tank bioleach plant to treat sulphidic gold concentrate to enhance the gold recovery was commissioned in 1986 at Fairview Mine in South Africa. The largest operations are carried out in reactors having a size of 900 m^3 which process 1000 t of concentrate are at Sansu, Ghana (Africa). The slurry density of 15–20 % is used for refractory gold plants. The bacteria primarily oxidise the sulphide coating covering the gold microparticles in the ores and in the concentrates. The microorganisms which participate in this process belong to the *Acidithiobacillus* and *Leptospirillum* genera. Further thermophilic archaea are being bioprospected for their possible use in this process. In the primary reactor the slurry residence time is 2–3 days, wherein most microbial growth occurs. The primary reactor overflows to a series of smaller secondary reactors, thereby increasing efficiency of sulphide oxidation by reducing short circuiting of sulphide particles. The total residence time in the circuit is around 4–6 days. Without pretreatment 30–50 % of gold is recovered, while after bio-oxidation the recovery of the gold is enhanced to 95 %.

5.6 Microbially Enhanced Oil Recovery (MEOR)

The global demand for crude oil is often exceeding the existing production in many industrialised countries, which leads to reliance of these industries on imports. Microbially enhanced oil recovery is a unique process and an economically attractive method to enhance the recovery of oils from wells.

Enhanced oil recovery (EOR) generally relies upon the use of chemical or thermal energy for recovering the crude oil that is trapped in pores of the reservoir after primary and secondary (water flood) crude oil production has ceased from the oil well. Chemicals used in EOR processes include surfactants which reduce the interfacial tension between the oil and water and oil and rock interfaces.

Microbes contribute in enhanced oil recovery which can produce biosurfactants and biopolymers on the interfacial surface; microorganisms can also grow in the reservoir rock pore to produce gases and surfactants and other chemicals to recover trapped oil in the reservoirs, and finally they may plug high-permeability channels in reservoir rock to increase the sweep efficiency of the recovery process.

For carrying out in situ MEOR, it is necessary to use microbial cultures that can survive and grow at the temperatures, pressures and salinities present in the reservoir. Microorganisms produce several compounds that have potential for enhanced oil recovery, including carbon dioxide, acids and alcohols. Carbon dioxide may increase reservoir pressure and decrease the viscosity and gravity of the crude oil, allowing it to move more freely to the producing wells. Scleroglucan and xanthan gum possess similar rheological properties. The bacterial fermentation of polysaccharides leads to the production of gases like carbon dioxide and hydrogen which can contribute to in situ repressurisation of a pressure-depleted petroleum reservoir. These gases may dissolve in crude oil and reduce its viscosity.

Table 5.2 Applications of microorganisms used in enhanced oil recovery (EOR)

Microorganisms	Applications in MEOR	Microbial product
Leuconostoc, *Xanthomonas* and *Bacillus*	Selective plugging and wettability alteration	Biomass
Arthrobacter, *Bacillus*, *Pseudomonas* and *Acinetobacter*	Emulsification and de-emulsification through reduction of interfacial tension	Surfactants
Bacillus, *Brevibacterium*, *Leuconostoc*, *Xanthomonas*	Injectivity profile and viscosity modification, selective plugging	Polymers
Clostridium, *Zymomonas*, *Klebsiella*	Rock dissolution for better permeability, oil viscosity reduction	Solvents
Clostridium, *Enterobacter*, *mixed acidogens*	Permeability increase, emulsification	Acids
Clostridium, *Enterobacter*, *Methanobacterium*	Increased pressure, oil swelling, IFT and viscosity reduction	Gases

Zobell (1946) patented the process of secondary recovery of petroleum using anaerobic, hydrocarbon-utilising and sulphate-reducing bacteria such as *Desulfovibrio* species in situ. Several microorganisms have been isolated which produce biopolymers and emulsifiers. One isolate was found to grow in 10 % salt concentration, over a pH range of 4.6–9.00 and temperatures up to 50 °C in the presence of crude oil. The first field test of MEOR was carried out in Arkansas in 1954. In this trial, 2 % solution of beet molasses in fresh water was injected during a 6-month period, along with 18,200-gal containers of broth containing *Clostridium acetobutylicum*. Fresh water breakthrough occurred in the production well 70 days after the injection, and fermentation products (short-chain fatty acids, CO_2 and traces of ethanol, 1-butanol and acetone) and sugars appeared after 80–90 days. The production of oil increased from 0.6 to 2.1 bbl/day without production of any hydrogen gas. Table 5.2 summarises the different microorganisms used in MEOR and their metabolic products.

Several field trials of MEOR have been carried out till 2003, and more than 400 MEOR tests have been conducted in the USA alone as compared to other field tests carried out in the rest of the world. The MEOR field applications are conducted as a single-well treatment and full-field treatment. The successful field trials were single-well treatments in the USA, China, Romania, India, Russia and Argentina where incremental oil varied from no impact to 204 %.

5.7 Summary

The role of microorganisms is tremendous not only in removing the pollutants but also in bioconcentrating essential metals and minerals from the ores as well as recovering the oil stuck in stripped oil wells, thus making the process more environment friendly and economical.

Selected Reading

Acevedo F (2000) The use of reactors in biomining process. Electron J Biotechnol 3(3). http://www.ejb.org/content/vol3/issue3/full/4

Al-Sulaimani H, Joshi S, Al-Wahaibi Y, Al-Bahry S, Elshafie A, Al-Bemani A (2011) Microbial biotechnology for enhancing oil recovery: current developments and future prospects. Biotechnol Bioinf Bioeng 1(2):147–158

Crawford RL (1995) The microbiology and treatment of nitroaromatic compounds. Curr Opin Biotechnol 6:329–336

Jain RK, Kapur M, Labana S, Lal B, Sarma PM, Bhattacharya D, Thakur IS (2005) Microbial diversity: application of microorganisms for the biodegradation of xenobiotics. Curr Sci 89:101–112

Jimenez N, Vinas M, Bayona JM, Albaiges J, Solanas AM (2007) The prestige oil spill: bacterial community dynamics during a field biostimulation assay. Appl Microbiol Biotechnol 77:935–945

Lazar I, Petrisor IG, Yen TF (2007) Microbial enhanced oil recovery (MEOR). Pet Sci Technol 25:1353–1366

Lloyd JR, Renshaw JC (2005) Bioremediation of radioactive waste: radionuclide–microbe interactions in laboratory and field-scale studies. Curr Opin Biotechnol 16:254–260

Pritchard PH, Mueller JG, Rogers JC, Kremer IV, Glaser JA (1992) Oil spill bioremediation: experiences, lessons, and results from the Exxon Valdez Oil spill in Alaska. Biodegradation 3:315–335

Sharma S (2012) Bioremediation: features, strategies and applications. Asian J Pharm Life Sci 2:202–213

Siddiqui MH, Kumar A, Kesari KK, Arif JM (2009) Biomining – a useful approach toward metal extraction. Am-Eurasian J Agron 2(2):84–88

Sinha RK, Valani D, Sinha S, Singh S, Herat S (2009) Bioremediation of contaminated sites: a low-cost nature's biotechnology for environmental clean up by versatile microbes, plants and earthworms In: Solid waste management and environmental remediation. Nova Science, New York. ISBN: 978-1-60741-761-3

Swanell RPJ, Lee K, McDonagh M (1996) Field evaluations of marine oil spill bioremediations. Microbiol Rev 60:342–365

Varsha YM, Naga Deepthi CH, Chenna S (2011) An emphasis on xenobiotic degradation in environmental cleanup. J Bioremediat Biodegrad S11:001. doi:10.4172/2155-6199.S11-001

Wang Xiang-Qi, Majumdar AS (2007) Biomining and its application in mining industry. Technology Review Report Number: M3TC-TRR-07003. Minerals, Metals and Materials Technology Centre (M3TC), National University of Singapore, Singapore

Zobell CE (1946) Bacterial process for treatment of fluid bearing earth formulations. US Patent No. 2,413,278, 24 December

6 Microbes in the Food Industry

Role of Microbes in Food Processing, Fermented Beverages and Fermented Foods

6.1 Introduction

The development of fermented food is one of the oldest technologies known to man since the dawn of civilisation. Methods of fermentation of milk product like yogurt have been described in ancient scripture like the Vedas in India and the Bible. Records of fermentation of meats, vegetables and milk have also been found as early as 6000 BC (Fox et al 1993). With the discovery of microbes by Antoni van Leeuwenhoek and development of the science of microbiology in the 1850s, the biological basis of fermentation was understood for the first time. Adam (1990) defines food fermentation as a form of energy-yielding microbial metabolism in which an organic substrate usually a carbohydrate is incompletely oxidised and an organic carbohydrate acts as an electron acceptor. Thus, by this definition the production of ethanol by yeasts or organic acids by lactic acid bacteria is considered as fermentation. Thus, fermented foods are the foods which are produced or preserved by the action of microorganisms. Originally, food fermentations were carried with the primary purpose to achieve preservation effect for long-term use. With the development of alternative technologies for food preservation and long-term storage, there is no longer pressing need of food to be preserved by fermentation. Fermented foods are currently being manufactured because of their unique flavour, aroma and texture attributed which are generally relished by the consumer. Today, fermented foods are an integral part of our staple diet.

6.2 Fermented Foods

The substrates used for the commercial production of fermented products are milk, vegetables like cabbage and cucumber, meat, oriental fermented foods and bakery products.

6.2.1 Milk Products

The fermented milk products basically comprise of cheese, yogurt and fermented milks. Cheese is a concentrated form of the milk protein casein and milk fat. There are over 400 different varieties of cheese which can be clubbed into 20 distinct types (Jay 1996). Similarly, there exist an extensive list of types of yogurts and fermented milks. The primary requirement for cheese making is milk, and the variation of cheese is based on the quality and type of milk being used. The milk is pasteurised and a starter culture is added. The starter cultures can be mesophilic or thermophilic based on the type of cheese being manufactured (Table 6.1). For cheddar cheese, *Streptomyces lactis* and *Streptococcus cremoris* are generally used which are mesophiles, while *Lactobacillus bulgaricus* and *Streptococcus thermophilus* are used as starter cultures as they can sustain high milk temperatures around 132 °F

S. Saxena, *Applied Microbiology*,
DOI 10.1007/978-81-322-2259-0_6, © Springer India 2015

Table 6.1 Starter culture for cheese manufacture

Organism type	Organism name	Cheese type
Mesophiles	*Lactococcus lactis*	Cottage cheese
	Lactococcus cremoris	Cheddar cheese
	Leuconostoc mesenteroides ssp. *cremoris*	Cream cheese
		Continental varieties
	Lactococcus lactis ssp. *lactis* biovar *diacetylactis*	Blue cheese
		Baby Swiss
Thermophiles	*Streptococcus thermophilus*	Mozzarella
	Lactobacillus delbrueckii ssp. bulgaricus	Grana
	Lactobacillus helveticus	Swiss

which are used for the preparation of Swiss and Italian cheese. The bacteria produce small amount of acid which helps in clotting of milk required for cheese making. Subsequently, the curd is formed which entraps fat and water. A suitable coagulant is then added to the curd which splits the colloidal casein into carbohydrate-rich peptide fraction and insoluble paracasein that precipitates in the presence of calcium ions.

Fungi which are used in producing coagulants for cheese making are *Mucor miehei*, *Mucor pusillus* and *Endothia parasitica*. After the formation of coagulum, it is cut into pieces or cubes so that there is loss of whey (syneresis). The cubes of coagulum/curd are suspended in whey and heated at a given temperature (37–38 °C) for approximately 30 min in the case of cheddar cheese. This process helps in the control of acid production because the starter culture suppresses growth of spoilage bacteria and influences the structure of the curd. Finally, the whey is removed from the coagulum. This is generally carried out by filling the curd/whey mixture to perforated mould (in case of camembert/Brie cheese) where the curd is allowed to press by its own weight. Semi-continuous pressing systems like Casomatic® are used which comprise of cylindrical columns to which whey/curd mix is dispensed, pressing button to pre-press the curd beneath the whey and perforated bands which enable whey drainage followed by curd cutting/moulding system. The curd particles are knot into a cohesive mass depending upon the temperature, pressure and time. Sodium chloride is applied to the curd in several ways; dry salt may be sprinkled on loose curds as in manufacture of cheddar cheese or rubbed on the cheese surface. Salt contributes to the development of flavour, texture and appearance of the cheese and controls the production of lactic acid. Ripening of the cheese involves the finished cheese to be placed in controlled temperature and relative humidity (4 °C/85 % humidity for cheddar cheese) for 3 months to 1 year depending upon the type of cheese.

Yogurt is the fermented milk product which consists of two lactose-fermenting organisms *Lactobacillus bulgaricus* and *Streptococcus thermophilus*. Yogurt is generally prepared from standardised whole milk, partially defatted milk, condensed skim milk cream and non-fat dry milk. *L. bulgaricus* and *S. thermophilus* are used in 1:1 ratio as starter culture for yogurt manufacture. *Streptococcus* grows first followed by *Lactobacillus bulgaricus* which provides its aroma and flavour. Attempts have been made to prevent syneresis from yogurt by including a slime-producing strain *Streptococcus filant* or *Streptococcus lactis* var. *hollandicus*.

Kefir is an acid and alcoholic fermented milk which is commonly produced and consumed in Russia. The alcohol content of kefir is approximately 1 %. Koumiss is prepared from the mare's milk and contains alcohol which causes mild intoxication. The alcohol content of koumiss varies from 1 to 2.5 %, while the titratable acidity varies from 0.7 to 1.8 % lactic acid. Mare's milk is low in casein content and does not curdle like cow's milk; it is a greyish white wholesome drink. The starter culture of koumiss consists of *Lactobacillus bulgaricus* and lactose-fermenting *Torulopsis holmii*.

6.2.2 Fermented Vegetables

Lactic acid fermentation or pickling is one of the important methods of food conservation. The pickled foods generally comprise of cucumbers, olives, various peppers and green tomatoes which serve as appetisers or are consumed as substantial part of the meal. Other vegetables which are less frequently pickled are carrots, cauliflower, celery, okra, onion and sweet and hot peppers.

Sauerkraut is a product which is produced as a result of lactic acid fermentation of the shredded cabbages, its literal meaning in sour cabbage. The concept behind sauerkraut fermentation is initiated by *Leuconostoc mesenteroides* and continued by *Lactobacillus brevis* and *Lactobacillus plantarum*.

6.2.3 Fermented Meat Preparations

Fermented sausages are produced as a result of by lactic fermentation of a mixture of comminuted meat mixed with fat, salt, curing agents (nitrate/nitrite), sugar and spices, and these represent traditional foods of central and southern Europe.

Fermentation temperatures vary according to the individual product, but they are generally less than 22 °C for dry- and mould-ripened sausages and 22–26 °C for semidry varieties. European fermented sausages formulated with nitrite are produced with added starter culture, generally consisting of lactic acid bacteria (lactobacilli and pediococci) and catalase-positive cocci (*S. carnosus*, *Micrococcus varians*). Yeasts and moulds that are available as starters include *Debaryomyces hansenii*, *Candida famata* and *Penicillium nalgiovense* and *P. chrysogenum*, respectively.

6.2.4 Traditional Fermented Food

Soy sauce (or shoyu) is a condiment widely used in Japan for cooking and seasoning of Japanese food. There are five main types of soy sauce in Japan, each with its own distinctive colour, flavour and use. The characteristic aroma and flavour of soy sauce is attributed to the enzymatic activities of yeasts, *Tetragenococcus halophilus* and some *Lactobacillus* species.

'Idli' is a fermented steamed cake of rice and dehulled blackgram dhal produced in India. LAB such as *Lc. mesenteroides*, *Lb. delbrueckii*, *P. cerevisiae*, *E. faecalis* and *L. lactis* are responsible for pH reduction and may increase the thiamine and riboflavin content. Nan is a leavened flat sourdough bread with a central pocket now prepared worldwide. *Saccharomyces*, yeasts and lactic acid bacteria are basically involved in the fermentation process. Philippines' 'puto' is also a fermented product similar to idli. It consists of a steamed rice cake prepared from year-old rice grains which are soaked, ground with water and allowed to undergo a natural acid and gas fermentation. The acid is partially neutralised with sodium hydroxide and does not contain pulses as in idli.

Tempeh is a soyabean-based fermentation product of Indonesia that contains over 40 % protein. Tempeh is a meat substitute that is used in soups or sliced, salted, deep fired in coconut oil and consumed. Lactic acid bacteria including *Lb. casei* and *Lactococcus* species dominate the fermentation.

'Ogi' is a fine paste-like sour gruel eaten in Nigeria resulting from the submerged fermentation of cereals. It is consumed as a breakfast cereal by adults and is an important traditional weaning food of infants. A starter culture has been used to produce an improved version of ogi called DogiK. The starter strains are lactobacilli isolated from local fermented foods possessing strong antibacterial activity.

6.2.5 Bakery Products

Sourdough breads are made with starters containing yeasts such as *Saccharomyces* spp. and

Torulopsis and homo-fermentative and heterofermentative lactic acid bacteria. Heterofermentative strains such as *Lb. sanfrancisco*, *Lb. brevis* and *Lb. fermentum* are responsible for the characteristic sensory qualities of such breads. More than 20 types of yeast are found in sourdoughs. *S. cerevisiae* is frequently present (or added) due to the use of baker's yeast.

The sourdoughs have been classified into three groups: type I, type II and type III. Type I sourdoughs are traditional doughs maintained by the continuous propagation at ambient temperature (20–30 °C). *Lactobacillus (Lb) sanfranciscensis* and *Lb. pontis* are the dominant LAB in these sourdoughs. Bakers' yeast is used for the leavening of type II sourdoughs. This is essential since type II doughs are a less time-consuming, one-stage fermentation process at temperatures exceeding 30 °C. The dominant strains in industrial processes of type II doughs are mostly *Lb. panis*, *Lb. pontis*, *Lb. reuteri*, *Lb. johnsonii*, *Lb. sanfranciscensis*, *Lb. fermentum*, *Lb. delbrueckii*, *Lb. acidophilus*, *Lactococcus lactis*, *Lb. brevis* and *Lb. amylovorus*. The third type of sourdough is basically consist of dried preparations which are made by traditional sourdough fermentation with subsequent water evaporation by freeze-drying, roller/spray drying or drying in a fluidised bed reactor. The type III sourdoughs are the most convenient way to introduce superior bread taste into modern bakery industry.

6.3 Fermented Beverages

Alcoholic beverages have been produced by humans since thousands of years. These drinks were basically based on fermentation of cereals and fruits by yeasts. However, with the advancement in microbiology and biological sciences, the technologies of production have improved and these have been identified with specific names as wine, beer and whisky.

6.3.1 Wine

Wines are primarily prepared from the fermentation of grapes majorly but can be also produced by some other fruits. The five basic components of wine making involve harvesting, crushing and pressing (mush formation), fermentation, clarification, aging and bottling. Wines are named after the type of grapes or the geographic area or specific village where they were first produced. Burgundy, Bordeaux, Champagne and Alsace are important wines of France. *Saccharomyces ellipsoideus* is the common yeast used for the preparation of table wines. These have an alcohol content of 10–12 %.

Champagne is a sparkling wine in which the alcohol content reaches up to 20 % as it undergoes double fermentation. They have a natural effervescence; others are made effervescent by bubbling them with carbon dioxide. The starter culture for yeast fermentation is *Saccharomyces bayanus* which is also known as Premier Cuvee. Cognac is the distilled product of wine also known as wine brandy.

6.3.2 Beer

Beer is one of the oldest beverages humans have produced, dating back to at least the fifth millennium BC (prior even to writing), and recorded in the written history of Ancient Egypt and Mesopotamia. Beer is prepared from malted barley and malted wheat. Sometimes a mixture of starch sources can be used, such as rice. The main ingredients of beer are water, malted barley, hops and yeast. Other ingredients, such as flavouring or sources of sugar, are called adjuncts and are commonly used; common adjuncts are corn, rice and sugar. The process of making beer is called brewing. It includes breaking the starch in the grains into a sugary liquid, called wort, and fermenting the sugars in the wort into alcohol and carbon dioxide by yeasts. Beers tend to fall in one of the two large families: ale (using top fermenting yeast) or lager (using the bottom fermenting yeasts). *Saccharomyces cerevisiae* is the strain used for manufacturing ale, while *Saccharomyces uvarum* is known as the lager's yeast. On average, beer's alcohol content is between 4 and 6 % alcohol by volume, although it can be as low as 2 % and as high as 14 % under ordinary circumstances. Lagers are the most commonly consumed beer in the world. They are of Central European origin, taking their name from the German lagern which means 'to store'.

Other fermented beverages include the mead, prepared by fermenting honey; cider, a fermented apple juice while perry being a fermented pear juice.

6.3.3 Whiskey

Whiskey is the name for a broad category of alcoholic beverages distilled from grains that are subsequently aged in oak casks. The grains used to make various types of whisky include barley, malted barley, rye, malted rye, wheat and maize/corn. The alcoholic content in whiskey ranges from 40 to 50 %. The distillate must age for at least 3 years to be called Scotch whisky.

6.3.4 Kombucha

Kombucha tea is a fermented tea beverage produced by fermenting sugared black tea with tea fungus (kombucha). The fermentation of kombucha tea is done by symbiotic association between acetic acid bacteria and yeasts for a period of 14 days. Kombucha is composed of two portions: a floating cellulose pellicle layer and the sour liquid broth. Kombucha has been consumed in Asia for over two millennia and is a popular beverage among traditional fermented foods across the world.

6.4 Summary

Today, microorganisms play a tremendous role in the development of fermented food products by serving as starter culture as well as maintaining the aroma and nutrition of the product. They have a significant effect on the organoleptic properties of the end product.

Selected Reading

Adams MR (1990) Topical aspects of fermented foods. Trends Food Sci Technol 1:141–144

Fox PF, Lucey JA, Cogan TM (1993) Cheese: an overview. In: Fox PF (ed) Cheese: chemistry, physics and microbiology, 2nd edn. Chapman & Hall, London, pp 1–36

Jay JM (1996) Modern food microbiology, 5th edn. Chapman & Hall, New York

7 Microbes in Production of Commodity Chemicals

Ethanol, Acrylamide, Citric Acid, Adipic Acid, 1, 2-Propanediol and Penicillin

7.1 Introduction

Commodity chemicals are inexpensive, have larger demands and are produced and sold in bulk. They generally are intermediates involved in the syntheses of high end products (Table 7.1). Initially the chemical industry was dependent on nonrenewable resources for virtually all commodity chemicals. The cost of the feedstocks for commodity chemicals is directly associated with the cost of the petroleum and hence represents 50–75 % of the manufacturing cost of the commodity chemicals. However, considering the enhanced cost of the petroleum and natural gas resources as well as their possible exhaustion in the future due to continuous industrial demand, newer alternatives are being explored. One of the major technologies being explored by the industries in the USA, Europe and Japan is conversion of biomass into commodity chemicals using microbial interventions. Biomass generally comprises of crop and forest product wastes and municipal and agricultural wastes. Technologically it is possible to produce all the commodity chemicals from biomass feedstocks like starch and cellulose.

Microbes offer to be the best manipulative systems which could be exploited for customised synthesis of commodity chemicals, thereby decreasing the reliance on petroleum for production of chemical feedstocks. Microbes which could be exploited in this process could be natural isolates as well as genetically/metabolically engineered to produce the desired product. The commodity chemicals which are being produced from biomass include ethanol, acetone, citric acid, propanoic acid, fumaric acid, butanol and 2,3-butanediol.

7.2 Commercial Production of Ethanol

Previously ethanol was being made from ethylene derived from petroleum sources. Presently the ethanol/alcohol is being produced by fermentation through the conversion of biomass. Ethanol is a renewable energy which is being produced by fermentation of sugars and is being used as a blending agent up to 15 %v/v in petroleum in many countries of the world. Ethanol-blended petroleum for automobiles can significantly reduce the use of petroleum as well as bring down the emission of greenhouse gases. Brazil is one of the largest producers of motor grade fuel ethanol (MGFE).

The common fermentation substrates for the production of ethanol are corn starch, molasses, sugarcane juice, cassava starch and other fermentable carbohydrates. There are a variety of microbes which convert these substrates into ethanol (Table 7.2). Corn starch is generally used as a raw material for the commercial production of motor grade fuel ethanol. The broad process of production of motor grade fuel ethanol comprises grinding, cooking, fermentation, distillation and dehydration (Fig. 7.1).

S. Saxena, *Applied Microbiology*,
DOI 10.1007/978-81-322-2259-0_7, © Springer India 2015

General assumption of ethanol production is based on the amount of the fermentable sugar available for the process. It has been estimated that approximately 18–23 kg of ethanol is produced from 45 kg of fermentable sugar, i.e. glucose. For starchy material, the yield is about the same, i.e. between 40 and 50 % based on the dry weight of the carbohydrate. Commonly the substrates used for ethanol production are directly fermentable or starchy materials which could be easily hydrolysed to fermentable sugars. Recently technologies are being developed to use the cellulosic as well as lignocelluloses biomass by pretreating them so that they are easily hydrolysed for use as a fermentable substrate.

During the commercial production of MGFE, corn is generally ground by dry milling process into proper consistency for efficient conversion of corn starch into ethanol. The purpose basically is to prepare the grain for efficient and rapid

Table 7.1 Important commodity chemicals and their uses

Commodity chemicals	Major uses
Ethanol	Detergent, solubiliser, cosmetics, solvent, fuel
Acetic anhydride	Cellulose esters
Adipic acid	Nylon
Cyclohexane	Nylon, caprolactam
Isopropanol	Acetone, solvents
Propylene oxide	Propylene glycol, urethanes
Butadiene	Rubber
Acrylonitrile	Polymers

Table 7.2 Ethanol-producing microorganisms

Class	Organism name	Carbon source
Yeasts	*Saccharomyces cerevisiae*	Glucose
	Schizosaccharomyces pombe	Xylulose
	Kluyveromyces lactis	Xylulose
Fungi	*Pachysolen tannophilus*	Glucose, xylose
	Mucor indicus	Glucose
Bacteria	*Zymomonas mobilis*	Glucose
	Thermobacteroides saccharolyticum	Glucose, xylose
	Thermoanaerobacter ethanolicus	Glucose, xylose
	Clostridium thermohydrosulfuricum	Glucose, xylose

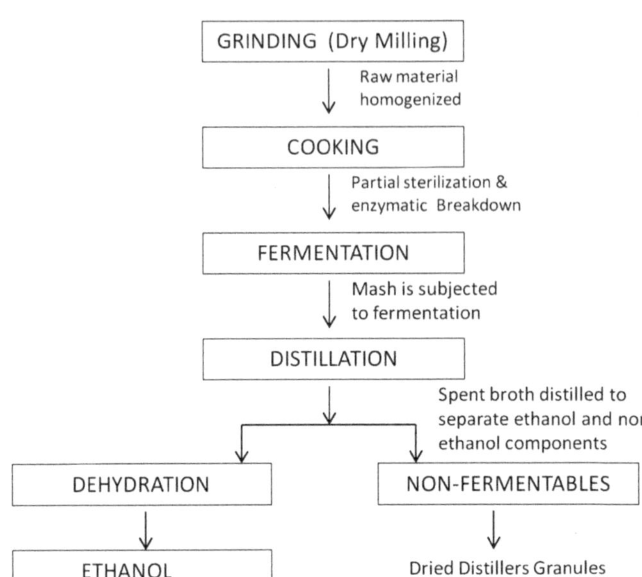

Fig. 7.1 Broad processes for the production of motor grade fuel ethanol

introduction of water and enzymes to achieve a homogenous mixture which could be easily pumped.

To have 25 % solids in the fermentation substrate, 1.149 kg of ground corn is mixed with 2.85 l of water at 60 °C and stirred continuously in a homogeniser held at a temperature ranging between 80 and 90 °C for 4–8 h and then mixed with two different enzymes. Approximately 3.3 ml of α-amylase (145,000 amylase units/ml) per kg of ground corn is mixed to hydrolyse the starch (i.e. amylase) and reduce the viscosity. This process is referred to as liquefaction and also prevents starch retrogradation. The slurry is then autoclaved at 121 °C for 20 min. Subsequently the sterilised slurry is held at 85 °C for 1 h and then mixed with 6.7 ml of amylase (saccharification step) which is just before fermentation. This process is known as saccharification and results in the formation of mash, the final fermentable substrate. Mash is cooled at room temperature and the water lost is made up with sterile water. Antibiotic lactoside is added 5 μg/ml to prevent bacterial contamination.

The pH value for fermentation is between 4.5 and 5.5 which are typically encountered in fuel ethanol plants. In the fermentation medium, urea is added at 0.016 % of the weight of mash as a nitrogen source. The fermentation was carried out at a temperature of 30 °C. Yeast growth and ethanol formation are generally inhibited by solutions having high osmotic pressure and accumulating high concentrations of ethanol. In batch fermentation for MGFE production, the inoculum size is six to eight million cells per ml or 0.24 kg of yeast solids per kilolitre. A five- to tenfold multiplication is generally expected during the batch process which approximately gives a yield of 1.2–2.2 kg of yeast solids per kg of mash. *Saccharomyces cerevisiae* requires biotin for enhancing the rate of fermentation. The total batch fermentation time ranges between 48 and 72 h. The alcohol produced during the fermentation process ranges between 6 and 8 % by volume. Twenty-five kilograms of corn on fermentation usually yields 8.5–10 l of ethanol and about 7.5 kg of distillers dried grains in a batch fermentation process.

The production of MGFE from the fermented beers is similar to those found in beverage spirits industry. Previously the dehydration step was achieved by azeotropic distillation, but currently molecular sieve dehydration utilising integrated pressure swing adsorption (PSA) technology is being used which is an energy-efficient process when compared to combined distillation and dehydration. Molecular sieves are manufactured from materials such as potassium aluminosilicates and are hard, granular, spherical or cylindrical extrudates. Their grading is done according to the nominal diameter of myriad internal pores that provide access to the interstitial free volume found in the microcrystalline structure. The grade used for ethanol dehydration is type 3Å, which refers to average diameter of the interstitial passageways is 3 Angstroms (Å). As the water molecule has a diameter of less than 3 Å and the ethanol molecule has an average diameter of more than 3 Å, the water molecules are retained in the interstitial spaces and the ethanol molecules move out. Thus, ethanol is also recovered from the azeotropic concentrations. The molecular sieving process generally occurs in the liquid or vapour phase. This dehydration step produces a final product that is nominally 100 % ethanol (200 proof, ethanol). After the centrifugation, the non-fermentable solids, i.e. unfermented grains, are concentrated thin stillage with 10–12 % moisture. These are known as Dried Distillers Granules Soluble (DDGS). DDGS has been used in poultry diets and in livestock diets for the improvement in the yield of milk and biomass.

7.3 Industrial Production of Acrylamide

Acrylamide ($CH_2{=}CHCONH_2$) is a commodity chemical which is used as a starting material for different polymeric substances like polyacrylamide and similar polymers which find applications as thickening, binding, strengthening and flocculating agent in industrial applications. The current annual global demand for acrylamide is ~250,000 metric tonnes. Conventionally acrylamide was produced using acrylonitrile as a

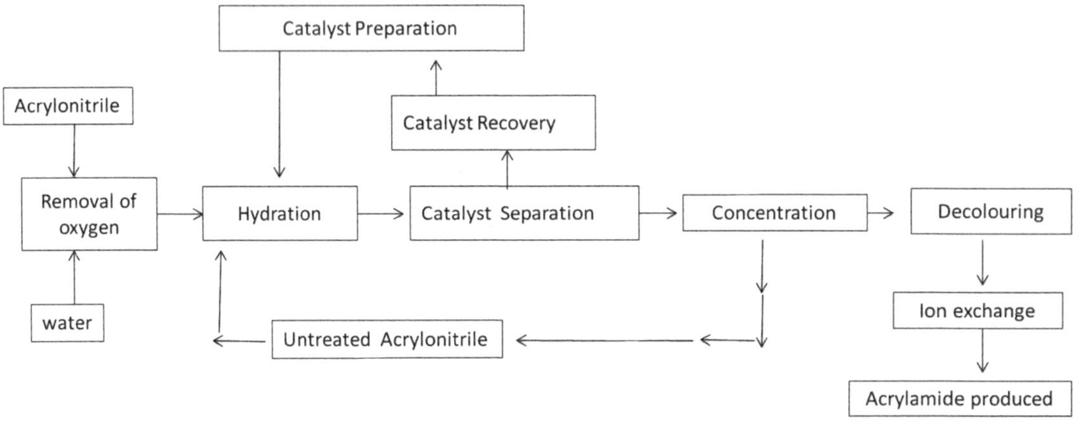

Fig 7.2 Conventional (chemical) process of acrylamide production

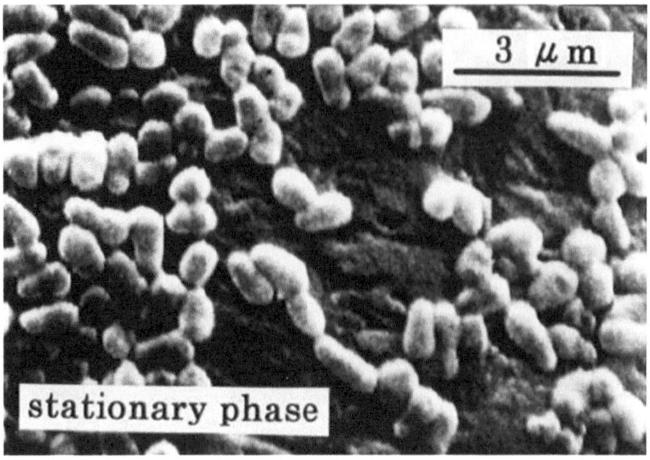

Fig. 7.3 Electron micrograph of *Rhodococcus* species N-774 (Taken from Dia-Nitrix Co., Ltd website)

starting material by passing it over a Raney copper catalyst at about 100 °C (Fig. 7.2). However, the major drawback of the process is the complex nature of the preparative procedure for the catalyst, difficulties in regenerating the used catalyst and problems associated with separating impurities from acrylamide. The impurities are ethylene cyanohydrin, β-hydroxypropionamide and nitrilotrispropinamide which may interfere with the chain reactions. The Nitto Chemical Company in Japan started the commercial acrylamide production using an enzymatic process which was jointly developed by researchers at Kyoto University and Nitto Chemical Industry (now Mitsubishi Rayon). The process was based on the biocatalyst nitrile hydratase which was discovered during studies on degradation of nitrile compounds by microorganisms and catalyses the hydration reaction of nitrile to amide (Asano et al. 1980).

During the extensive screening process, some microorganisms under resting conditions were able to accumulate acrylamide when incubated with acrylonitrile. *Rhodococcus* sp. N-774 was the first strain that was put into commercial use to produce acrylamide from acrylonitrile (Fig. 7.3). The nitrile hydratase enzyme is produced in this strain without an inducer. The resting cells of *Rhodococcus* sp. N-774 and *Pseudomonas chlororaphis* B23 were incubated with acrylonitrile provided that acrylonitrile was added gradually to avoid inhibition of nitrile hydratase activity, thereby accumulating approx. 400 g of acrylamide per litre at 10 °C. Ninety-nine percent conversion

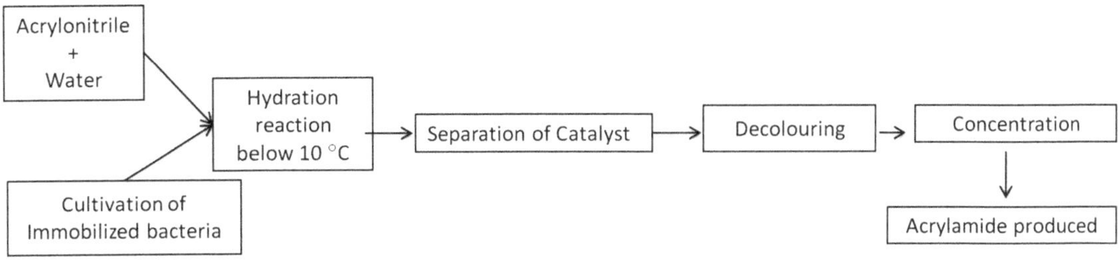

Fig. 7.4 Process outline for microbial production of acrylamide (Catalyst here refers to the immobilised bacteria producing nitrile hydratase)

Table 7.3 Performance of microorganisms producing nitrile hydratase for acrylamide production

	Reaction conditions			Reaction performance		
	Temp (°C)	pH	Acrylonitrile conc. (%)	Conversion of acrylonitrile (%)	Selectivity of acrylamide (%)	Acrylamide concentration at outlet of reactor (%)
Rhodococcus sp. N-774	0–5	7.5–8.5	1.5–2.0	99.9	99.9	20
Pseudomonas chlororaphis B23	0–5	7.5–8.5	1.5–2.0	99.97	99.98	30
Rhodococcus rhodochrous J1	0–15	6.5–8.0	1.5–2.0	99.97	99.98	50

of acrylonitrile into acrylamide was achieved without the formation of impurities (Fig. 7.4). *Rhodococcus rhodochrous* J1 strain produced cobalt containing nitrile hydratase which could produce 700 g/l acrylamide in the reaction solution. Mitsubishi Rayon's acrylamide production using bacterial nitrile hydratase produces approximately 30,000 tpa, consumes less energy and steers clear of heavy metal problems in the wastewater (Table 7.3).

7.4 Industrial Production of Citric Acid

Citric acid is a naturally occurring organic acid which is found in citrus fruits, pears and pineapples. This tricarboxylic acid is a critical intermediate of metabolism in plants and animals as it is the first product formed in the aerobic respiration, i.e. citric acid cycle or Krebs cycle. Carl Wilhelm Scheele isolated and crystallised citric acid from lemon juice in 1784. Citric acid is commodity chemical and is available in anhydrous or monohydrate forms. The commercial success of citric acid is attributed to its use in industry. Approximately 70 % of the citric acid produced is used in food and beverage industry with carbonated products having a market size of approximately 50 %. The pharmaceuticals consume approximately 12–14 %, and the remaining 16–18 % is used for other industrial applications like metal cleaning and detergent markets, in preparation of blue print paper, in passivation of stainless steel, etc. The annual global production of citric acid is to the tune of over 1.4 million tonnes.

The citric acid is currently being produced by fermentation. Microorganisms that can produce citric acid were first observed by C. Whemer (1893). He found that the mould *Penicillium glaucum* could accumulate significant quantities of citric acid when grown on sugar solution. It was James N. Currie (1917) who found that an isolate of *Aspergillus niger* produced better yields of citric acid and is still considered the organism of choice for industrial production of citric acid (Table 7.4). Currie joined Pfizer and in 1923, Pfizer started the commercial production of citric acid. Currently the major producers of

Table 7.4 Microorganisms producing citric acid

Fungi	*Aspergillus niger*
	A. aculeatus
	A. carbonarius
	A. awamori
	A. foetidus
	A. fonsecaeus
	A. phoenicis
	Penicillium janthinellum
Yeasts	*Candida tropicalis*
	C. oleophila
	C. guilliermondii
	C. citroformans
	Hansenula anomala
	Yarrowia lipolytica
Bacteria	*Arthrobacter paraffinens*
	Bacillus licheniformis
	Corynebacterium ssp.
	Brevibacterium flavum
	Bacillus subtilis

citric acid are Archer Daniels Midland (ADM), USA; Cargill, USA; Tate & Lyle, UK; DSM, Netherlands; Jungbunzlauer, Switzerland; Gadot Biochemical Industries, Israel; and Anhui BBCA Biochemical Co. Ltd., China.

The basic fermentation processes used in industry for production of citric acid by *Aspergillus niger* are (1) surface fermentation, (2) submerged fermentation and (3) solid substrate or the Koji fermentation.

7.4.1 Citric Acid Production by Surface Fermentation

This was the very first method adopted for the industrial manufacture of citric acid. The process is still used in small and medium scale industries as the installation is cost-effective and the process operations are cost- and energy effective. The raw material used for production of citric acid comprise of molasses, hydrolyzed corn starch or other inexpensive sugary solutions. Dextrose or beet molasses is generally preferred raw material wherein it is diluted to 15–20 % with dilute sulphuric acid and the pH is adjusted between 5.5 and 6.5. Inorganic nitrogen is provided by ammonium sulphate, ammonium nitrate, sodium nitrate, potassium nitrate and urea. Phosphorus is the third major constituent of the fermentation medium, and its concentration ranges between 0.1 and 2 % based on the type of strain used for carrying out the fermentation process.

Shallow trays made of high purity aluminium or stainless steel with a capacity of 200–1,000 l are used for the surface fermentation process. These trays are stacked in a rack under aseptic conditions. Media is pumped through the trays aseptically and then spore inoculation is carried out into the liquid medium or via air. Aeration is important for the process of fermentation as well as removal of heat from the aseptic system. The heat generation during the fermentation is at the rate of 1 kJ/h/m^3, but the surface and the medium temperature is maintained in the range of 28–30 °C throughout the fermentation by air circulation in the fermentation chamber. Air flow at initial fermentation stage is low but increases after 12 h, and when the growth is maximal, it is supplied at a rate of 10 m^3. The air is humidified between 40 and 60 % to prevent the loss of moisture from the surface of the medium and is passed through bacteriological filter prior to entering the fermentation chamber. The fermentation duration is between 8 and 15 days. After the fermentation process, the tray contents are separated into crude fermentation fluid and mycelial mats are washed to remove the impregnated citric acid. The productivity of citric acid is 1 kg per sq. m/day.

7.4.2 Submerged Fermentation for Citric Acid Production

In industrialised countries, the submerged process is the choice method since it is less labour intensive, uses less space compared to surface fermentation and gives higher production rate. Submerged process is generally carried out in stirred tank fermentation though air lift fermenters with higher aspect ratio are also being used industrially. The reactors are designed using high stainless steel grade due to low pH being developed during the fermentation process.

At times, two-stage fermentation process is adopted in which sufficient inoculum is developed by using the growth medium in the first stage and in the second stage the whole biomass is transferred to the production medium. However, the temperature for inoculum production and that for citric acid production is the same at 30 °C. Low aeration rates are generally used since oxygenation is toxic hence low aeration rate of 0.1 vvm is generally used at the beginning of the fermentation process which is slowly enhanced to 1 vvm as the growth proceeds. Fed-batch processes are generally preferred over continuous process for citric acid production by submerged fermentation. The raw material generally used is molasses amended with nitrogen source like urea, peptone, ammonium nitrate and potassium dihydrogen phosphate as phosphate source. The duration of the fermentation is 3–5 days and the fermentation mother liquor is drained off, mycelium washed from which citric acid is extracted.

7.4.3 Solid-Substrate Fermentation for Citric Acid Production

Solid state fermentation process or Koji process was first developed in Japan. The process is characterised by the development of organism with low water activity environment on insoluble material which served as nutrient as well as solid support. A variety of agro-industrial wastes have been used for citric acid production. These include wheat bran, rice bran, coffee husk and pineapple waste husk. Solid substrate fermentation is carried out in trays or horizontal drum bioreactor.

Solid state fermentation is carried out in trays ($0.0045\ m^3$) where the moist and pre-inoculated substrate (10^6–10^7 spores/g of dry substrate) was distributed in order to have thickness of 6 cm (0.45 Kg of dry substrate). The trays are placed in a room with controlled temperature of 28 °C and humidity of about 97 %. Fermentation is carried out for 120 h.

In horizontal drum bioreactor (Fig. 7.5), 2 kg of substrate with initial moisture of 60 % and is placed inside the drum. The drum is made using steel 360 with 32 cm diameter and 30 cm length (internal volume of $0.024\ m^3$) which consisted of shovel coupled to a motor axle which is rotated with a controlled speed. The HD reactor is rotated three to four times a day. After 20 h of fermentation, saturated air is passed continually into the drum in order to control substrate temperature and moisture. The air flow is maintained at 5 L/min. Fermentation is carried out for 144 h. *A. niger* is the common organism which is generally employed for solid substrate fermentation.

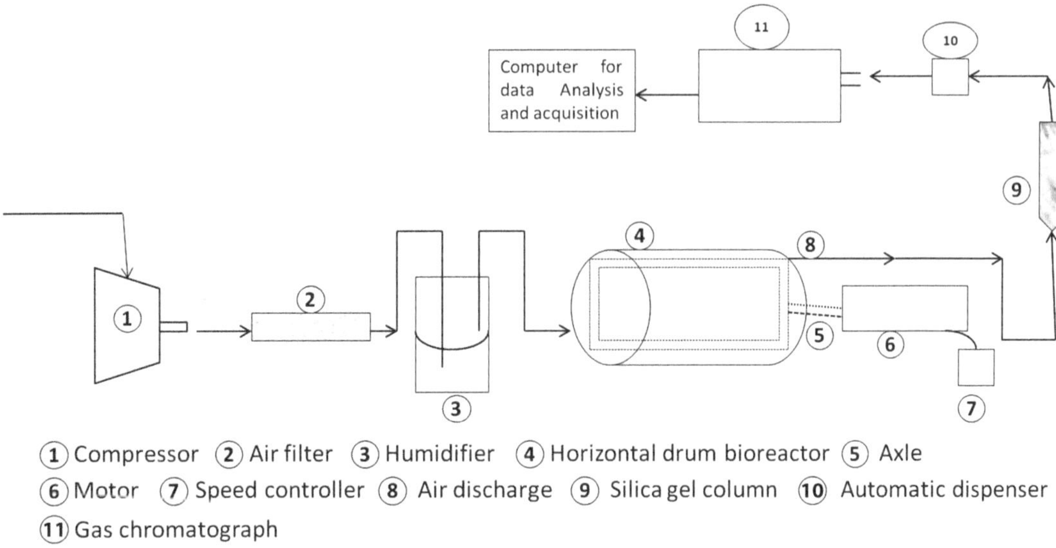

Fig. 7.5 Solid state fermentation for citric acid production using horizontal drum bioreactor

7.4.4 Recovery of Citric Acid

The process of citric acid recovery is same for surface and submerged fermentation wherein the mother liquor containing citric acid is mixed with mycelial washes so as to recover the impregnated citric acid. Broadly the three procedures adopted are (1) precipitation, (2) extraction and (3) adsorption. Precipitation is the first method which is a conventional process of mixing the filtered mother liquor with calcium oxide (hydrated lime) in the ratio of 2:1 at a temperature of 50 °C for 20 min to achieve 100 % precipitation of citric acid. The citric acid is converted into tricalcium citrate tetrahydrate. The precipitated calcium citrate is filtered off and then washed several times with deionised water. The precipitate is finally recovered by filtration and then treated with sulphuric acid which results in the formation of calcium sulphate (gypsum) which can be filtered off and the mother liquor containing citric acid is obtained. Further this mother liquor of citric acid is treated with activated charcoal and passed through cation and anion exchangers. Finally, the liquor is concentrated under vacuum crystallisers between 20 and 25 °C wherein the citric acid crystallises as citric acid monohydrate. Anhydrous citric acid is obtained when the crystallisation temperature is higher than 36.5 °C. The second process of recovery of citric acid is solvent extraction method which involves the use of tridecylamine or triisononylamine with water insoluble ester, ketone or alcohol. The third process which has been recommended by the US Food and Drug Administration uses a mixture of n-octyl alcohol, synthetic isoparaffin petroleum hydrocarbons and tridodecylamine used for the recovery of citric acid from fermented liquors.

7.5 Microbial Production of Adipic Acid

Adipic acid is a dibasic acid which was essentially used as a chemical feedstock for the production of Nylon 6, 6. Apart from its role in the production of nylon, today adipic acid sports a position of commodity chemical due to its versatile use in adhesives, coatings, hydraulic fluids, flue gas desulfurisation scrubber additive, cleaning additive, soil conditioners, glass protection agents, polymer/plasticiser additives, leather tanning, personal care emollients and chemical intermediates. A green method has been developed for the synthesis of the dibasic acid, adipic acid. In this process the dextrose is converted into cis, *cis*-muconic acid which is subsequently by catalytic hydrogenation leads to the formation of adipic acid (Fig. 7.6). This process of synthesis is not known and has been engineered in *E. coli* to develop an ecofriendly process so as to reduce the greenhouse effects due to the production of nitrous acid. More recently, Verdezyne Inc. has engineered a yeast for the commercial production of adipic acid. It has set up a fermentation plant for adipic acid production at Carlsbad, California, in 2011.

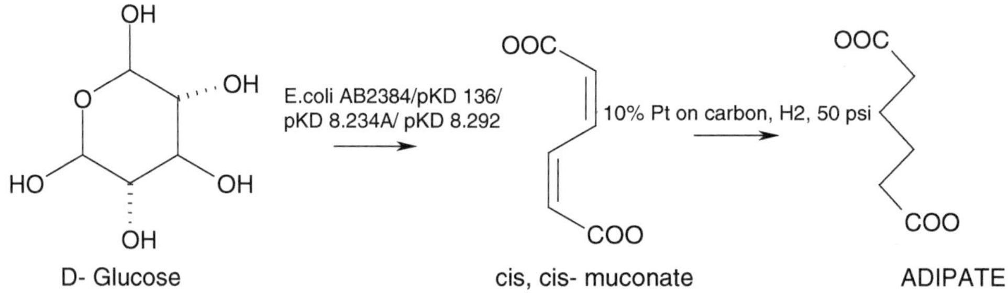

Fig. 7.6 Microbial production of adipic acid from d-glucose via cis, *cis*-muconate

7.6 Microbial Production of 1, 2-Propanediol

1,2-Propanediol (PDO) is major commodity chemical which is used to carry out polycondensations to produce biodegradable plastics and polymer resins. Apart from this, PDO is also being used for the production of non-ionic detergents, as anti-freezing agent, de-icing agent in cosmetics and liquid detergents. It is also used as a feed additive in dog food after being recognised as GRAS by USFDA. The global consumption of 1, 2-propanediol is over 1.5 million metric tonnes, and the biggest producers are Dow and Lyondell. Microbial production of 1, 2-propanediol was first reported from *Clostridium thermobutyricum* (Enebo 1954). The other microorganisms which metabolised sugars, viz. fucose, rhamnose, glucose, xylose to produce 1, 2-propanediol, are *Salmonella typhimurium*, *Klebsiella pneumonia*, *Bacteroides ruminicola*, *Clostridium sphenoides*, *Clostridium thermosaccharolyticum* and *Thermoanaerobacterium thermosaccharolyticum* HG-8. *Thermoanaerobacterium thermosaccharolyticum* exhibits a unique feature of producing enantiomerically pure (R)-1, 2-propanediol and hence been explored for developing a fermentative mass production process.

Attempts are also being carried out to develop recombinant microbial strains for cost-effective production of 1, 2-propanediol from renewable resources. One of the strategies involves development of a recombinant bug which could convert glucose to glycerol and glycerol to 1, 2-propanediol and subsequently optimise the fermentation process. Recently, Cargill and Ashland Inc. have started a joint venture for production of propylene glycol from glycerol coming out of biodiesel plant industry. Cargill has announced the process wherein carbohydrates would be converted into 1, 2-propanediol by *E. coli* or *T. thermosaccharolyticum*.

7.7 Penicillin as a Commodity Chemical

Penicillin was discovered as an antibiotic and marked the golden era of antibiotic drug discovery and development. However, since the late 1990s, it is sporting a status of a commodity chemical for the production of semi-synthetic penicillins. The key factors which led penicillin to become a commodity chemical were (1) addition of precursors in the fermentation medium to enhance the yield, (2) development of high penicillin-yielding strains by strain improvement programmes (see Chap. 10) and (3) development of appropriate industrial equipments for the production as well as isolation of penicillin in bulk quantities. Another important reason for penicillin being converted into a commodity chemical was antibiotic resistance encountered by the pathogenic bacteria and hence newer versions of more potent antibiotics were required which could overcome antibiotic resistance. The discovery of the enzyme penicillin acylase helped in isolation of the β-lactam structure known as 6-aminopenicillanic acid (6-APA) which could be used as a template for development of semi-synthetic penicillins with better antimicrobial activity against drug-resistant bacteria as well as broader spectrum of activity against pathogenic microbes (please refer to Chap. 8). This process of enzymatic splitting of penicillin has been referred to as biosplitting (Fig. 7.7). The global production of penicillins is approximately

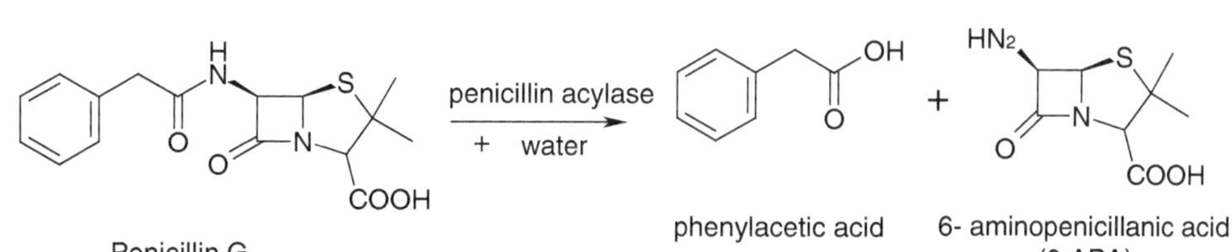

Fig. 7.7 Biosplitting of penicillin G to 6-APA by penicillin acylase

75,000–85,000 tonnes annually. Majority of the penicillin being produced is used for the development of semi-synthetic penicillins, and a very limited amount is directly used as an antibiotic.

7.7.1 Production of Penicillin

Currently the penicillin is produced by high-yielding strains of *Penicillium chrysogenum* which have been developed using classical mutagenesis and recombinant DNA technology. Commercially the production is carried out in a Fed-batch reactor having a tank volume of 20,000–60,000 gal. The carbon source is glucose which is provided as molasses, while the nitrogen source is corn steep liquor. The pH of the fermentation medium is maintained between 6.4 and 6.8, and the temperature during the fermentation process is maintained throughout at 25 °C. Membrane filters are used for providing air during the fermentation process. Thirty percent dissolved oxygen is critical for the production of penicillin. Phenoxyacetic acid and phenylacetic acid are used as precursors for production of penicillin V and penicillin G, respectively. There are three-staged seed fermenters for the production of the inoculum and the fermentation time is 120–200 h for penicillin production. The production of penicillin is monitored by HPLC during the fermentation process. The production of penicillin, i.e. titer achieved in the above process, ranges between 40 and 60 g/l.

7.7.2 Recovery and Purification of Penicillin

Whole broth is filtered through a membrane filter which separates the insolubles from the aqueous liquor. The mycelium is rewashed with distilled water to recover entrapped penicillin in the mycelial cake due to capillary force. The aqueous liquor/mother liquor obtained is adjusted to a pH range of 2–2.5, and then penicillin is extracted in butyl acetate. The butyl acetate fraction is re-extracted with a buffer of pH 6.0 to yield penicillin-rich buffer solution. This penicillin-rich solution is re-extracted with butyl acetate to yield a solvent solution consisting of a high potency of penicillin. This is further concentrated using multiple back extractions using buffer and solvent at different pH using countercurrent contractors, leading to a considerable penicillin concentration in early stages of recovery.

Subsequently, the pigment and broth impurities are removed by activated charcoal. The penicillin is re-crystallised by addition of potassium acetate and isolated as crystalline potassium salt. Additional carbon treatment and solvent washes ensure highly purified penicillin V.

7.8 Summary

Microbes are playing an important role in development of greener industrial processes for the production of commodity chemicals. Thus, reliance of petrochemical for harnessing them is progressively reducing with parallel development of microbially catalysed processes, thereby saving energy and environment. Production of ethanol, acrylamide and adipic acid through microbial process clearly indicates the shift in the paradigm.

Selected Reading

Asano Y, Tani Y, Yamada H (1980) A new enzyme "nitrile hydratase" which degrades acetonitrile in combination with amidase. Agric Biol Chem 44:2251–2252

Cooney CL (1983) Prospects for chemicals and fuels production by fermentation. Basic biology of new developments in biotechnology. Plenum Press, New York

Draths KM, Frost JW (1994) Environmentally compatible synthesis of adipic acid from D-glucose. J Am Chem Soc 116:399–400

Elander RP (2003) Industrial production of β-lactam antibiotics. Appl Microbiol Biotechnol 61:385–392

Enebo L (1954) Studies in cellulose decomposition by an anaerobic thermophilic bacterium and two associated non-cellulolytic species. Viktor Pettersons Bokindustrie Akuebolag, Stockholm

Karaffa L, Kubicek CP (2003) *Aspergillus niger* citric acid accumulation: do we understand this well working black box? Appl Microb Biotechnol 61:189–196

Nagasawa T, Yamada H (1995) Microbial production of commodity chemicals. Pure Appl Chem 67(7):1241–1256

Saxena RK, Pinki A, Saurabh S, Jasmine I, Lata A (2010) Microbial production and application of 1, 2-propanediol. Ind J Microb 50(1):2–11

Soccol CR, Vandenberghe LPS, Rodrigues C, Pandey A (2006) New perspectives for citric acid production and application. Food Technol Biotechnol 44(2):141–149

Watanabe I (1987) Acrylamide production method using immobilized nitrilase containing microbial cells. Meth Enzymol 136:523–530

8 Microbes in Production of Fine Chemicals (Antibiotics, Drugs, Vitamins, and Amino Acids)

8.1 Introduction

Fine chemicals are single pure substances that are produced in small to medium quantities and have a high value (>US$10/kg). They are synthesized via multi-step batch chemical or biotechnological processes. Fine chemicals consist of organic aromatic compounds, organic amines, proteogenic/non-proteogenic amino acids, carbohydrates, heteroaromatic compounds, and saturated and unsaturated fatty acids, uses for which are found in the pharmaceutical, specialty chemical, agricultural, and cosmeceutical industries (Pollack 2007) (Fig. 8.1).

Conventionally, fine chemicals were synthesized via traditional multi-step chemical processes that involved a high consumption of raw materials and resulted in a high level of by-products and wastes. The global demand for greener, ecofriendly, less energy intensive industrial processes has resulted in the harnessing of the potential of microbes for fine chemical production using genetic engineering tools and techniques. Microbial systems are attractive, as they produce and offer a wide range of molecules while using low levels of energy resources compared with energy-intensive chemical processes. Microbes provide novel biosynthetic routes to already known products and are also a source of entirely new products.

8.2 Pharmaceutical Fine Chemicals

Fine chemicals in pharmaceuticals are broadly classified as an 'active pharmaceutical ingredient' (API) and as 'pivotal', 'critical' or 'basic' intermediates.

8.2.1 Antibiosis and Antibiotics

The notion that microbes inhibit other microbes dates back to 1877 when Louis Pasteur observed that *Bacillus anthracis* lost its virulence in the presence of some aerobic organisms. In 1877, Louis Pasteur and Robert Koch observed that an airborne bacillus inhibited the growth of *B. anthracis*. This phenomenon was named 'antibiosis', meaning 'against life', by French bacteriologist Jean Paul Vuillemin. The American microbiologist Selman A. Waksman (1947) later introduced the term 'antibiotics' for those substances that mediated the phenomenon of antibiosis.

As soon as leading microbiologists of the late nineteenth century, such as Louis Pasteur and Robert Koch, implicated microbes as the cause of infections, humankind's search for agents that could control these infections began. This marked the beginning of the chemotherapeutic era, the

S. Saxena, *Applied Microbiology*,
DOI 10.1007/978-81-322-2259-0_8, © Springer India 2015

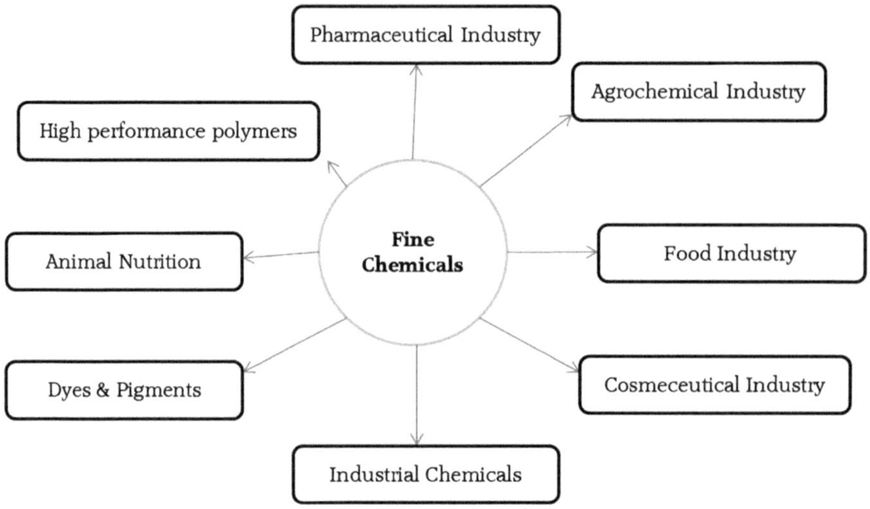

Fig. 8.1 Application areas of fine chemicals

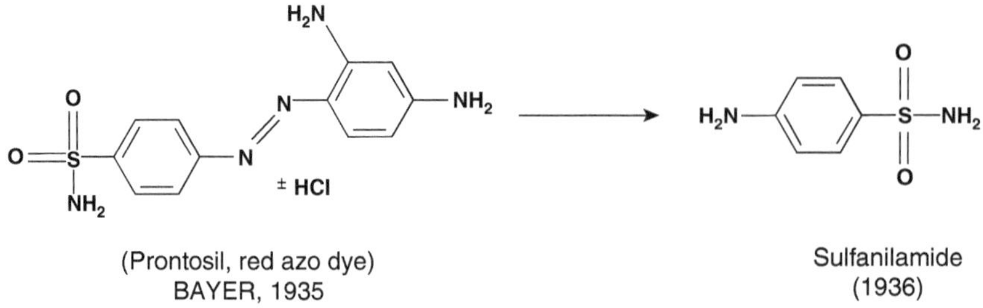

Fig. 8.2 Conversion of prontosil into sulphanilamide

foundations of which were laid by Paul Ehrlich. He coined the term 'chemotherapy' and popularized the concept of 'magic bullets'. Ehrlich established the role of salvarsan (dihydroxydiaminoarsenobenzenedihydrochloride) in 1909 as a modality for the treatment of syphilis while he was working on dyes that stained specific chemicals. Similarly exploring the action of dyes, Gerhard Domagk (1895–1964), a scientist at the Bayer wing of the IG Farbenindustrie consortium, tested a number of dyes synthesized by his colleagues, Fritz Mietzsch and Josef Klarer. Prontosil red was identified as a compound that bound very strongly with fibers; it was hoped it would do the same with bacteria. Initially, it was found to recover a mouse from severe hemolytic streptococcal infection; however, when tested in vitro it did not exhibit any activity. Further investigations revealed that the active compound dissociated from prontosil in vivo and was identified as sulphanilamide. Thus, the first broad spectrum antibacterial chemotherapeutic class 'sulphonamide' was identified (Fig. 8.2). In the early 1940s, several companies used sulphonamide derivatives until one formulation, consisting of diethylene glycol and sulfanilamide, killed over 100 people in the USA. The USA then empowered the Food and Drug Administration to regulate the licensing and approval of new drugs for commercialization.

8.2.2 Discovery of Penicillin: Beginning of the Antibiotic Era

The golden era of antibiotics began with the serendipitous discovery of penicillin by Sir Alexander Fleming while working at St. Mary's Hospital, London. A plate of *Staphylococcus* was found to have been contaminated by a blue-green mold (*Penicillium*) while he was on vacation. On his return, Fleming observed that colonies in the vicinity of the mold were dissolved. He subsequently grew the *Penicillium* in a broth and concluded that it could destroy a variety of pathogenic bacteria using a method known as the agar cup method, which was the benchmark for the development of current in vitro antimicrobial susceptibility-testing methods. Ernst B. Chain and Professor Howard Florey at the Department of Pathology, Oxford University, were testing the bactericidal activity of different substances. While going through Fleming's paper, they set their target on the purification of penicillin and obtained the culture from Alexander Fleming.

Norman Heatley aided in the culture, purification, and assay of the crude and pure penicillin. Mice with staphylococcal and streptococcal infections were injected with purified penicillin molecule; they survived, indicating the stability of the antibiotic. Penicillin production was first attempted using Fleming's original strain, *Penicillium notatum*, using surface fermentation methods (Bennett and Chung 2001). However, Florey and his team started using submerged fermentation attempts using corn steep liquor and other sources to enhance the production of penicillin. In the quest for a better strain that could further enhance production under deep submerged vat fermentation, they happened upon a moldy cantaloupe from which the industrial strain of *Penicillium chrysogenum* was isolated and improved in due course for the commercial production of penicillin through the efforts of Heatley, Florey, Chain, and colleagues. Fleming's discovery of a chemotherapeutic agent from a microbial source paved the way for the commercial production of penicillin and marked the beginning of the golden era of antibiotics.

After the discovery of penicillin, screening programs were initiated in academia to explore the potential of microbial secondary metabolites as anti-infectives. Soil was the prime resource for hunting microbes and screening their antimicrobial activity. Streptomycin, another important anti-bacterial, was isolated from *Streptomyces griseus*, which was a soil-inhabiting filamentous bacterium, i.e. actinomycetes. Streptomycin was discovered in 1943 by Professor Selman A. Waksman from Rutgers University, USA, and his student Albert Schatz et al. (1973). It was effective in fighting bacterial infections such as tuberculosis, whooping cough, and typhoid. Thus, actinomycetes became the second most preferred organisms for screening new antibacterial drugs. Bacteria, especially *Pseudomonas* and *Bacillus* spp., were also screened to explore their antibiotic-producing potential. Today, antibacterial antibiotics are classified into different functional classes of compounds and have been studied for their mechanistic action in inhibiting pathogenic bacteria (Fig. 8.3).

8.2.3 Antibiotics Discovered from Fungi

Fungi belonging to *Penicillium* and *Acremonium* spp. predominantly synthesize antibiotics that possess the β-lactam group and are therefore referred to as β-lactam antibiotics. Penicillins are produced by *Penicillium* spp. and are characterized by a four-member β-lactam ring along with a five-member thiazolidine ring (Fig. 8.4a). Penicillins G and V are the stable form of penicillins developed for intravenous and oral use, respectively (Fig. 8.4b, c). *Penicillium griseofulvum* produces an antifungal antibiotic griseofulvin (Fig. 8.5a), initially identified as the curling factor. It was discovered by Harold Raistrick and his team from the London School of Hygiene and Tropical Medicine in 1939. The properties of griseofulvin were recognized when it was re-isolated from *Penicillium janczewski*. It is fungistatic in nature and used for treating dermatophytic fungal infections like ringworm. The commercial production of griseofulvin was undertaken by

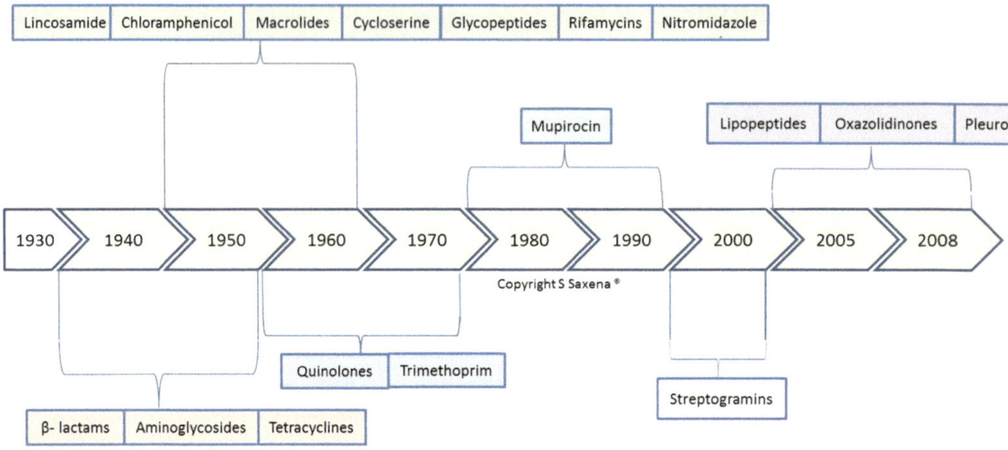

Fig. 8.3 Timeline for the discovery of antibacterial antibiotics

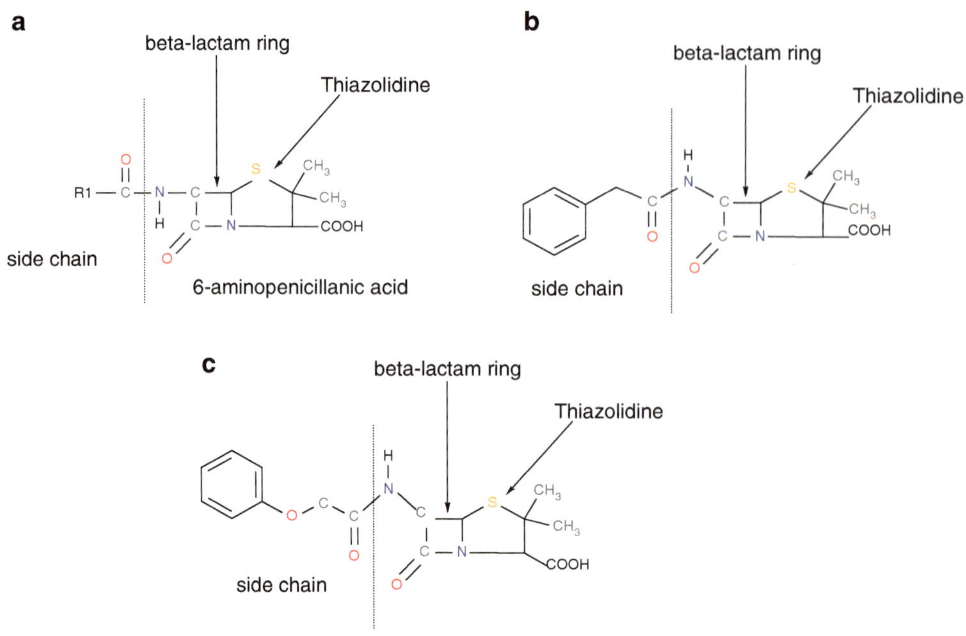

Fig. 8.4 General structure of (**a**) penicillin, (**b**) penicillin G, and (**c**) penicillin V

Glaxo using the strain *Penicillium patulum* via the submerged fermentation process.

Cephalosporins are produced by *Acremonium* spp. (previously known as *Cephalosporium* spp.). Cephalosporins were initially isolated from a culture of *Cephalosporium acremonium* isolated from sewers in Sardinia by Professor Giuseppe Brotzu in the University of Cagliari in 1945. The fungal extract was able to effectively kill *Salmonella typhi*, but a lack of funds meant that Brotzu could not conduct detailed isolation and characterization of the secondary metabolite/

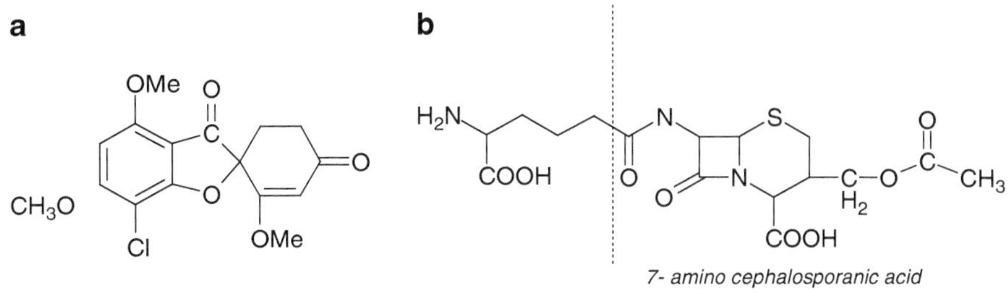

Fig. 8.5 (**a**) Griseofulvin and (**b**) cephalosporin C

antibiotic. The actual isolation of Cephalosporin was carried out by Edward Abraham, a student of Alexander Fleming, to whom the culture was passed by Brotzu (Murray 1981). Initially, cephalosporin P was isolated from *C. acremonium* and found to be steroidal in nature. It was active against Gram-positive microorganisms only. Cephalosporin C (Fig. 8.5b) was isolated in 1953 by Abraham and it resembled penicillin in that it had an acyl side chain attached to a double-ring nucleus. As well as the four-member β-lactam ring, it possessed a six-member dihydrothiazine ring. This nucleus was named '7-aminocephalosporanic acid' (7-ACA). The side chain of the cephalosporin C was later modified by Eli Lilly for the development of the first stable antibiotic cephalothin.

Penicillins and cephalosporins constitute the broad β-lactam class of antibiotics. Development of resistance to penicillins and poor activity against Gram-negative bacteria has motivated research to find improved penicillins with a broader spectrum of activity. The following semi-synthetic penicillins were introduced by Beecham Research Labs: methicillin (1960), ampicillin (1961), and cloxacillin (1962). Semi-synthetic penicillin was synthesized using 6-aminopenicillanic acid (6-APA) as the starting point. 6-APA was initially obtained as a fermentation by-product along with penicillin G or V in very low yields. After screening various microorganisms, an enzyme from *Streptomyces lavendulae* was found to be capable of carrying deacetylation of penicillin V for the manufacture of 6-APA. Similarly, cephalosporin C acylase-producing microorganisms have been exploited for the conversion of cephalosporin C to produce β-lactam nuclei of 7-aminodeacetoxy cephalosporanic acid (7-ACA) in the production of semi-synthetic cephalosporins. D-amino oxidase converts the cephalosporin C into 7 β-(5-carboxy-5-oxopentaamido)-cephalosporanic acid followed by the auto-conversion to glutaryl-7-ACA. Cephalosporin acylase subsequently deacylates glutaryl-7-ACA to 7-ACA. Today, semi-synthetic penicillins and cephalosporins are being synthesized from penicillin and cephalosporin for use against infections caused by various pathogenic bacteria (Tables 8.1 and 8.2).

Strobilurins A and B are a class of antifungal antibiotics that were initially isolated from *Strobilurus tenacellus* (a fungus that grows on pinecones) (Anke et al. 1977). Strobilurin A has been produced from *Cyphellopsis* sp. and has served as a template for the development of stable azoxystrobin. In 1996, this was introduced as a broad spectrum fungicide of fungal diseases on 84 different crops in 71 countries, representing over 400 crop/disease systems.

Caspofungin (MK-0991) (1, 3)-β-D-glucan synthase inhibitor is an echinocandin class antifungal drug marketed as Cancidas (Merck) in Europe and Caspofungin MSD (Merck) in the USA. Caspofungin is a water-soluble amphipathic lipopeptide that is semi-synthetically derived from pneumocandinB0, a fermentation product from *Glarea lozoyensis* (Deresinski and Stevens 2003). Papulacandins are a family of antibiotics isolated from *Papularia sphaerosperma* (Pers.) *Hoehnel* and strongly inhibit the growth of

Table 8.1 Semisynthetic penicillins developed using a 6-APA core structure

Class of semi-synthetic penicillins (penams)	Semi-synthetic penicillin name	Year introduced	Antimicrobial activity
Aminopenicillins	Ampicillin	1961	Broad spectrum[a]
	Amoxicillin	1972	
Carboxypenicillins	Carbenicillin	1967	Broad spectrum
	Ticarcillin	1975	
	Temocillin	1984	
Ureidopenicillins	Piperacillin	1980	Broad spectrum
β- Lactamase resistant	Methicillin	1960	Narrow spectrum[b]
	Oxacillin	1962	
	Cloxacillin	1962	

[a]Broad spectrum indicates it is effective against Gram-positive and Gram-negative bacteria
[b]Narrow spectrum indicates only Gram-positive bacteria

Table 8.2 Semi-synthetic cephalosporins developed using a 7-ACA core structure

Generation of semi-synthetic cephalosporins	Cephalosporin name	Year introduced	Antimicrobial activity
First generation	Cefalexin	1970	Good antimicrobial activity against Gram-positive bacteria but limited activity against Gram-negative β- Lactamase sensitive
	Cefazolin	1971	
	Cefradine	1972	
	Cefadroxil	1977	
Second generation	Cefoxitin	1977	Increased binding affinity to penicillin-binding proteins (PBPs) and increased antimicrobial activity
	Cefmandole	1977	
	Cefuroxime	1977	
	Cefaclor	1979	
	Cefmetazole	1980	
	Cefotiam	1981	
	Cefotetan	1984	
	Cefprozil	1992	
Third generation	Cefotaxime	1980	Improved activity against *Enterobacteriaceae* associated with hospital-acquired infections; some agents are also active against *Pseudomonas aeruginosa*
	Cefsulodin	1980	
	Cefoperazone	1981	
	Ceftazidime	1983	
	Cefpiramide	1985	
	Cefixime	1987	
	Cefpodoxime	1989	
	Cefdinir	1991	
Fourth generation	Cefpirome	1992	Have greater activity against Gram-negative bacteria
	Cefepime	1994	
Fifth generation	Ceftobiprole	2008	Only β-lactam antibiotics that are effective against methicillin-resistant *Staphylococcus aureus*
	Ceftaroline fosamil	2010	

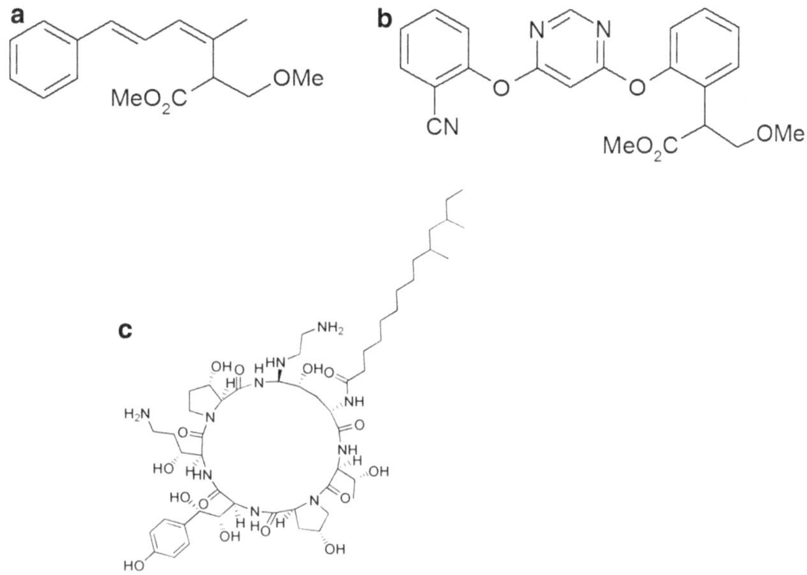

Fig. 8.6 Antifungal antibiotics produced by fungi: (**a**) strobilurin A, (**b**) azoxystrobin, and (**c**) caspofungin

Candida albicans. These compounds could not be further developed into antibiotics due to a limited efficacy in animal model studies despite medicinal chemistry efforts (Fig. 8.6).

8.2.4 Actinomycetes in Antibiotic Discovery

Of about 20,000 antibiotics produced via microorganisms, 45 % of antibiotics come from actinomycetes, 80 % of which come from a single genus *Streptomyces*. Streptomycin was the first aminoglycoside class of antibiotic discovered by Selman A. Waksman and his graduate student Albert Schatz at Rutgers University from *S. griseus* for tuberculosis. Several trials of this drug were carried out at different centers, including Mayo Clinic, to establish its antibacterial potential. Waksman and Schatz were granted US patent "Streptomycin and process of preparation" on 21 September 1948. Streptomycin (Fig. 8.7a) was licensed to Merck and Waksman, who received the Nobel Prize in Medicine in 1952 for this significant discovery. The other antibiotics discovered by Waksman and his group are neomycin, grisein, streptothricin, candidin, and candicidin. *Streptomyces* sp. has been an important source of major classes of antibacterial drugs, namely tetracyclins, aminoglycosides, macrolides, chloramphenicol (acetamide), and β-lactams (Table 8.3). David Gottlieb discovered chloramphenicol (Fig. 8.7b) from *Streptomyces venezuelae* and brought it into clinical practice under the trade name Chloromycetin. In 1945, Benjamin Minge Duggar, working at Lederle laboratories under the supervision of Y. Subbarao, discovered the first tetracycline class of antibiotic, chlortetracycline (Aureomycin) (Fig. 8.7c), from *Streptomyces aureofaciens*. This is an amphotericin B polyene class antibiotic isolated from *Streptomyces nodosus* and used for the treatment of fungal infections caused by *Candida* sp. and *Aspergillus fumigatus*.

β-lactamase is an enzyme produced by microorganisms to resist the β-lactam antibiotics, i.e. penicillins and cephalosporins.

However, several inhibitors of β-lactamases have been reported from *Streptomyces* species. These compounds, when formulated with β-lactam antibiotics, re-potentiate them to overcome resistance encountered in the pathogenic bacteria

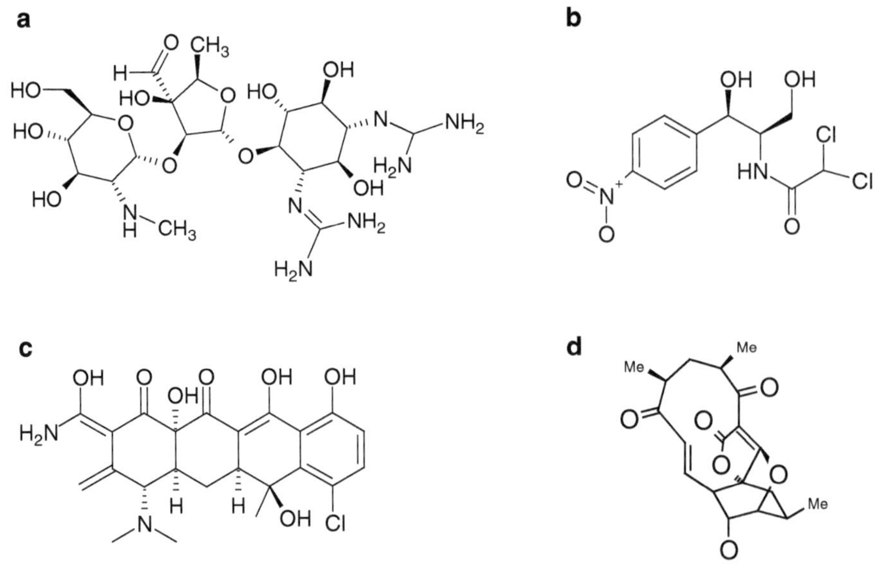

Fig. 8.7 Antibiotics from streptomyces: (**a**) streptomycin, (**b**) chloramphenicol, (**c**) chlortetracycline, and (**d**) abyssiomicins

Table 8.3 Antibiotics produced by actinomycetes

Microorganism	Antibiotic	Antimicrobial activity	Antibiotic class	Year
Streptomyces griseus	Streptomycin	Antibacterial	Aminoglycoside	1944
Streptomyces fradiae	Neomycin	Antibacterial	Aminoglycoside	1949
Streptomyces kanamyceticus	Kanamycin	Antibacterial	Aminoglycoside	1957
Streptomyces fradiae	Framycetin	Antibacterial	Aminoglycoside	
Micromonospora purpurea	Gentamicin	Antibacterial	Aminoglycoside	1964
Streptomyces tenebrarius	Tobramycin	Antibacterial	Aminoglycoside	1975
Streptomyces krestomuceticus	Parmomycin	Anti-protozoal	Aminoglycoside	1950
Streptomyces spectabilis	Spectinomycin	Antibacterial	Aminocyclitol	1961
Streptomyces rimosus	Tetracycline	Antibacterial	Tetracycline	1950
Streptomyces venezuelae	Chloramphenicol	Antibacterial	Acetamide	1949
Streptomyces orientalis	Vancomycin	Antibacterial	Glycopeptide	1955
Actinoplanin teichomyceticus	Teicoplanin	Antibacterial	Glycopeptide	1988
Actinoplanes sp. *ATCC33706*	Ramoplanin	Antibacterial	Glycolipodepsipeptide	
Streptomyces roseosporus	Daptomycin	Antibacterial	Lipopeptide	
Streptomyces erythraea	Erythromycin	Antibacterial	Macrolide	1952
Saccharopolyspora erythraea	Telithromycin	Antibacterial	Semi-synthetic macrolide	
Amycolatopsis rifamycinica	Rifamycin	Anti-tubercular	Ansamycins	
Streptomyces cattleya	Thienamycin (Imipenem)	Antibacterial	β-lactam (carbacephem)	1976
Streptomyces noursei	Nystatin	Antifungal	Polyene	1950
Streptomyces nodosus	Amphotericin B	Antifungal	Polyene	
Streptomyces natalensis	Natamycin	Antifungal	Macrolide polyene	

Staphylococcus aureus, *Escherichia coli*, etc. *Streptomyces clavuligerus* yielded a β-lactamase inhibitor that was isolated and characterized as clavulanic acid by scientists from Beecham who then patented it in 1981. Merck, in their screening program, isolated the thienamycin family of carbapenem β-lactams from *Streptomyces cattleya* (Williamson et al. 1985). However, due to dimerization, the thienamycin was degraded and a derivative N-formi-imidoyl thienamycin was developed. This was commercially known as Imipenem by Merck scientists and is classified under β-lactams (carbacephams). Marine actinomycetes recently have become an interesting resource for untapped secondary metabolites. Abyssomicin C (Fig. 8.7d) has been isolated from *Verrucosispora* strain AB-18-032 from the Sea of Japan. This polycyclic polyketide antibiotic inhibits para-aminobenzoic acid (PABA) biosynthesis in the fatty acid pathway. Salinosporamide A is a β-lactone-ϒ-lactam proteasome inhibitor produced by the novel marine actinomycete *Salinispora tropica* (Feling et al. 2003).

8.2.5 Antibiotics Discovered from Bacteria

Bacillus species has also been a prolific producer of secondary metabolites possessing antibiotic properties. Approximately 800 different secondary metabolites possessing antibiotic properties against bacteria or fungi have been documented to date. The types of antibiotics produced by bacteria include the β-lactams, phenazines, and fatty acid derivatives apart from peptides. *Chromobacterium violaceum* produces a simple monobactam (monocyclic β-lactam) SQ 26,180 that has been derivatized into aztreonam by the Squibb Institute of Medical Research. Aztreonam (Fig. 8.8a) is active against Gram-negative bacteria and highly effective against *E. coli* and *Klebsiella pneumoniae*.

Monobactams SQ 28,502 and SQ 28,503 have also been isolated from fermentations of Gram-negative bacterium *Flexibacter* spp. SC 11479 (ATCC 35103). They exhibit weak antimicrobial activity but possess potential resistance to the action of β-lactamases (Cooper et al. 1983). Zwittermicin A (Fig. 8.8b) is a fungistatic polyol class of antibiotic that has been isolated from culture broth of *Bacillus cereus* UW85 and protects alfalfa seedlings from damping off disease caused by *Phytophthora medicaginis*. The antibiotic has also been found to possess a protective action against *Phytophthora nicotiana*, *Pythium* spp., and *Sclerotinia minor* (Silo Suh et al. 1994).

Andrimid (Fig. 8.8c) is a natural peptide antibiotic isolated from marine *Pseudomonas fluorescens* isolated from a tunicate in Alaska. It was first isolated from the strain of *Enterobacter* sp. that was an intracellular symbiont in the brown grasshopper and exhibited potent activity against the blight pathogen *Xanthomonas campestris* pv. *orate* (Fradenhagen et al. 1987). Andrimid exhibits a minimum inhibitory concentration (MIC) in the range of 2–8 μg/ml against methicillin-, gentamicin-, and ciprofloxacin-resistant *S. aureus*.

Bacitracin, a cyclic polypeptide, has been formed by *Bacillus subtilis* var. *Tracy* in 1945. It is active against Gram-positive microorganisms. Colistin is also a polypeptide antibiotic produced by *Bacillus polymyxa* var. *colistinus* and is composed of cyclic polypeptides colistin A and colistin B. It is very active against Gram-negative bacilli and is used to treat multi-drug-resistant (MDR) infection caused by *Pseudomonas aeruginosa*. In 1939, Rene Dubos discovered gramicidin, a mixture of 80 % gramicidin A, 6 % gramicidin B, and 14 % gramicidin C from *Bacillus brevis* isolated from soil. Gramicidin is used topically since, when administered internally, it is responsible for hemolysis rather than inhibiting bacteria. Gramicidin S was discovered by Russian microbiologist Georgyi Frantsevitch Gause and his wife Maria Brazhnikova in 1942. It is a cyclodecapeptide formally written as *cyclo* (-Val-Orn-Leu-D-Phe-Pro-)$_2$. Gramicidin S is also used topically. Mycobacillin is an antifungal cyclic peptide antibiotic produced by *B. subtilis B3* (Majumdar and Bose 1958).

Pseudomonic acid A is an antibiotic that was discovered from *P. fluorescens* NCIMB10586 (Fuller et al. 1971) and exhibited activity against streptococci and staphylococci and against Gram negative bacteria such as *Haemophilus influenza* and *Neisseria gonorrhoea*. Beecham Pharmaceuticals developed it as the topical

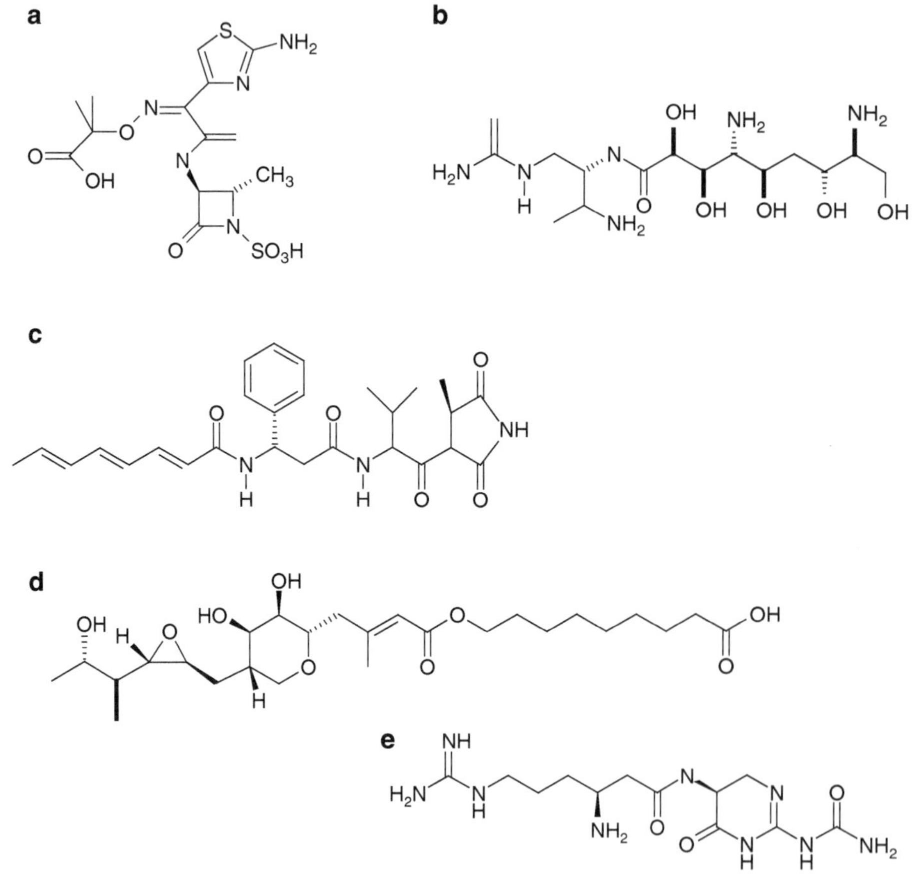

Fig. 8.8 Antibiotics synthesized by bacteria: (**a**) aztreonam, (**b**) zwittermycin A, (**c**) andrimid, (**d**) mupirocin, and (**e**) TAN-1057A

antibiotic mupirocin (Fig. 8.8d). It has been used for topical disinfection of skin and soft tissue infections caused by methicillin-resistant *S. aureus* (MRSA). When administered internally, mupirocin binds very strongly to serum protein (95 %), thereby reducing its efficacy against systemic infections. Hence, it has not been approved for systemic use (Sutherland et al. 1985).

A few years ago, a strain of *Flexibacter* sp. PK-74 isolated in Japan exhibited potential activity against MRSA (Katayama et al. 1993). Elucidation of the bioactive compound led to the isolation of the ot tetrahydropyrimidine class of antibiotics, designated as TAN-1057 A-D. These are broad spectrum antibiotics with an MIC range between 6.25 and 12.5 μg/ml. TAN-1057A (Fig. 8.8e) has been found to be more effective than vancomycin and imipenem in a murine *S. aureus* sepsis model while studying the in vivo efficacy of the compound.

Myxobacteria are Gram-negative obligate aerobic, chemotrophic δ-proteobacteria that predominantly inhabit soil. Their unique feature is the ability to develop fruiting bodies during starvation periods. They are considered to be one of the best producers of novel antibiotics after actinomycetes. Approximately 100 core structures have been discovered to date, many of which are antibacterial in nature. *Myxococcus flavus*, *Myxococcus virescens*, and *Cystobacter velatus*

Table 8.4 Lantibiotics produced by bacteria[a]

Microorganism	Antibiotic	Use	Reference
Lactobacillus lactis	Nisin A	Gastrointestinal infections; skin infections	Piper et al. (2009)
Lactobacillus lactis	Nisin F	Skin infections	Kwaadsteniet et al. (2010)
Microbispora ATCC PTA 5024	Microbisporicin	Hospital-acquired infections; gastrointestinal infections	Jabes et al. (2011)
Staphylococcus hominis	Hominicin	Hospital-acquired infections	Kim et al. (2010)
Staphylococcus gallinarum T	Gallidermin	Skin and cutaneous infections	Mansoroi et al. (2010)
Bacillus sp. strain HIL, Y-85,54728	Mersacidin	Respiratory tract infections	Kruszewska et al. (2004)
Streptococcus salivarinus K12	Salivaricin A	Gastrointestinal tract infections	Burton et al. (2006)
Lactococcus lactis DPC3147	Lacticin 3147	Gastrointestinal infections; skin infections	Piper et al. (2012)
		Hospital-acquired infections	
Bacillus thuringensis DPC 6431	Thuricin CD	Gastrointestinal tract infections	Rea et al. (2010)
Streptococcus mutans	Mutacin 1140	Gastrointestinal tract infections	Smith et al. (2008)

[a]Data taken from Bactibase

are myxobacteria that have produced myxothiazol, myxalamid, and stigmatellin, respectively, which possess bioactivity against several fungi. Soraphen from *Sorangium cellulosum* has been found to possess potent anti-fungal activity as it acts as an inhibitor of the enzyme acetyl-CoA-carboxylase. *Myxococcus stipitatus* was found to produce melithiazol A, which possesses high antifungal activity (Sasse et al. 1999). The other potent anti-fungal agent was haliangicin, isolated from a marine myxobacteria *Haliangum luteum* (AJ-13395). More recently, thuggacins have been isolated from *Sorangium cellulosum* So ce895. Thuggacins are a macrolide class of antibiotics that possess excellent antibacterial properties against Gram-positive bacteria such as *Micrococcus luteus*, *Corynebacterium* sp., and *Mycobacterium* sp. (Irschik et al. 2007).

A variety of peptide antibiotics have also been isolated from bacteria, and they have been referred to as lantibiotics, which essentially means lanthionine-containing antibiotics. Lantibiotics contain polycyclic thioether amino acids as well as the unsaturated amino acid dehydroalanine and 2-aminoisobutyric acid. Varieties of lantibiotics have been discovered from bacteria and find applications as preservatives and as medicines (Table 8.4).

8.2.6 Microorganisms Producing Other Pharmaceutically Active Metabolites

Microbes, apart from producing antibiotics, are also prolific producers of structurally diverse compounds that possess anti-hypercholesterolemic, anti-diabetic, immunosuppressant, and anti-cancer properties. Enzyme inhibitors are molecules that inhibit or decrease the activity of an enzyme under in vitro or in vivo conditions. The inhibitor can also lead to a complete loss of activity of an enzyme. Enzyme inhibitors also serve as therapeutic interventions in the management of metabolism-associated disorders or diseases. Hence, the discovery and development of enzyme inhibitors is an important area of research in biochemistry and pharmacology disciplines during the drug discovery and development process. Hence, these have

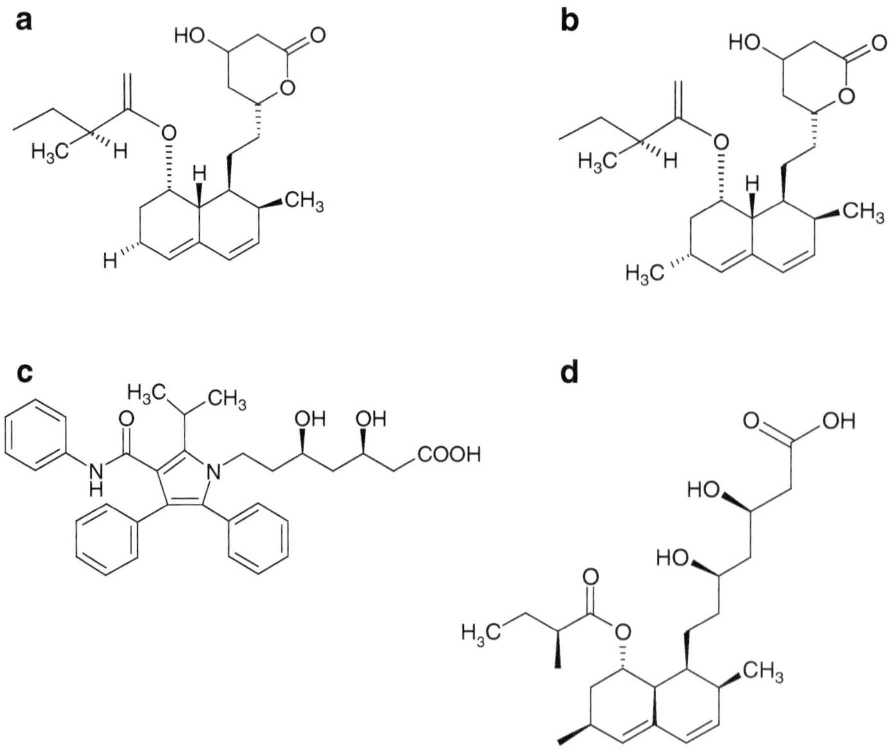

Fig. 8.9 Anti-lipidemic agents produced by microbes: (**a**) mevastatin, (**b**) lovastatin, (**c**) atorvastatin, and (**d**) pravastatin

been developed into drugs for the treatment of specific diseases. Some microbial products have also been used for developing drugs for Alzheimer's dementia.

8.2.6.1 Anti-hypercholesterolemic Agents from Microbes

These drugs are also referred to as anti-lipidemic drugs or statins and are prescribed to patients with hypercholesterolemia, i.e. excess cholesterol production and deposition. Endo et al. (1976) first isolated mevastatin (ML-236B) (Fig. 8.9a) from a strain of *Penicillium citrinum* and later from *Penicillium brevicompactum*. It is a hexahydro-naphthalene skeleton substituted with a p-hydroxy-δ-lactone moiety that, on reaction with alkali, could be converted into a water-soluble open acid. Mevastatin inhibited hydroxymethylglutaryl co-enzyme A (HMG-CoA) reductase, the enzyme responsible for conversion of mevalonate into cholesterol. The acidic form of mevastatin exhibited more potent inhibition, as observed from its K_i values. Merck researchers isolated another analog of mevastatin called lovastatin (previously known as mevinolin or monacolin K) from *Aspergillus terreus*. This new agent proved to be slightly more active in inhibiting HMG-CoA reductase than the parent compound. Lovastatin (Fig. 8.9b) exhibited a drastic reduction in low-density lipoprotein (LDL)-cholesterol levels in a population that primarily included patients with primary and severe hypercholesterolemia.

Currently, commercial drugs for the treatment of hyperlipidaemia are structures derived from mevastatin and lovastatin. These include simvastatin, which is a synthetic derivative of mevastatin marketed under the brand name Zocor® by

Merck. Pravastatin (CS-514) (Fig. 8.9d) was isolated from *Nocardia autotrophica* by researchers at Sankyo Pharma Inc., Japan, and is currently being marketed by Bristol-Myers Squibb. Atorvastatin (Fig. 8.9c) is a synthetic compound that is a very potent statin drug developed by Pfizer and marketed as Lipitor®.

8.2.6.2 Microbially Produced Anti-diabetic Agents

Acarbose, a complex oligosaccharide, has been isolated from neglected genera of actinomycetes *Actinoplanes* sp. by scientists at Bayer, Germany, while screening inhibitors of oligosaccharide- and polysaccharide-degrading enzymes to treat non-insulin-dependent diabetes (type II diabetes). These inhibitors were generally expected to restrict the food-induced increase in blood sugar levels by retarding the rate of disintegration of the complex in the intestine and delaying their absorption. Acarbose has been found to inhibit the α-glucosidase type of enzymes present in the intestine, thereby blocking the breakdown and absorption of oligosaccharide and polysaccharides. Bayer launched acarbose under the trade name Glucobay in 1990 (Hanefeld et al. 1991).

Valiolamine is an aminocyclitol that was isolated from *Streptomyces hygroscopicus* subsp. *limoneus* and has been found to exhibit potential α-glucosidase inhibitory activity (Kamada et al. 1984). Valiolamine is converted into valiolone by oxidative deamination; it is then used to prepare voglibose (US patent 61505684). Voglibose has proved beneficial for the treatment of postprandial hyperglycemia in diabetic patients. In Japan, voglibose, BASEN®, was introduced in 1993 for diabetes and obesity by Takeda Pharmaceutical Company. Voglibose was introduced by Ranbaxy Laboratories as VOLIX® and by Zydus Cadila as Prevog®.

Streptomyces toxytricini produces lipstatin, a pancreatic lipase inhibitor, and interferes with the gastrointestinal absorption of fat to combat obesity and diabetes (Weibel et al. 1987). The commercial product is tetrahydrolipstatin, known as orlistat (US20030149095 A1). Orlistat is marketed under the trade names Xenical® by Roche, Alli® by GlaxoSmithKline, Lipcut® by Biocon, and Cobese® by Ranbaxy.

8.2.6.3 Immunosuppressants Produced by Microorganisms

Immunosuppressive agents are substances that inhibit or prevent the activity of the immune system. They are basically used in the transplantation of organs or tissues to prevent rejection, in the treatment of autoimmune disorders or diseases like arthritis that have autoimmune origins, and also in the treatment of non-autoimmune inflammatory conditions (Fig. 8.10).

The discovery of cyclosporine from the fungus *Tolypocladium inflatum* marked the beginning of immunopharmacology. Sandoz scientists isolated the fungus from a soil sample from Hardangervidda, Norway, in 1969 (Borel et al. 1976). Cyclosporine comprises 11 amino acid residues arranged in a cyclic fashion and one amino acid, D-amino acid. Cyclosporine selectively regulates T cells without exhibiting excessive toxicity and is widely used in organ and tissue transplantation surgery. Tacrolimus (FK-506) is a macrolide class of natural product isolated from *Streptomyces tsukubaensis* No. 9993 in 1984 and is used in allogeneic organ

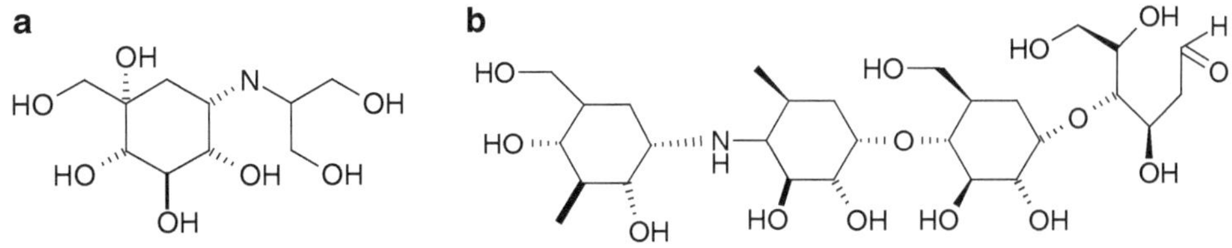

Fig 8.10 Anti-diabetic agents: (**a**) voglibose and (**b**) acarbose

transplantation surgery (Kino et al. 1987). It is 100 times more potent than cyclosporine and approved by the US FDA initially for liver transplantation; in 1994, approval was extended to kidney and heart transplantation. Tacrolimus ointment is also used topically for the treatment of inflammatory skin diseases (psoriasis) and atopic dermatitis. Tacrolimus is commercially sold as Prograf® and Advagraf®.

Rapamycin (Sirolimus) has been used as an immunosuppressant to prevent organ transplant rejection. It was first isolated from *S. hygroscopicus* from Rapa Nui/Easter Island (a part of Chile) in 1975. Rapamycin was approved in 1999 for bone marrow and kidney transplantation (Huang et al. 2003). Mycophenolic acid from *Penicillium brevicompactum* has a selective anti-proliferative effect on lymphocytes that rely on the de novo synthesis of purine. Mycophenolic acid as mycophenolate mofetil has been clinically developed and approved for the prevention of acute renal allograft rejection when given in combination with cyclosporine and steroids.

8.2.6.4 Anti-cancer/Anti-tumor Agents of Microbial Origin

Actinomycin-D, one of the first natural metabolites used for tumor treatment, was first isolated from *Streptomyces antibioticus*. It is still used in the treatment of Wilms tumor in children. Bleomycins are glycopeptide antibiotics. Bleomycin A2 was isolated from the fermentation of *Streptomyces verticillus* by H. Umezawa in 1966 and approved by the US FDA in 1973 for clinical use (Umezawa et al. 1966). It is marketed under the trade name Blenoxane. *Streptomyces achromogenes* produces streptozotocin, which exhibits selective toxicity against pancreatic β cells. It was approved by the US FDA in 1982 as an anti-tumor drug for pancreatic β cells.

The anthracyclin class of anti-tumor agents are the most clinically efficacious agents. Scientists at Farmitalia Research Laboratories isolated *Streptomyces peucetius* from a soil sample from an area in Castel de Monte. It produced the anthracycline daunomycin, which exhibited activity against murine tumors. The organism was mutated with N-nitroso-N-methylurethane, and the mutant produced a red colored antibiotic adriamycin (doxorubicin) (Fig. 8.11a). Adriamycin is used for the treatment of leukemia and Hodgkin's lymphoma. Doxil is the pegylated liposome encapsulated form of doxorubicin. 4-demethoxy daunorubicin (idarubicin) (Fig. 8.11b) has been clinically tested and exhibits powerful anti-leukemia activity with reduced cardiotoxicity. It is commercially known as Zavedos®. Epirubicin® is marketed as Ellence®. Rapamycin (sirolimus) also possesses anti-proliferative action and could be used as a chemotherapeutic agent. Geldanamycin, a macrocyclic polyketide isolated from *S. hygroscopicus* var. *geldanus*, possesses potent anti-tumor potential and reduced hepatotoxicity and has been derivatized as 17-N-allylamino-17-demethoxygeldanamycin (Fig. 8.11c). This compound has successfully cleared phase I clinical trials and entered into phase II clinical trials (Solit et al. 2007).

Mitomycins are isolates from *Streptomyces caespitosus* and were approved by the US FDA in 1974 for the clinical treatment of lung, breast, colorectal, and anal cancers and melanomas. It is also used as a chemotherapeutic agent in glaucoma surgery. Deoxycoformycin (Fig. 8.11d), also known as pentostatin, is produced by *S. antibioticus*. In 1993, the US FDA approved pentostatin for hairy cell leukemia, acute lymphoblastic leukemia, and pro-lymphocytic leukemia. Salinosporamide A is a β-lactone γ-lactam isolated from the fermentation broth of *Salinispora tropica* marine actinomycetes and exhibits a potential cytotoxicity against NCI 60 tumor cell lines. Nereus Pharmaceuticals is carrying out phase I clinical trials.

8.2.6.5 Drugs for Alzheimer's Dementia

Streptomyces griseofuscus produces phytostigmine, which improves memory function in the brain of healthy humans as well as in those with Alzheimer's dementia.

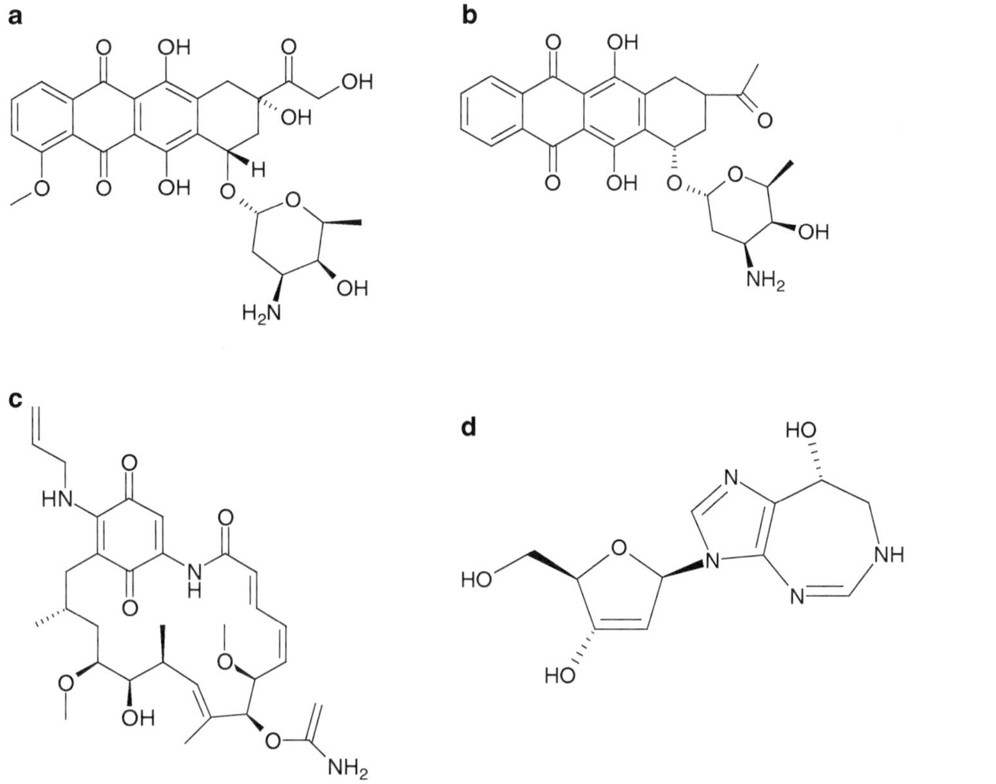

Fig. 8.11 Anti-cancer agents from microbes: (**a**) doxorubicin, (**b**) idarubicin, (**c**) 17-N-allylamino-17-demethoxygeldanamycin, and (**d**) deoxycoformycin

8.2.7 Endophytic Microbes as Sources of Putative Phytochemicals

Endophytic microorganisms comprise unicellular bacteria, fungi, and actinomycetes, which spend part of their complete life cycle colonizing in healthy plant tissues inter- or intra-cellularly. Almost all vascular plants on earth harbor endophytic microbes.

That endophytes could biosynthesize plant compounds or phytochemicals was first comprehended by Stierle et al. 1993. *Taxomyces andreanae* produces the multi-billion dollar anti-cancer compound Taxol® (generic name paclitaxel), initially isolated by Wani and Wall from the yew tree *Taxus brevifolia*. Since this discovery, supply has been a major issue due to the high demand for treatment of prostate, ovarian, breast, and lung cancer. Many approaches (e.g. plant cell culture technology or chemical synthesis for paclitaxel production) have been developed, but cost-effective bulk production is still not achievable, resulting in the high cost of the drug. The discovery of *T. andreanae* from *T. brevifolia* propelled the screening of taxol production from *Taxus* species and other plants to develop a fermentative method for paclitaxel production (Table 8.5). The identification and isolation of taxol-producing fungi has paved the way for its fermentative production as well as the possibilities of strain improvement using classical and modern methods that can further enhance the production, as with penicillin.

Podophyllotoxin is an aryltetralin lignan synthesized by *Podophyllum* species and is a highly

Table 8.5 Taxol-producing endophytic microorganisms

Host plant	Fungal isolate	Paclitaxel yield (μg/L)	Reference
Taxus brevifola	*Taxomyces andreanae*	0.024–0.5	Stierle et al. (1993)
Taxus mairei	*Colletotrichum gloeosporioides*	–	
	Tubercularia species TF5	185.4	Wang et al. (2000)
Taxus wallichiana	*Pestalatiopsis microspora Ne-32*	0.5	Strobel et al. (1996a)
Taxus chinensis	*Metarrhizium anisoplae H27*	846	Liu et al. (2009)
	Fusarium solani Tax-3	163.35	Deng et al. (2009)
	Mucor rouxianus DA10		Miao et al. (2009)
	Fusarium mairei UH23	286.4	Dai et al. (2008)
Taxus cuspidata	*Aspergillus niger var. taxi HD86-9*	273.6	Zhao et al. (2009)
	Nodulisporum sylviformae HQD33	468.62	Zhao et al. (2011)
	Botrytis sp. *HD181-23*	206.34	Zhao et al. (2008)
	Alternaria sp. *Ja-69*	0.16	Strobel et al. (1996a, b)
Taxus baccata	*Botryodiplodia theobromae BT115*	280.5	Venkatachalam et al. (2008)
	Kitasatospora sp.		Caruso et al. (2000)
Taxodium distichum	*Pestalatiopsis microspora Cp-4*	1.49	Li et al. (1996)
Taxus media	*Cladosporium cladosporioides MD2*	800	Zhang et al. (2009)
Taxus celebica	*Fusarium solani*	1.6	Chakravarthi et al. (2008)
Torryea grandiflora	*Periconia* sp. No. 2026	0.03–0.83	Li et al. (1998)
Aegle marmelos	*Bartalinia robillardoides*	187.6	Gangadevi and Muthumary (2008)
Morinda citrifolia	*Lasiodiplodia theobromae*	245	Pandi et al. (2011)
Ginkgo biloba	*Phoma betae*	795	Kumaran et al. (2012)
Cupresses sp.	*Phyllosticta spinarum* no. 625	235	Kumaran et al. (2008)
Podocarpus sp.	*Aspergillus fumigatus EPTP-1*	557.8	Sun et al. (2008)
Terminalia arjuna	*Chaetomella raphigera*	79.6	Gangadevi and Muthumary (2009)
Wollemia nobilis	*Pestalatiopsis guepinii*	–	Strobel et al. (1997)

valued precursor of the clinical drugs etoposide and teniposide. The availability of podophyllotoxin from natural sources is far less for the commercial production of etoposide and teniposide. Synthetic approaches for podophyllotoxin synthesis are commercially unattractive. *Trametes hirsuta* and *Phialocephala fortinii* are isolated from *Podophyllum hexandrum* and *Podophyllum peltatum*, respectively (Puri et al. 2006; Eyberger et al. 2006). *Fusarium solani* has recently been isolated from the roots of *P. hexandrum* (Nadeem et al. 2012). Screening of endophytic flora for the production of podophyllotoxin is being carried out from podophyllum and related species (Table 8.6).

Camptothecin is a pentacyclic quinolone alkaloid first isolated by Wall et al. (1966) from *Camptotheca acuminate*, a Chinese plant. Camptothecin mainly targets the intra-nuclear enzyme DNA topoisomerase, which is required for the winding and relaxation of DNA during replication and transcription. Camptothecin also inhibits the replication of human immunodeficiency virus. The first endophytic fungus that produced camptothecin was *Entrophospora infrequens* obtained by *Nothapodytes foetida* (Puri et al. 2005). Other endophytes producing camptothecin include *F. solani* and *Neurospora* species (Table 8.7). Initially, vincristine

Table 8.6 Endophytic microbes producing podophyllotoxin

Host plant	Fungal isolate	Podophyllotoxin yield	Reference
Podophyllum hexandrum	*Trametes hirsuta*		Puri et al. (2006)
Juniperus communis	*Aspergillus fumigatus*		Kusari et al. (2009)
Juniperus recurva	*Fusarium oxysporum JRE1*	28 μg/g	Kour et al. (2008)
Podophyllum peltatum	*Phialocephala fortinii*	0.5–189 μg/L	Eyberger et al. (2006)
Sinopodophyllum hexandrum	*Alternaria neesex*	2.4 μg/L	Cao et al. (2007)
Podophyllum hexandrum	*Fusarium solani P1*	29 μg/g	Nadeem et al. (2012)
Sinopodophyllum hexandrum	*Cephalosporium species*		Liu et al. (2010)

Table 8.7 Camptothecin-producing endophytic microorganisms

Host plant	Fungal isolate	Reference
Nothapodytes foetida	*Entrophosphora infrequens*	Amna et al. (2006)
	Neurospora species	Rehman et al. (2008)
	Nodulisporium species	Rehman et al. (2009)
Nothapodytes nimmoniana	*Botryosphaeria parva*	Gurudatt et al. (2010)
Capmtotheca acuminate	*Fusarium solani*	Kusari et al. (2009)
	Xylaria sp. *M20*	Liu et al. (2010)
Apodytes dimidata	*Fusarium solani*	Shweta et al. (2010)
Juniperus communis	*Aspergillus fumigatus*	Kusari et al. (2009)

Table 8.8 Vinca alkaloids producing endophytic microorganisms

Host plant	Fungal isolate	Compound	Reference
Catharanthus roseus	*Alternaria species*	Vinblastine	Guo et al. (1998)
Catharanthus roseus	*Fusarium oxysporum*	Vincristine	Zhang et al. (2000)
Catharanthus roseus	*Unidentified*	Vincristine	Yang et al. (2004)

(Oncovin®) and vinblastine (Velbe®) were isolated from the Madagascar periwinkle (*Catharanthus roseus*) and were the major drugs for leukemia and lymphoma in 1978. These alkaloids have been found to naturally occur in plants only in trace amounts. Approximately 500 kg of leaves provides 1 g of pure vincristine. The market price of these low-volume fine chemicals ranges between $US1 and 3.5 million per kg. Thus, there is global interest in finding alternative sources of these compounds. Endophytic microbes/fungi offer immense possibilities of producing these putative phytochemicals via fermentative processes. To date, only a few organisms producing vinca alkaloids have been reported (Table 8.8).

Huperzia serrata is a Chinese medicinal plant that produces huperzine A, a cholinesterase inhibitor for the treatment of Alzheimer's dementia and for further memory degradation. *Acremonium* sp. 2F09P03B is the first endophytic fungus isolated from *H. serrata* producing huperzine A (Li et al. 2007). *P. chrysogenum*, an endophyte from *Lycopodium serratum* could also produce huperzine A, as much as 4.761 mg/l in liquid culture (Zhou et al. 2009). Huperzine A-producing endophytic fungi, *Blastomyces* sp. (HA15) and *Botrytis* sp. (HA23), were also isolated from *Phlegmariurus cryptomerianus* (Ju et al. 2009).

Thus, endophytic microorganisms serve as an abundant and novel resource for producing the compounds or their analog that have originated in plants and possess medicinal or industrial value and generated interest among researchers for their exploration and exploitation for basic and applied research.

8.3 Engineering Microbes in the Production of Plant Products

Some natural products produced by plants have immense importance as drugs; however, the quantity recovered from plant sources is very meagre compared with the medicinal doses per patient per year required. Total chemical synthesis of these complex structures produced by plants is impractical because of the complexity of the reactions and the cost-intensive process. Using molecular biology tools, reconstruction of biosynthetic pathways in heterologous microorganisms is a promising strategy to provide sufficient quantities of desired natural products by using inexpensive renewable resources.

8.3.1 Isoprenoid Biosynthesis Engineering

The common precursor for biosynthesis of isoprenoids is isopentyldiphosphate (IPP). The most notable isoprenoids with medicinal value are Taxol®, artemisinin, and lycopene.

Taxol® synthesis through microbes: The biosynthesis of taxol utilizes 19 enzyme-catalyzed steps starting from the precursor geranylgeranyl diphosphate (GGPP). Taxadiene synthase catalyses the committed step, yielding taxadiene (pentamethyl tricyclopentadecane). Taxadiene is subsequently oxygenated by several cytochrome P450-dependent monooxygenases to generate taxol. *Saccharomyces cerevisiae* (yeast) produces a limited amount of GGPP and no taxadiene. *S. cerevisiae* produces a large amount of farnesyl pyrophosphate (FPP) for the biosynthesis of steroids, hence transgene-encoding taxadiene synthase from *Taxus chinensis* was introduced along with *T. chinensis* GGPP synthase controlled by phosphoglycerate kinase (PGK) promoters. This resulted in approximately 204 μg/l of taxadiene.

The complex feedback regulation of HMG-CoA reductase is the principal regulatory target. The N-terminal regulatory domain from yeast HMG-CoA reductase coenzyme-1 was truncated in combination with *T. chinensis* GGPP synthase, and *T. chinensis* synthase resulted in taxadiene production of 50 %, i.e. approximately 306 μg/l of taxadiene. Steroid biosynthesis by *Saccharomyces* was inhibited to increase taxoid biosynthesis by generating a mutant allele *upc* 2.1, which could uptake endogenous steroids. The *T. chinensis* GGPP synthase gene competes with squalene synthase for farnesyl diphosphate, hence it was replaced by the *Sulfolobus acidocaldarius* GGPP synthase gene, which has the ability to synthesize GGPP via the sequential addition of dimethylallyl diphosphate (DMAPP) building blocks (Ohnuma et al. 1994). *S. cerevisiae* with *T. chinensis* taxadiene synthase truncated HMG-CoA reductase, *upc* 2.1 transcription allele factor gene and *S. acidocaldarius* GGPP synthase gene resulting in very high production of taxadiene of circa 8.7 mg/l, a 40-fold increase (Engels et al. 2008) (Fig. 8.12).

Microbial artemisinin synthesis: Malaria is a tropical disease prevalent in Africa, Southeast Asia, and Central and South America. It threatens more than one-third of the total global population and kills approximately two million people annually. The malarial parasite *Plasmodium vivax* responsible for infection spreads through mosquito bites. Despite the use of quinine-derived anti-malarial agents, MDR strains of *P. vivax* have evolved and have been responsible for the global increase in mortality rate.

Artemisinin is a sesquiterpene lactone endoperoxide produced by the plant *Artemisia annua*, also known as quinghaosu and first identified by Chinese scientists as a potent drug capable of killing drug-resistant *P. vivax* (Liu et al. 1979). The isolation and extraction of artemisinin is too expensive and is an environmentally unfriendly process. Chemical synthesis has a poor yield and is also an expensive process for the manufacture of artemisinin. The World Health Organization (WHO)

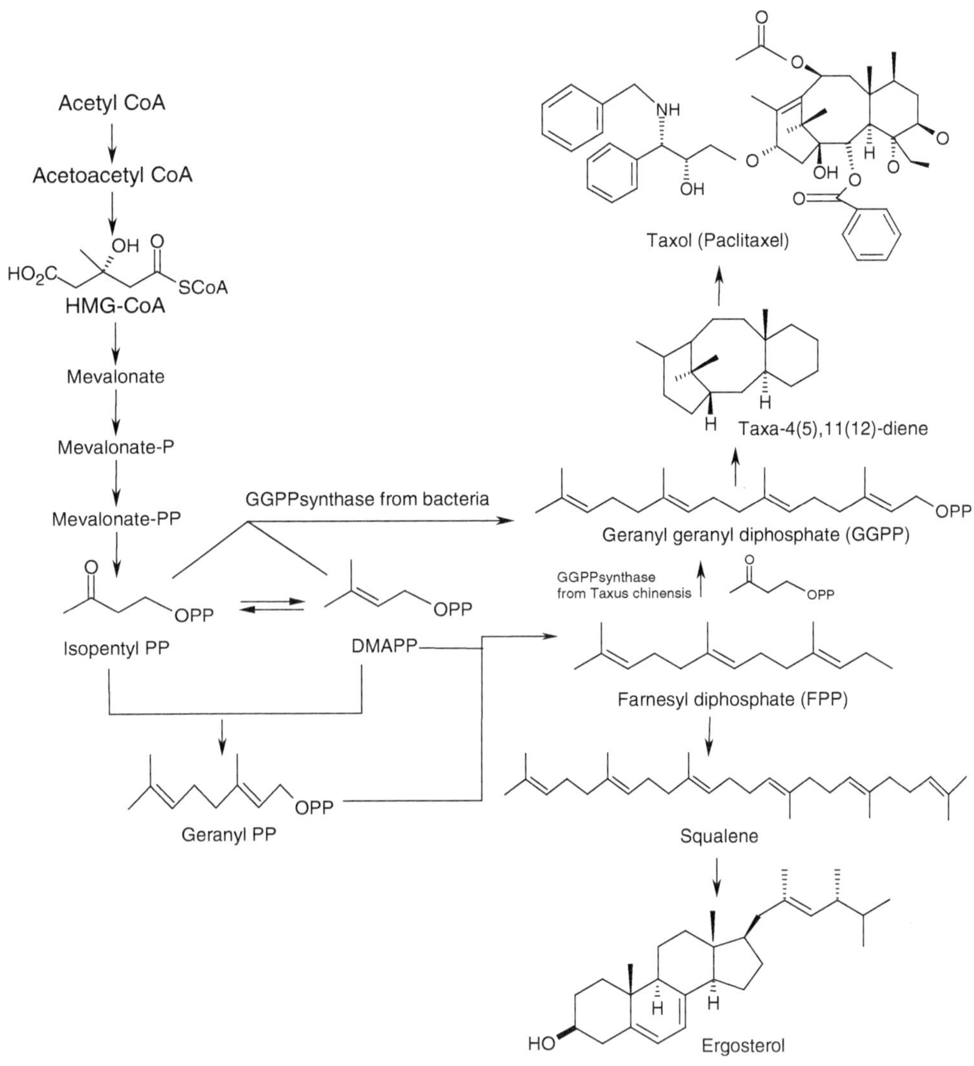

Fig. 8.12 Biosynthesis of taxol through *Saccharomyces cerevisiae*

recommended artemisinin-based combination therapies (ACT) for the prevention and onward transmission of MDR malaria. ACT comprises artesunate, artemether, and dihydroartemisinin. Endemic regions require over 100 million courses of ACT every year, a total that cannot be met because artemisinin is produced in trace amounts in the plants (Mutabingwa 2005). *A. annua* is a labor-intensive crop predominantly cultivated in Southeast Asia and China. The time from planting to harvesting and extraction of artemisinin is around 12–18 months. To meet the demand and to make the process cost effective, a need arose to shift the manufacture/biosynthesis of artemisinin from plants to microbes.

Professor Jay Keasling's group at the University of Berkeley, and scientists from Amyris Biotechnologies, developed a synthetic microbial process for the production of artemisinic acid, the precursor of artemisinin. They inserted the plant biosynthetic pathway into the microbe. His group created a new metabolic pathway comprising genes from bacteria, yeast, and plants, thereby creating a platform for copious supply of artemisinic acid (Chang and Keasling 2006).

The yeast *S. cerevisiae* EPY224 was engineered in three steps: (1) FPP biosynthetic pathway to increase FPP production and decrease it for sterols; (2) introduce *amorphadiene synthase* (ADS) gene from *A. annua* to current FPP producer to convert FPP to amorphadiene; (3) clone the novel cytochrome P450, which performs the three-step oxidation of amorphadiene to artemisinic acid from *A. annua* and expressing in the amorphadiene producing clone (Ro et al. 2006) (Fig. 8.13).

E. coli has also been engineered with an engineered mevalonate pathway, amorphadiene synthase, and a novel cytochrome P450 monooxygenase (CYP71AV1) from *A. annua*. Thus, well characterized biosynthetic pathways of plant metabolites could help in mimicking them in microorganisms for cost-effective and copious manufacture. Advances in systems biology and synthetic biology have opened avenues in performing metabolic engineering at the whole cell level, enabling the efficient production of drugs and drug precursors via engineered microbes. Artemisinin and taxol production through engineered microbes has already been achieved, but other plant products are also being attempted using heterologous expression systems.

8.4 Microbial Synthesis of Vitamins

Vitamins are defined as micronutrients that are required by all organisms in trace quantities but that cannot be synthesized by mammals and are instead synthesized by microorganisms or plants. They are also used as food/feed additives and as therapeutic agents. Today, many processed foods, feeds, pharmaceuticals, cosmetics, and chemicals contain extraneously added vitamins or vitamin-related compounds. The production of vitamins is through chemical synthesis or through extraction processes. However, these processes are energy intensive and generate waste, the disposal of which is also a cost-intensive process. Nowadays, biotechnological processes are being explored for the production of vitamins and are competing with the current chemical processes.

Vitamins are broadly classified in two groups: fat soluble and water soluble.

8.4.1 Vitamin E

Vitamin E or tocochromanol comprises a group of fat-soluble compounds, i.e. consists of α-, β-, δ-, and γ-tocopherols and α-, β-, δ-, and γ-tocotrienols. α-tocopherol plays a major role in humans in the prevention of light-induced pathologies of the skin, eyes, and degenerative disorders such as atherosclerosis, cardiovascular diseases, and cancer. α-tocopherol is also a major constituent of cosmetics, sunscreens, and for the food preservatives.

Tocopherols from natural sources are used for human applications and have a high market price of around US$20/kg, while the chemically synthesized α-tocopherol acetate is generally used for fortification of animal feed and is relatively cheap at US$11/kg. Presently, α-tocopherol is being obtained via chemical synthesis and chemical extraction of vegetable oils. Higher plants as sources of tocopherol are not lucrative as the α-tocopherol content is low (Valentin and Qi 2005). Photosynthetic microorganisms are known to accumulate detectable amounts of tocopherols. *Spirulina platensis*, *Dunaliella tertiolecta*, *Synechocystis* sp., species of *Chlorella*, *Chlamydomonas*, and *Ochromonas*, and *Euglena gracilis* are examples of microalgae that accumulate detectable amounts of tocopherols (Powls and Redfearn 1967; Carballo-Cardenas et al. 2003; Taketomi et al. 1983; Hughes and Tove 1982; Ogbonna et al. 1998; Vismura et al. 2003).

The model system for genetic engineering for over-production of tocopherols is a cyanobacterium, *Synechocystis* sp. strain PCC 6803 in which the nirA promoter is very useful for engineering the tocopherol metabolic pathway (Qi et al. 2005). To date, there are no reports on the genetic modification of good tocopherol producers such as *E. gracilis* as carried out in *Synechocystis* sp. in which expression of p-hydroxyphenylpyruvate dioxygenase is induced, thereby increasing the total tocopherol content five times (Qi et al. 2005). With the knowledge of higher plants and

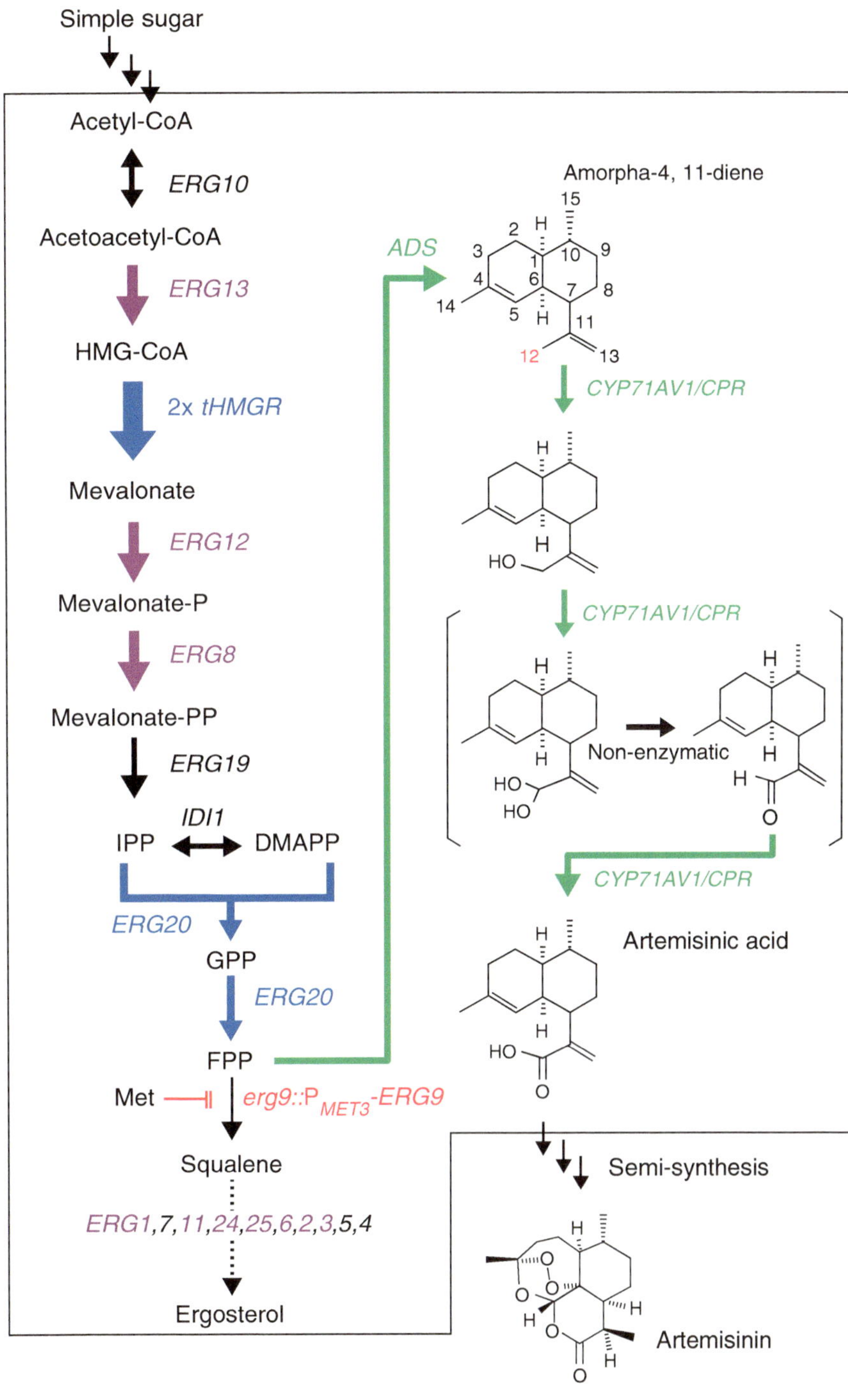

Fig. 8.13 Schematic representation of the engineered biosynthetic pathway of artemisinin production in *Saccharomyces cerevisiae* EPY224 (Adapted from Ro et al. Nature 440, doi: 10.1038/nature04640)

Synechocystis sp., it would be possible to develop genetically modified *Euglena* species that would accumulate high concentrations of tocopherols.

8.4.2 Vitamin K

Vitamin K is a quinone comprising two broad types: K1 and K2. Vitamin K1 is known as phylloquinone and is synthesized by plants, whereas vitamin K2 or menaquinone (MK) is synthesized by bacteria. The clinical roles of K vitamins are in the prevention of bone loss and bone fractures in humans, antioxidant activity, and reducing the effects of Alzheimer's disease (Cheung et al. 2008; Li et al. 2003; Allison 2001). The major role of vitamin K is in blood coagulation (Lurie et al. 2010) and they are increasingly drawing attention as nutritional supplements for humans. The commercial production of vitamin K is via extraction from natural resources or chemical synthesis. Microbial production of vitamin K has been reported with *Flavobacterium* sp., *B. subtilis*, and *Propionibacterium freudenreichii*, but a model microorganism for vitamin K biosynthesis has not been developed for its commercial production. Recently, a gut *E. coli* strain was found to produce naphthoquinone-type MK-8 under anaerobic conditions and has been taken up for systemic metabolic engineering for the production of vitamin K (Kong and Lee 2011).

8.4.3 β-Carotene (Provitamin A)

Carotenoids are tetraterpenoid pigments present in the chloroplast and chromoplasts of plants, photosynthetic bacteria, fungi, and microalgae. β-carotene is the precursor of vitamin A in humans and functions as an antioxidant, thereby having a protective effect against cancer. Approximately 50 of the 600 carotenoids known serve as precursor to vitamin A. The recommended dietary intake of carotenoids for humans is between 3 and 6 mg/day to ensure the beneficial concentration of retinol equivalents in the blood.

Microorganisms contribute to approximately 15 % of the total industrial production. *Blakeslea trispora* has been used after intensive classical strain improvement by random mutagenesis and selection of an isolate producing 7 g/l of β-carotene in large-scale fermentations. The fermentation spent broth is filtered and biomass is extracted with solvent to concentrate β-carotene, which undergoes a crystallization process to obtain the pure product. *Dunaliella salina* is a halophilic green microalgae that accumulates β-carotene in the oil globules in the chloroplast inter-thylakoid spaces, thereby protecting them against photo-inhibition and photo-destruction. Commercial β-carotene production uses large shallow, high-salinity ponds to grow *Dunaliella*. Nature Beta Technologies is carrying out the production, employing open channels in the form of oblong raceways agitated by paddle wheels (Table 8.9).

Several microorganisms are able to produce β-carotene, but biotechnological production (synthesis of β-carotene through homologous or heterologous production) becomes a commercially attractive proposition. Red yeast *Xanthophyllomyces dendrorhous* (formerly *Phaffia rhodozyma*) is mainly a producer of astaxanthin but also accumulates β-carotene as an intermediate in the astaxanthin biosynthesis

Table 8.9 Commercial producers of β-carotene by *Dunaliella*

Company	Location	Culture area	β-carotene production (tpy[a])	Culture system
Cyanotech	Hawaii	Not applicable	Not applicable	Raceways ponds
Inner Mongolia Biological Engg.	China	Not applicable	Not applicable	Raceways ponds
Nature Beta Technologies	Israel	5	3–4	Raceways ponds
Tianjin Lantai Biotechnology	China	Not applicable	Not applicable	Raceways ponds
Parry Agro Industries Limited	India	Not applicable	Not applicable	Raceways ponds

[a]*tpy* tons per year

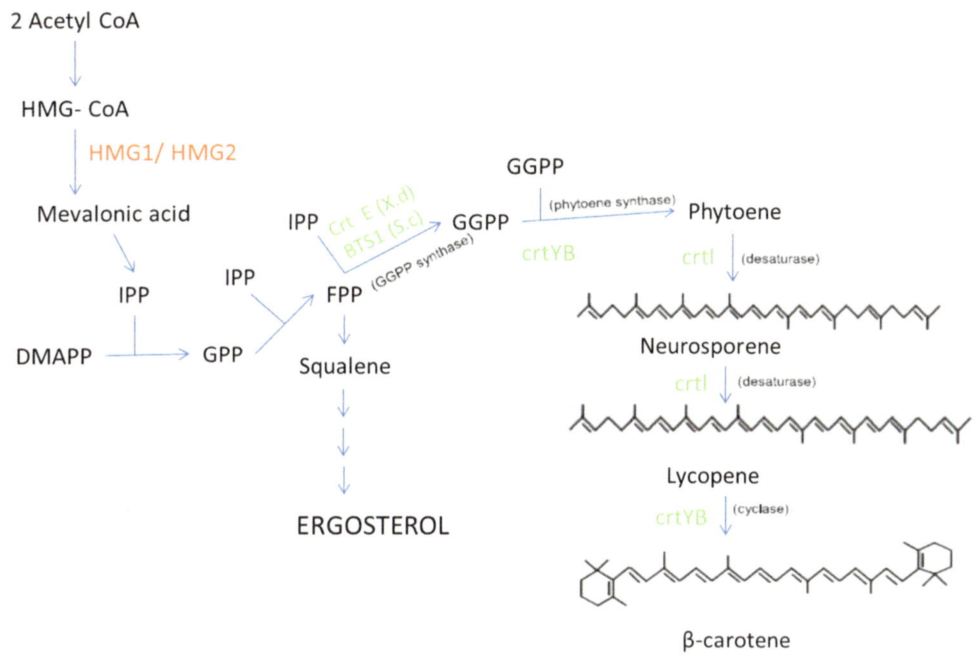

Fig. 8.14 β-carotene biosynthesis pathway of *Saccharomyces cerevisiae* transformant

pathway. To ascertain that *S. cerevisiae* can serve as an efficient host for the production of carotenoids, particularly β-carotene, carotenogenic genes from *X. dendrorhous* were cloned and expressed in *S. cerevisiae*. The over-expression of the bifunctional gene *crtYB*, which encodes for phytoene synthase and lycopene cyclase and *crtI*, which encodes for phytoene desaturase from *X. dendrorhous*, were capable of carotenoid production. Improvement in carotenoid production was further achieved by over-expression of homologous GGPP synthase from *S. cerevisiae* encoded as BTS1. A subsequent increase in carotenoid production was achieved by combined over-expression of *crtE* (heterologous GGPP synthase) from *X. dendrorhous* with *crtYB* and *crtI* and introduction of an additional copy of a truncated 3-hydroxy-3-methylglutaryl-coenzyme A reductase gene (*tHMG1*) into carotenoid-producing cells (Veerwal et al. 2007) (Fig. 8.14).

The final transformant was an *S. cerevisiae* strain capable of producing high levels of β-carotene, i.e. up to 5.9 mg/g (dry weight), which consisted of an additional copy of *crtI* and *tHMG1* into carotenoid-producing yeast cells. Development of this strain has opened up avenues for β-carotene production via a light- and biomass-independent process as encountered in *D. salina*.

8.4.4 Vitamin B$_2$

Vitamin B$_2$ is also known as riboflavin (7, 8-dimethyl-10-(D-1')-ribityl) isoalloxazine. The riboflavin molecule is divided into two parts: ribose and lumichrome (aromatic structure) (Fig. 8.15). It plays a very important role as a precursor to flavin mononucleotide (FMN, riboflavin-5′-monophosphate) and flavin adenine dinucleotide (FAD), which function as coenzymes to a variety of enzyme catalyzed reactions in the intermediate metabolism of human beings and other living organisms. Deficiency of riboflavin or ariboflavinosis is characterized by cracked red lips, inflammation of the lining of mouth and tongue, mouth ulcers, cracks at the corners of the mouth (angular cheilitis), photophobia, and scrotal dermatitis. The recommended dietary allowance of riboflavin is 2 mg/day.

Commercial riboflavin is produced via chemical and biochemical synthesis. However, biochemical/fermentative synthesis has become more popular for the production of riboflavin with the discovery and development of microbes producing riboflavin (Table 8.10). The first commercial fermentation was established using *Clostridium butylicum* grown on grain, mashes, and whey. It was replaced by *Eremothecium ashbyii* in 1940 and subsequently shifted to *Ashbya gossypii* in 1946 (Perlman 1979). A wild strain of *A. gossypii* (Fig. 8.16) produced 2 mg/g of fungal biomass possibly for photoprotection of the spores. The wild strains have been optimized for (1) culture conditions, i.e. growth and production phase; (2) selection of anti-metabolite mutant, e.g. limiting reaction of isocitrate lyase; and (3) over-expression of *RIB* genes and integration of additional copies of *RIB 3* genes, thereby developing the over-producing strains through homologous recombination. Merck employed the over-producing strain *A. gossypii* for commercial production, with a yield of 15 g/L. BASF (Germany) started commercial production using *A. gossypii* in 1990. Industrial production of riboflavin is to the tune of 6,000–7,000 tons per year by BASF, Hoffmann-La Roche (Switzerland), and Hubei Guangji Pharmaceutical Co., Ltd (China). Currently, *A. gossypii* and genetically engineered *B. subtilis* expressing *RIB* genes are being commercially used for the industrial production of riboflavin. Perkins et al. (1999) developed the riboflavin strain containing multiple copies of modified *B. subtilis* riboflavin biosynthetic operon (rib operon) integrated at two different sites in the *B. subtilis* chromosome. The strain has also been engineered to deregulate the purine metabolism pathway. Roche (Japan) has started industrial use of recombinant *B. subtilis*, which uses a single-step conversion of glucose into riboflavin by fed batch fermentation. Vitamin B_2 is recovered from the fermentation broth by centrifugation after inactivation of the microorganisms by heat. The cell mass is removed and riboflavin is recovered by evaporation and vacuum drying.

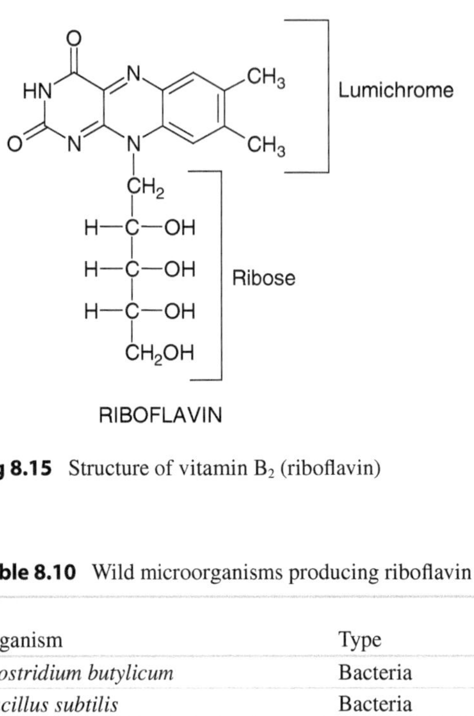

Fig 8.15 Structure of vitamin B_2 (riboflavin)

8.4.5 Vitamin B_{12}

Vitamin B_{12} is also known as cobalamin and performs a key role in the normal function of the brain and nervous system and in the formation of blood. It is biochemically related to cobalt. The

Table 8.10 Wild microorganisms producing riboflavin (vitamin B_2) after medium optimization

Organism	Type	Culture time	Riboflavin concentration (g/l)
Clostridium butylicum	Bacteria	–	0.1
Bacillus subtilis	Bacteria	18 min	0.08
Candida flareri	Yeast	–	0.6
Aspergillus terreus	Fungi	16 days	1.0
Eremothecium ashbyii	Fungi	7 days	3.3
Ashbya gossypii	Fungi	8 days	5.5

Adapted from Li et al. (2001)

physiological forms of cobalamin are methylcobalamin and adenosylcobalamin. Deficiency of vitamin B_{12} leads to pernicious anemia and can also cause symptoms of mania and psychosis. Bacteria and archaea are only capable of synthesizing vitamin B_{12} as they possess the required enzymes for the synthesis of the complex structure (Fig. 8.17). Cobalamins consist of a 15-member planar corrin ring with a central cobalt ion, a dimethyl benzimidazole group as the lower ligand, and an adenosyl, methyl, hydroxy, or cyano group as the upper ligand.

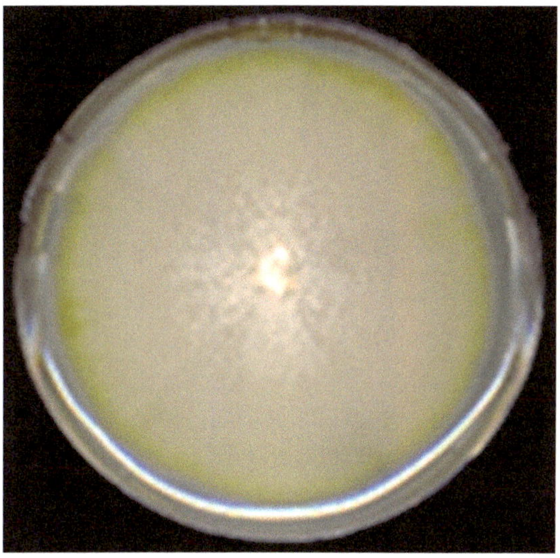

Fig. 8.16 Riboflavin producer: *Ashbya gossypii*

A host of microorganisms produce vitamin B_{12}: *Aerobacter*, *Agrobacterium*, *Alcaligenes*, *Azotobacter*, *Bacillus*, *Clostridium*, *Corynebacterium*, *Flavobacterium*, *Micromonospora*, *Mycobacterium*, *Nocardia*, *Propionibacterium*, *Protaminobacter*, *Proteus*, *Pseudomonas*, *Rhizobium*, *Salmonella*, *Serratia*, *Streptomyces*, *Streptococcus*, and *Xanthomonas*. The biosynthesis of vitamin B_{12} is via both anaerobic and aerobic pathways in bacteria. *Pseudomonas denitrificans* exhibits the aerobic pathway of vitamin B_{12} synthesis, while an anaerobic pathway is followed in *Propionibacteria* or *Salmonella typhimurium*. Some prominent microbes producing vitamin B_{12} are given in Table 8.11.

P. denitrificans is being used in industrial processes by the main B_{12}-producing companies such as Merck and Aventis Pharma. Scientists at

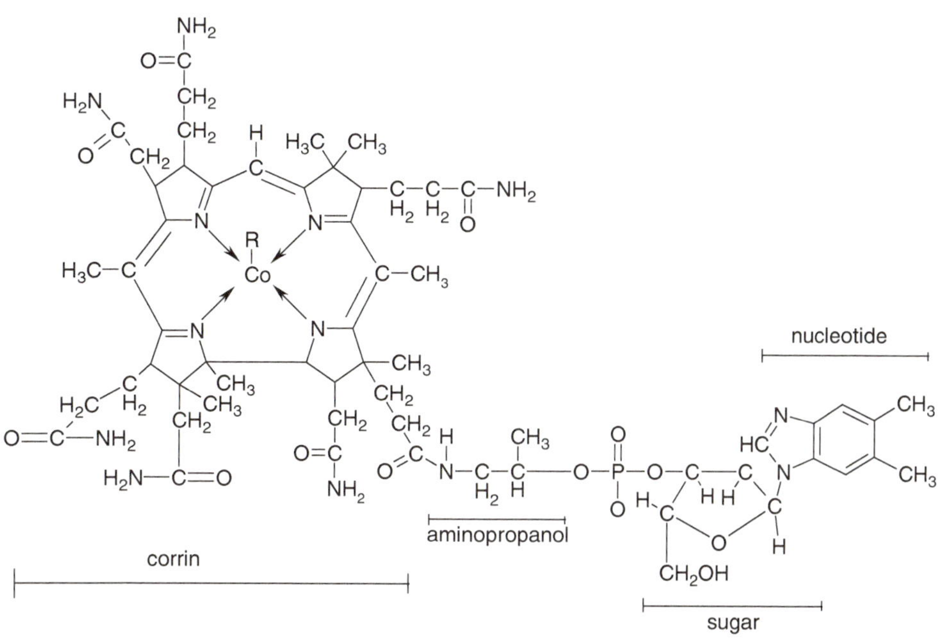

Fig 8.17 Structure of vitamin B_{12}

Table 8.11 Microorganisms producing vitamin B_{12}

Organism	Fermentation condition	Main carbon source used	Vitamin B_{12} production (mg/l)
Propionibacterium freudenreichii	Anaerobiosis, 5,6-dimethyl benzimidazole	Glucose	206
Propionibacterium shermanii	5,6-dimethyl benzimidazole	Glucose	60.0
Pseudomonas denitrificans	Aerobiosis, betaine	Sucrose	60.0
Nocardia rugosa	Aerobiosis	Glucose	18
Rhizobium cobalaminogenum	Aerobiosis		16.5
Streptomyces olivaceus	5,6-dimethyl benzimidazole	Glucose	6.0
Nocardia gardneri	Aerobiosis	Hexadecane	4.5
Butyribacterium methylotrophicum	Anaerobiosis	Methanol	3.6
Arthrobacter hyalinus	5,6-dimethyl benzimidazole	Isopropanol	1.1

Adapted from Martens et al. (2002)

Merck improved the efficiency of *P. denitrificans*, thereby enhancing the production 30-fold compared with the wild type isolates through genetic manipulations. Blanche et al. (1998) at Rhone Poulenc (now Aventis) used random mutagenesis for over 10 years to improve the production of vitamin B_{12} by *P. denitrificans* to bring it to 6 g/l. A genetic construct via cloning of the 22 *cob* genes has been used to further enhance the productivity and yield on raw materials over strains solely obtained by classical strain improvement and optimized for industrial production. However, the optimization study of genetically engineered *P. denitrificans* for vitamin B_{12} synthesis is still ongoing.

The contents of the fermentors are concentrated/spray dried. They are subsequently heated and extracted by methanol to get the dissolved vitamin B_{12}. This vitamin solution is clarified through activated carbon or aluminum oxide and then treated with 0.1 % potassium cyanide, thereby converting it into cyanocobalamin in the presence of sodium nitrate and heat. The vitamin solution is subsequently clarified via filtration treatment with zinc chloride, and then precipitated out with the addition of tannic acid or cresol to give 80 % pure vitamin, which is generally used in animal food. For pharmaceutical purposes, the clarified solution is re-extracted with organic solvents like carbon tetrachloride water and butanol. Pure vitamin B_{12} can be obtained by crystallization after the addition of organic solvents such as phenol and water.

8.5 Production of Amino Acids

Amino acids are the building blocks of proteins. They possess nutritional values, have medicinal actions, and flavor-enhancing properties and therefore find applications in the food and feed industries, in pharmaceutical formulations, as additives in the cosmetic industry, and also in the manufacturing of biopolymers. The present global demand of amino acids is approximately 2 million metric tons. Amino acids can be produced via four general methods: (1) extraction from natural protein hydrolysates; (2) chemical synthesis; (3) biochemical or enzymatic synthesis; or (4) fermentative production. The selection of the production method of each amino acid is based on the cost effectiveness of the process.

The use of monosodium L-glutamate (MSG) as a flavor enhancer by K. Ikeda propelled the development of amino acid production as an industrial process. The amino acid was present in the seaweed 'knobu' used as a traditional seasoning material during cooking in Japan. Ajinomoto Co., Inc. (Japan) pioneered the extraction of MSG from acid hydrolysates of wheat gluten or defatted soy bean. The microbial production of L-glutamic acid as MSG by fermentation was first carried out in 1957 by Kinoshita and his colleagues at Kyowa Hakko Kogyo Co. who discovered a bacterium *Corynebacterium glutamicum*, which could efficiently assimilate cheap sugar and ammonia for the production of L-glutamate (Kinoshita 1985). This discovery resulted in drastic production cost reductions for MSG when compared with the traditional process of protein hydrolysates extraction or chemical synthesis.

Table 8.12 Microorganisms producing L-Glutamate

Microorganism	Type	Production (g/L)
Brevibacterium flavuum ATCC 13826	W	15
Brevibacterium roseum ATCC 13825	W	14.8
Brevibacterium lactofermentum ATCC 13869	W	14.3
Corynebacterium acetoacidophilum ATCC 13870	W	7.3
Corynebacterium hydrocarboclastus M-104	W	6.3
Corynebacterium glutamicum (first isolate)	W	30
Brevibacterium lactofermentum AJ3611	M	21
Brevibacterium lactofermentum ATCC 13869	M	52
Brevibacterium flavuum AJ3612	M	52
Brevibacterium flavuum ATCC14067	M	50

Source: Vijaylaxmi and Sarva Mangala (2011)

M mutant, *W* wild type

8.5.1 Ajinomoto Process of Fermentative Production of L-Glutamate

Different strains with improved L-glutamate production have been developed from *C. glutamicum* and *Brevibacterium divaricatum*, with detailed investigations carried out on their biosynthetic properties (Kimura 2003). *Corynebacteria* and *Brevibacteria* are Gram-positive, non-spore forming, non-motile, and biotin-requiring microorganisms that can be classified as (1) wild types; (2) auxotrophic mutants; and (3) genetically modified strains (Table 8.12). The fermentative production process is very simple and involves the charging of a sterile fermentation tank with culture medium containing a suitable carbon source such as sugarcane syrup, nitrogen, sulphur, or phosphorus along with trace elements. The seed culture is prepared in the pre-fermenter (seed fermentor) and then added to fermentation tank aseptically and stirred under defined pH, temperature, and aeration. The carbon source is utilized by the culture, and L-GLUTAMIC acid is released and accumulated in the fermentation broth. To recover it from the fermentation broth, it is converted into a sodium salt that is then recovered via crystallization in the recovery section of the industrial plant (Fig. 8.18). Approximately 1.5 million tons of MSG are being produced by this process, making it the leading amino acid in terms of production capacity and demand.

The microbial culture collection of Ajinomoto includes over 21,000 strains comprising 5,000 bacteria, 6,000 yeasts, and 10,000 unknown isolates. Thus, we expect that other commercial producers of amino acids also have diverse microbial collections for the development of novel industrial strains for amino acid production. Thermo-tolerant strains of *C. glutamicum* are preferred over mesophilic strains since temperature maintenance becomes a cost-intensive factor in the L-glutamate production process. Hence, thermo-tolerant strains that can grow at up to 40 °C have been isolated and developed for the industrial production of L-glutamate. The preferred carbon sources are cane molasses, beet molasses, and corn starch hydrolysates. However, in Thailand and Indonesia, tapioca hydrolysates have been used for the commercial production of glutamic acid. Ammonium salts serve as the nitrogen source. The production of L-glutamic acid is carried out in an agitated tank fermentor or airlift fermentor of 50–500 m^3 capacity. The first seed fermenter for the production of inoculums is 1–2 m^3 followed by a second seed fermenter with a capacity of between 10 and 20 m^3. The fed batch process is generally preferred over

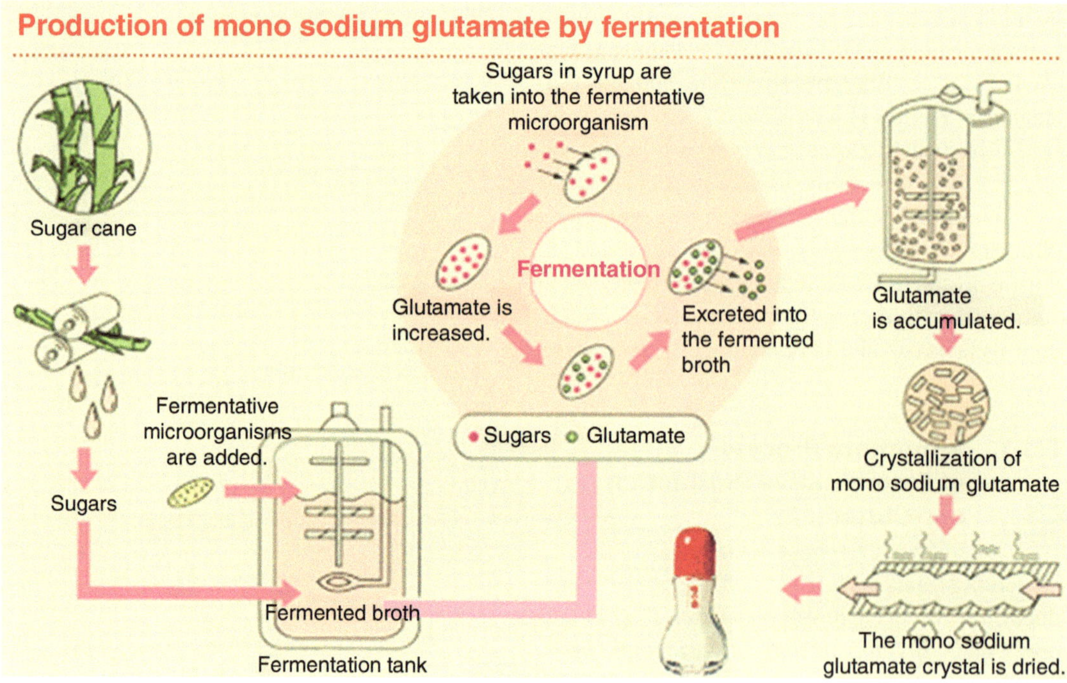

Fig. 8.18 Monosodium glutamate production by fermentation: Ajinomoto process (Source: Ajinomoto Inc.)

the batch process since productivity is enhanced because of a reduction in fermentation time.

Glutamate is recovered as an ammonium salt through ion exchange chromatography wherein the amino acid gets bound to the resin while the ammonia is released and recovered by distillation for reuse. Subsequently, the ion exchange resin holding L-glutamate is washed through a sodium hydroxide solution to recover the L-glutamate as MSG, which is subsequently recovered via crystallization and converted into a food grade product by decolorization and re-crystallization.

8.5.2 Fermentative Production of L-Lysine

L-Lysine is an essential amino acid generally found in all naturally occurring proteins. Lysine is an additive in animal feed for monogastric diets to provide adequate nutrition to livestock. As a fine chemical, it is utilized in humans as a medicine and in cosmetics. In the pharmaceutical industry, it is used as an ingredient for infusion solutions. The annual production of L-lysine is over 700,000 tons per annum globally. *C. glutamicum* has been found to be a producer of L-lysine. Homoserine auxotroph of *C. glutamicum* ATCC13287 was used for the production of L-lysine by Kinoshita and colleagues at Kyowa Hakko Kogyo Co. The strain yielded 44 g lysine per liter, with a conversion efficiency of 26 % from sugar (g lysine/g sugar).

Homoserine is the starting point of making L-threonine, L-methionine, and L-isoleucine. The metabolic flux of individual amino acids occurs at a level of aspartyl kinase and homoserine dehydrogenase (HSD). Hence, mutants that block the synthesis of homoserine would over-produce L-lysine. HSD^{-} and HSD^{leaky} mutants over-produce L-lysine. HSD^{leaky} mutants generally express an HSD that is less effective at making homoserine so that L-threonine is not over-produced to feedback inhibit aspartyl kinase.

HSD⁻ mutants are homoserine auxotrophs that effectively lack the ability to make threonine, methionine, and isoleucine, thus eliminating the feedback inhibition of aspartyl kinase by threonine. Development of recombinant *Corynebacteria* has led to the development of strains that excrete 170 g/l of L-lysine (Hirao et al. 1989).

Ajinomoto Inc., ADM, BASF, Cheil Jedang, Degussa, Global Bio-Chem, and Kyowa Hakko Co. are the major producers of L-lysine. The major carbon sources used for industrial L-lysine production are cane molasses, beet molasses, sucrose, and dextrose. Ammonium sulphate or ammonia (gaseous or ammonia water) is used as a cheap nitrogen source. Ammonium sulphate provides the counter ion to neutralize the accumulating basic acid. Hence, L-lysine in fermentation broth is accumulated as sulphate. Tween 40 is used as an anti-foam agent in the fermentation process. Successive inoculum development strategies are adopted, as for the production of glutamic acid, to develop sufficient inoculums for 500 m^3 fermentation tank capacity. This is considered an ideal operation size considering the economy of the production process. After the fermentation duration, cell separation is carried out with vacuum filtration followed by ion-exchange methods to recover L-lysine, followed by the addition of HCl to recover the L-lysine, followed by evaporation and drying. This process provides 95 % pure L-lysine HCl, which is generally used in pharmaceutical preparations. An alkaline solution containing 50.7 % L-lysine can also be obtained after biomass separation, evaporation, and filtration. The fermentation broth can also be spray dried to produce granulated lysine sulphate after the separation of biomass.

There is a paradigm shift in the industry in terms of the commercial production of L-amino acids such as L-threonine, L-serine, and L-METHIONINE and aromatic amino acids such as L-tryptophan, L-phenylalanine, and L-tyrosine. Microbial production of these amino acids is being adopted over microbial enzymatic processes by developing auxotrophic mutants, analog-resistant mutants, and genetically engineered strains.

8.6 Microbes in the Production of Dyes and Pigments

Humans have been gifted with the ability to identify colors and correlate them with feelings. Red exudes warmth and enhances pulse rate and respiration, while blue and green are temperamentally cool and indicative of a peaceful environment and prosperity. The color of food materials reflects the quality and has sensory properties, while that of textiles provides personality attributes to individuals in different socio-geographic conditions. There exists a fine line of demarcation between pigments and dyes and the interchangeability of these terms. A pigment is insoluble in a given medium, while a dye is soluble. However, the common term for pigment and dyes is colors/colorants.

8.6.1 Microbial Pigments in the Textile Industry

A huge array of synthetic chemical colorants have been exploited by the textile industry for dyeing yarn, fabrics, wool, and clothes, thereby posing a potential threat to the environment and human health due to the toxicity of these synthetic chemicals. Natural colors are preferred over synthetic dyes as they are non-toxic, environmentally friendly, and least hazardous to living beings. These can be resourced from plants, animals, and microorganisms. Microbes appear to be a lucrative source of production of colored compounds since they proliferate rapidly and have the potential to be standardized for mass production in a bioreactor.

The first microbially resourced dye was indigo, which was initially obtained from the plant *Indigofera suffruticosa*, which was used in the textile industry for a long time. However, the process required massive cultivation of the plant to recover an appreciable amount of indigo for commercial purposes. To meet the growing demand, Ensley et al. (1983) developed an *E. coli* that could produce indigo in a bioreactor. *E. coli* already possesses an enzyme tryptophanase that converts L-tryptophan into indole. This organism was engineered and a naphthalene dioxygenase was transferred from a *Pseudomonas* species, thereby converting indole into indigo. Indigo is currently being produced via the microbial route.

As microbes produce a wide variety of colored compounds belonging to different groups, such as anthraquinones, carotenoids, flavonoids, quinines, and rubramine, studies are underway to assess the dyeing potential of these natural compounds to reduce the use of synthetic colorants in the textile industry. Fungi appear to be an interesting source of colors as they produce many pigments in their reproductive structure (spores). Many fungal species have been reported to produce brown to reddish-brown pigments. *Acrostalagmus* sp. (NRC 90) produces a brown dye with a reflectance at 485 nm and exhibits a good rate of washing and perspiration fastness. *Phymatotrichum* sp. (NRC 151) produces a reddish-brown dye that has the best properties in terms of washing and perspiration fastness and light stability (Atalla et al. 2011). Violacein type dye has also been isolated from *Chromobacterium violaceum* from wastewater from oil refineries in Malaysia and is being tested for textile dyeing. *Curvularia lunata*, *Alternaria alternata*, and *Trichoderma virens* have been found to produce pigments that can be used for dyeing textiles (Sharma et al. 2012).

8.6.2 Microbial Pigments in the Food Industry

The increased societal concern surrounding the negative impact of synthetic food dyes has led to a strong desire in the food industry to replace them with natural alternatives. Plant extracts have initially served as the source of natural pigments since the majority of them are consumed as fruits and vegetables. The source of a natural red color is paprika (seeds of *Capsicum annuum*, *Beta vulgaris* [*betanin*]); yellow has been sourced from saffron and marigold, and green from leafy vegetables.

However, with the increasing use of fermentation-based products such as gellan, curdlan, and xanthan in food products, pigments produced by microbes could also be used for coloring foods. β-carotene production has been reported from the fungus *B. trispora* and *Dunaliella* species for use as vitamins. Gist-brocades (now DSM) was the first company to produce β-carotene via a fermentative route. The fermentative production of β-carotene has European compliance E 160aii, which also includes the proportions of cis- and trans-isomers and is free of mycotoxins and other toxic metabolites. The fungal β-carotene was free of any genotoxicity. Carotenoids like canthaxanthin (4, 4-diketo-β-carotene) have been produced by *Bradyrhizobium* sp. and also by *Halobacterium* species (Hannibal 2000). Zeaxanthin has been produced by *Flavobacterium* species. Red yeast, *X. dendrorhous*, synthesizes astaxanthin and zeaxanthin as its main carotenoids (Roy et al. 2008). Commercial production of astaxanthin is being carried out via red yeast.

Monascus species is also a prolific producer of pigments that could be used for coloring foods. *M. ruber* was the first organism among the different *monascus* species known today and was isolated from linseed oil cakes and potatoes. It was named *monascus* since it possessed a single polyspore ascus. *Monascus purpureus* was isolated from red mould rice procured from markets in Java, Indonesia (Fabre et al. 1993). Presently, over 50 patents have been issued in Japan, Europe, and the USA for the use of *Monascus* pigments in the food industry (Table 8.13) (Fig. 8.19). *Penicillium aculeatum* and *Penicillium pinophilum* strains also produce monascus-like pigments or their amino acid derivatives. These strains do not co-produce citrinin or any other known mycotoxins. Further, these fungi are nontoxic to humans.

Polyketide naphthoquinone red pigments have been isolated from *Cordyceps unilateralis* BCC1869, and they possess structural and chemical similarities to the plant-derived commercial pigments shikonin and alkannin (Unagul et al. 2005). 3, 5, 8- trihydroxy -6-methoxy-2-(5-oxohexa-1, 3-dienyl)-1, 4-naphthoquinone is the major component of *C. unilateralis* and is very stable in light, heat, acid, and alkali solutions.

Table 8.13 Some microbial pigments as potential food colorants

Microorganism	Pigment	Color
Fungi		
Monascus ruber	Monascorubicin	Red
	Rubropunctatin	Red
	Monascin	Orange
	Ankaflavin	Orange
Epicoccum nigrum	Flavipin	Yellow
Penicillium herquei	Atrovenetin	Yellow
Penicillium oxalicum var. *armenica*	Arpink Red	Dark red
Penicillium purpurogenum	Purpurogenone	Orange yellow
	Mitrorubin	Yellow
Blakeslea trispora	β-carotene	Yellow
Mucor circinelloides	β-carotene	Yellow
Phycomyces blakesleeanus	β-carotene	Yellow
Algae		
Dunaliella species	β-carotene	Yellow
Haematococcus species	Astaxanthin	Orange-red
Yeasts		
Xanthophyllomyces dendrorhous	Astaxanthin	Orange-red
Rhodotorula glutiniis	β- carotene	Yellow
Bacteria		
Bradyrhizobium species	Canthaxanthin	Orangish pink
Halobacterium species	Canthaxanthin	Orangish pink
Flavobacterium species	Zeaxanthin	Yellow
Pseudomonas aeruginosa[a]	Pyocyanin	Blue
	Pyorubrin	Red

[a]A marine isolate

A soil isolate *Penicillium oxalicum* var. *armenica* CCM8242 produces an anthraquinone-type chromophore called 'arpink red'. This red colorant has been evaluated for its safety and toxicity and has been recommended for use in meat-based products with a concentration of 100 mg/kg, in alcoholic drinks at 200 mg/kg, and in milk products, including ice creams, at 150 mg/kg (Sardaryan 2002).

Prodigiosins are naturally occurring tripyrrole ring-containing red pigments produced by microorganisms. A variety of bacterial taxa like *Serratia rabidosa*, *Rugamonas rubra*, and *Streptoverticillium rubric reticulata* produce prodigiosin (Giri et al. 2004). It was first isolated from *Serratia marcescens* in 1929. They find potential use as colorants in the textile and food industry. Thus, microbes serve as cell factories for pigment production, which could be used as an extended color palette of natural food colorants.

8.7 Microbial Production of Flavors and Fragrances

Plants and animals have long been the natural sources of flavor and fragrances. The use of microbes to impart aroma and flavor began with fermentation products like beer, cheese, and wine. The harvesting of animals for fragrances has been made illegal, but previously musk deer were killed for the musk fragrance, which was found to be a cyclic ketone muscone. Currently, plant and microbial products are being used as flavors and fragrances and have a global market of over US$16 billion. The very first synthetic flavor and fragrance compounds were coumarin (1868) and vanillin (1874), respectively. Aromatic compounds like vanillin have a very large market due to their applications in the food, pharmaceutical, and cosmetic industries. The approximate global consumption of D-Vanillin is approximately 12,000 tons annually. Approximately 20 tons are extracted from vanilla beans and the rest are produced through biotechnological or synthetic routes. The stilbene dioxygenase process has been used for the conversion of isorhapotin to vanillin. However, the most promising synthetic route is the transformation of eugenol to coniferyl aldehyde or ferulic acid to vanillin. Recently, vanillin production has also been achieved through microbes, wherein a two-step process has been developed using *Aspergillus niger* CGMCCO774, which transforms the ferulic acid to vanillic acid, and *Pycnoporus cinnabarinus* or *Phanerochaete chrysosporium*, which convert vanillic acid into vanillin (500 mg/l). This has been further optimized to 1 g/l via the fermentation method (Priefert et al. 2001).

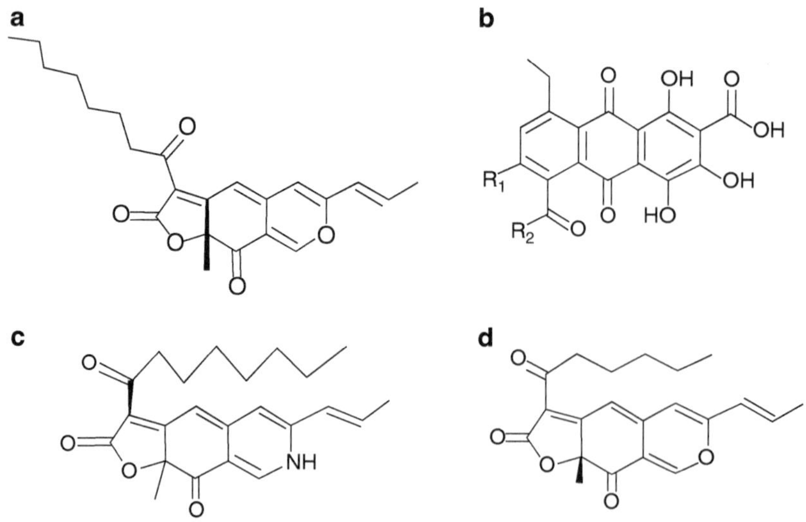

Fig. 8.19 Microbially produced pigments from *Monascus* and *Penicillium* species: (**a**) monascorubin, (**b**) arpink red basic structure, (**c**) monascorubramine, and (**d**) rubropunctatin

After vanillin, benzaldehyde is the second most important molecule for flavor and fragrance. Its market demand is 5,000 kg/year and price is approximately US$240/kg. Benzaldehyde is synthesized from *Pseudomonas putida* and white rot fungi *Polyporus tuberaster* and *P. chrysosporium*. The pervaporation process has been applied for the production of benzaldehyde recovery from the fermentation broth of white rot fungus *Bjerkandera adusta*.

Lactones are also potent pleasing compounds used in the flavor industry and are recognized as fruity, coconut, and buttery aromas. Microorganisms produce optically active lactones by the conversion of γ- and δ-keto acids. 6-pentyl-α-pyrone (6PP) is an unsaturated lactone that provides a characteristic coconut aroma. Perstraction has been employed for the online extraction of γ-decalactone (peach-apricot flavor) from yeasts *Sporidiobolus ruinen* or *Suillus salmonicolor*. The global annual production of γ-decalactone is approximately 10 tons. *Cladosporium suaveolens* and *Tyromyces sambuceus* efficiently produce γ-decalactone and δ-dodecalactones from ricinoleic and linoleic acid, respectively (Kapfer et al. 1989; Allegrone et al. 1991).

Pyrazines are heterocyclic nitrogen compounds that contribute significantly to the flavor of many foods. They generally are associated with the flavor of roasted or nutty products. Microorganisms have been recently reported to be the source of pyrazine, with the first being reported from tetramethylpyrazine from *B. subtilis* (Kosuge and Kamiya 1962). This gives a fermented soy bean flavor in soy sauce, natto, and miso. An auxotrophic mutant of *C. glutamicum* requiring leucine, isoleucine, valine, and pantothenate for growth has been reported to produce large amounts of tetramethylpyrazine.

Varieties of monoterpenes have been synthesized from microorganisms, the most prominent being the fungus *Ceratocyctis*. Monoterpenes are also produced by the fungus *Trametes odorata*, species of *Phellinus*. These are generally associated with wood rotting rather than flavor production. *Kluyveromyces lactis* has been reported to produce citronellol, linalool, and geraniol via submerged fermentation. β-ionone has been converted biologically through several fungi into tobacco flavorings and sclareolide. Sclareol was used as the precursor for generating amrox using *Cryptococcus* for perfumery applications (Cheetham 1993; Farbood et al. 1990).

2,5-dimethyl-4-hydroxy-3(2H)-furanone (Furaneol®) is an aroma exhibiting a strawberry flavor in dilute solutions and caramel-like flavors in concentrates. *Zygosaccharomyces rouxii* is able to form DMHF when specifically supplied with D-fructose-1, 6-bisphosphate, and glucose (Dahlen et al. 2001). Blackcurrant odorant, p-mentha-8-thiol-3-one, has been synthesized by incubating cysteine and pulegone with *Eubacterium limosum* ATCC10825 (Kerkeenar et al. 1993). Several aroma products, like ethyl acetate, propyl acetate, isobutyl acetate, isoamyl acetate, citronellol, and geraniol, are produced by the fungus *Ceratocystis moniliformis. Candida tropicalis* and *Torulopsis bombicola* convert the alkanes or fatty acids into musk fragrance. *Botryodiplodia theobromae*, the fungal plant pathogen, has been found to produce the jasmine fragrance, which is due to the formation of methyl (+) −7-isojasmonic acid from α-linolenic acid (C18:3; found in linseed oil).

Despite the production of a variety of aromas and flavors by microorganisms and their enzymes, their use at the industrial level is limited. This is attributed to their low fermentation yield and high downstream processing cost. Another important constraint is regulatory clearance of these microbial products as safe for food and flavor applications, which may take more time. BASF Germany has recently started the production of 4-decalactone, which has a peach aroma and is distributed by Fritzsche, Dodge, and Olcott.

8.8 Summary

Microbes have significantly contributed to the development of fine chemicals like antibiotics, amino acids, vitamins, dyes, pigments, flavors, and fragrances. Antibiotics and drugs were initially developed from microbes with the landmark discovery of penicillin by Alexander Fleming in 1929. A variety of microbial secondary metabolites are directly and indirectly used in the development of drugs like atorvastatin and pravastatin, which are used by people with hypercholesterolemia. They are also a source of novel anti-diabetic drugs like glucobay and voglibose. Orlistat, which is a lipase inhibitor isolated from actinomycete, has been used as a drug for combating obesity and diabetes.

Microbes are also responsible in evolving the science of immunopharmacology with the discovery of cyclosporine from the fungus *T. inflatum*. Other immunosuppressants of microbial origin are rapamycin and tacrolimus. Doxorubicin and geldanamycin from microbes have led to the development of chemotherapeutic agents to combat cancer. Recently, endophytic microorganisms have also been used as an under-explored resource for the discovery of new bioactive molecules with medicinal properties. These also provide immense opportunities in mimicking the plant biosynthetic pathways, thus opening avenues for fermentative production of plant medicinals. Recombinant DNA technology has helped in the de novo engineering of metabolic pathways for producing plant medicinals through microbial systems, thereby overcoming the problem of their limited production and short supply with respect to their increased global demand.

Vitamins are health supplements and have applications in the food, feed, and cosmetic industries. Microbes are increasingly being used for the production of vitamins E and K, β-carotene, and vitamins B_2 and B_{12} since the process is more ecofriendly than the chemical processes. Amino acids are increasingly finding applications as flavor enhancers, as additives in food and feeds, in intravenous infusions, and in the cosmetic industry. MSG (sodium salt of L-glutamate) has been used as a flavor enhancer and is being produced through fermentation carried out by *C. glutamicum*. The global market for MSG is approximately 1,700,000 MT, with Ajinomoto Inc., ADM, and Bayer as major global producers. The other important amino acids produced by microbes are L-lysine, L-threonine, L-serine, L-tryphtophan, and L-phenylalanine.

Dyes and pigments have wide applications in the textile and food industries. The production of indigo through recombinant *E. coli* was a landmark achievement in the production of plant pigments through microbes and paved the way for the use of microbes for the production of pigments and colors. Microorganisms produce various pigments like carotenoids, melanins, flavins, monascins, which could be helpful in the replacement of synthetic chemical dyes, which have already proved to be toxic and non-ecofriendly and unsafe for the end user. Several fungal and bacterial pigments are being isolated and tested for their use in the textile industry to the same standards as used for the synthetic dyes currently used.

In the food industry, microbially produced carotenoids have been recognised as GRAS (Generally Recognized as Safe) for human consumption. A variety of pigments produced by *Monascus ruber* have also found use in the food industry, and several patents have been granted on their use. More recently, arpink has been isolated from *Penicillium oxalicum* var. *armenica* and has been used for coloring meat products and seafoods with a bright red color.

More recently, microbes have been used for the production of flavors and fragrances, with a global market of US$16 billion. Vanillin, which was initially isolated from the vanilla plant is currently being produced through microbes due to its huge demand and application in the food, pharmaceutical, and cosmetic industries. Benzaldehydes, lactones, and pyrazines are the other major classes of compounds being produced through the microbial route for their use as flavors and fragrances. Regulatory clearance of the microbially produced flavors and fragrances is a time-consuming exercise and a major constraint in their commercialization. However, mechanism of the safety and regulatory clearance process is being optimized, with more and more products produced through the microbial route. A microbially produced peach aroma has been commercialized by BASF (Germany).

Thus, it could be inferred that microbes have revolutionized the fine chemical arena by providing alternative production methods through the fermentative route that are ecofriendly and cost effective and that can meet the growing global demands. The global emphasis is on the development of microbe-based processes to reduce the harmful effects of synthetic chemical processes to the environment as well as to society at large.

Selected Reading

Allegrone G, Barbeni M, Cardillo R, Fuganti C, Grasselli P, Miele A, Pisciotta A (1991) On the steric course of the microbial generation of (Z6)-gamma-dodecenolactone from (10R,S) 10-hydroxyoctadeca-(E8,Z12)-dienoic acid. Biotechnol Lett 13:765–768

Allison A (2001) The possible role of vitamin K deficiency in pathogenesis of Alzheimer's disease and in augmenting the brain damage associated with cardiovascular disease. Med Hypothesis 57(2):151–155

Amna T, Puri SC, Verma V, Sharma JP, Khajuria RK, Musarrat J, Spiteller M, Qazi GN (2006) Bioreactor studies on the endophytic fungus *Entrophospora infrequens* for the production of an anticancer alkaloid camptothecin. Can J Microbiol 52:189–196

Anke T, Oberwinkler F, Steglich W, Schramm G (1977) The strobilurins – new antifungal antibiotics from the basidiomycete *Strobilurus tenecellus*. J Antibiot 30:806–810

Attala MM, El-khrisy EAM, Youssef YA, Asem MA (2011) Production of textile reddish brown dyes by fungi. Malays J Microbiol 7(1):33–40

Bennett JW, Chung KT (2001) Alexander Fleming and the discovery of penicillin. Adv Appl Microbiol 49:163–184

Blanche F, Cameron B, Crouzet J, Debussche L, Levy-Schil S, Thibaut D (1998) Polypeptides involved in the biosynthesis of cobalamines and/or cobamides, DNA sequences coding for these polypeptides, and their preparation and use. Eur. Patent 0516647 B1

Borel JF, Feurer C, Gubler HU, Stahelin H (1976) Biological effects of cyclosporin A: a new anti-lymphocytic agents. Agents Action 6(4):458–475

Burton JP, Chilcott CN, Moore CJ, Speiser G, Tagg GR (2006) A preliminary study of the effect of probiotic *Streptococcus salivarinus* K12 in oral malodour parameters. J Appl Microbiol 100(4):754–764

Cao L, Huang J, Li J (2007) Fermentation conditions of *Sinopodophyllum hexandrum* endophytic fungus on production of podophyllotoxin. Food Ferment Ind 33:28–32

Carballo-Cardenas EC, Tuan PM, Janssen M, Wijffels RH (2003) Vitamin E (α-tocopherol) production by marine microalgae Dunaliella tertiolecta and Tetraselmis suecica in batch cultivation. Biomol Eng 20(4):139–147

Caruso M, Colombo AL, Crespi-Perellino A, Fedeli L, Pavesi A, Quaroni S, Saracchi M, Ventrella G (2000) Studies on a strain of Kitasatophora sp. paclitaxel producer. Ann Microbiol 50:89–102

Chakravarthi BVSK, Das P, Surendranath K, Karande AA, Jayabaskaran C (2008) Production of paclitaxel by *Fusarium solani* isolated from *Taxus celebica*. J Biosci 33:259–267

Chang MCY, Keasling JD (2006) Production of isoprenoid pharmaceuticals by engineered microbes. Nat Chem Biol 2:674–681

Cheetham PSJ (1993) The use of biotransformation for the production of flavours and fragrances. Trends Biotechnol 11:478–488

Cheung AU, Tile L, Lee Y, Tomlinson G, Hawker G, Scher J, Hu H, Vieth R, Thompson L, Jamal S, Josse R (2008) Vitamin K supplementation in post menopausal women with osteopenia (Ekcotrial): a randomized controlled trial. PLoS Med 5(10):1461–1472

Cooper R, Bush K, Principe PA, Trejo WH, Weels JS, Sykes RB (1983) Two new monobactam antibiotics produced by *Flexibacter* sp. I. Taxonomy, fermentation, isolation and biological properties. J Antibiot 36(10):1252–1257

Dahlen T, Hauck T, Wein M, Schwab W (2001) 2, 5-dimethyl-4-hydroxy-3(2H) - furanone as a secondary metabolite from D-fructose-1, 6-diphosphate metabolism by *Zygosacharomyces rouxii*. J Biosci Bioeng 91:352–358

Dai W, Tao W (2008) Preliminary study on fermentation conditions of taxol-producing endophytic fungus. Chem Ind Eng Prog 27:883–886

De Kwaadsteniet M, van Reenen C, Dicks L (2010) Evaluation of nisin F in the treatment of subcutaneous skin infections as monitored by using bioluminescent strain of *Staphylococcus aureus*. Probiotics Antimicrob Proteins 2(2):61–65

Deng BW, Liu KH, Chen WQ, Ding XW, Xie XC (2009) *Fusarium solani*, Tax-3, a new endophytic taxol-producing fungus from *Taxus chinensis*. World J Microbiol Biotechnol 25:139–143

Deresinski SC, Stevens DA (2003) Caspofungin. Clin Infect Dis 36:1445–1447

Ec C-C, Tuan PM, Janssen M, Wijffels RN (2003) Vitamin E (α-tocopherol) production by the marine microalgae *Dunaliella tertiolecta* and *Tetraselmis suecica* in batch cultivation. Biomol Eng 20(4–6):139–147

Endo A, Kuroda M, Trujita HY (1976) ML-236A and ML-236C: new inhibitors of cholesterol genesis produced by *Penicillium citrinum*. J Antibiot 29:1346–1348

Engels B, Dahm P, Jennewein S (2008) Metabolic engineering of taxadiene biosynthesis in yeast as a first step towards Taxol (Paclitaxel) production. Metab Eng 10:201–206

Ensley DB, Ratzkin JB, Osslund DT, Simon JM (1983) Expression of naphthalene oxidation genes in *Escherichia coli* results in the biosynthesis of indigo. Science 222:167–169

Eyberger AL, Dondapati R, Porter JR (2006) Endophyte fungal isolates from *Podophyllum peltatum* produces podophyllotoxin. J Nat Prod 69:1121–1124

Fabre CE, Santrre AL, Loret MD, Beberian R, Paralleux A, Gena G, Blanc PJ (1993) Production and food application of the red pigment of *Monascus ruber*. J Food Sci 58:1099–1103

Farbood MI, Morris JA, Sprecker MA, Beinkowski LJ, Miller KP, Vock MH, Hagedorn ML (1990) Flavoring with mixtures of lactones US Patent 4,960,597

Feling RH, Buchanan GO, Mincer TJ, Kauffman CA, Jensen PR, Fenical W (2003) Salinosporamide A: a highly cytotoxic proteasome inhibitor from novel microbial source, marine bacterium of new genus *Salinospora*. Angew Chem Int Engl 42:355–357

Floss HG, Lee S, Tornus I (2000) Valiolone, a method of preparing it and its use to prepare acarbose and voglibose. US Patent no. 6150568

Fradenhagen A, Tamura SY, Kenny PTM, Komura H, Naya Y, Nakanishi K, Sugiura J, Kita H (1987) Andrimid, a new peptide antibiotic produced by an intracellular bacterial symbiont from brown plant hopper. J Am Chem Soc 109(14):4409–4411

Fuller AT, Mellaros G, Woolford M, Banks GT, Barrow KD, Chain EB (1971) Pseudomonic acid: an antibiotic produced by *Pseudomonas fluorescens*. Nature 234:416–417

Gangadevi V, Muthumary J (2008) Taxol, an anticancer drug produced by endophytic fungus *Bartaliana robillardoides* Tassi, isolated from the medicinal plant *Aegle marmelos* Correa ex Roxb. World J Microbiol Biotechnol 24(5):712–724

Gangadevi V, Muthumary J (2009) A novel endophytic Taxol producing *Chaetomella raphigera* isolated from the medicinal plant *Terminalia arjuna*. Appl Biochem Biotechnol 158(3):675–684

Giri AV, Anandkumar N, Muthukumaran G, Pennathur G (2004) A novel medium for the enhanced cell growth and production of prodigiosin from *Serratia marcescens* isolated from soil. BMC Microbiol 4:1–10

Guo B, Li H, Zhang L (1998) Isolation of the fungus producing vinblastine. J Yunnan Univ (Nat Sci Ed) 20:214–215

Gurudatt PS, Priti V, Shweta S, Ramesha BT, Ravikanth G, Vasudeva R, Amna T, Deepika S, Ganeshaiah KN, UmaShaanker R, Puri S, Qazi N (2010) Camptothecin production and negative relationship between hyphal biomass and camptothecin content in endophytic strains of *Nothapodytes nimmoniana* Grahm (Icecinaceae). Curr Sci 98(8):1006–1010

Hanefeld M, Fischer S, Schulze J, Spengler M, Waagenar H, Schollberg K, Fuker K (1991) Therapeutic potentials of acarbose as a first line drug in NIDDM insufficiently treated with diet alone. Diabetes Care 14(8):732–737

Hannibal L, Lorquin J, D'ortoli NA, Chaintreuil C, Masson-Boivin C, Dreyfus B, Giruad E (2000) Isolation and characterization of canthaxanthin biosynthesis genes from the photosynthetic bacterium *Bradyrhizobium* sp. strain ORS278. J Bacteriol 182(13):3850–3853

Hirao T, Nakano T, Azuma T, Sugimoto M, Nakanishi T (1989) L-Lysine production in continuous culture of an L-Lysine hyperproducing mutant of *Corynebacterium glutamicum*. Appl Microbiol Biotechnol 32:269

Huang S, Bjornsti MA, Houghton PJ (2003) Rapamycins: mechanism of action and cellular resistance. Cancer Biol Ther 2:222–232

Hughes PE, Tove SB (1982) Occurrence of α-tocopherol quinone and α-tocopherol quinol in microorganisms. J Bacteriol 151:1397–1402

Irschik KH, Reichenbach H, Hofele G, Jansen R (2007) The thuggacins, novel antibacterial macrolides from

Sorangium cellulosum against selected gram positive bacteria. J Antibiot 60(12):733–738

Jabes D, Brunati C, Candiani GP, Rwa S, Romano G, Donadio S (2011) Efficacy of the new antibiotic NAI-107 in experimental infections induced by multidrug resistant gram positive pathogens. Antimicrob Agents Chemother 55(4):1671–1676

Ju Z, Wang J, Pan S (2009) Isolation and preliminary identification of the endophytic fungi which produce Hupzine A from four species in Hupziaceae and determination of Huperzine A by HPLC. Fudan Univ J (Med Sci Ed) 36:445–449

Kamada Y, Asano N, Yoshikawa T, Matsui K, Horii S, Fukase H (1984) Valiolamine – a new alpha glucosidase inhibiting aminocyclitol produced by *S. hygroscopicus*. J Antibiot 37(11):1301–1307

Kapfer GF, Berger RG, Draweti F (1989) Production of 4-decanolide by semicontinuous fermentation of *Tyromyces sambuceus*. Biotechnol Lett 11:561–566

Katayama N, Fukusami S, Funabashi Y, Iwahi T, Ono H (1993) TAN-1057 A-D, new antibiotics with potent antibacterial activity against methicillin resistant *Staphylococcus aureus*: taxonomy fermentation and biological activity. J Antibiot 46:606–613

Keri V, Csorvasi A, Aronhime J (2003) Preparation of orlistat and orlistat crystalline forms. US application no. US2003/149095 A1

Kerkeenar A, Schemedding DJM, Berg J (1993) Method of preparation of p-metha-8-thiol-3-one. US Patent 5182194

Kim PI, Sohng JK, Sung C, Joo HS, Kim EM, Yamaguchi T, Park D, Kim BG (2010) Characterization and structure identification of antimicrobial peptide Hominicin produced by *Staphylococcus hominis* MBBL-29. Biochem Biophys Res Commun 399(2):133–138

Kimura E (2003) Metabolic engineering of glutamate production. In: Scheper T, Faurie R, Thommel J (eds) Advances in biochemical engineering/biotechnology, vol 79. Springer, Berlin/Heidelberg/New York, pp 37–57

Kino T, Hatanaka H, Hashimoto M, Nishiyama M, Goto T, Okuhara M, Kohsaka M, Aoki H, Imanaka H (1987) FK-506, n novel immunosuppressant isolated from Streptomyces species I. Fermentation, isolation and physicochemical and biological characterization. J Antibiot 40(3):1249–1255

Kinoshita S (1985) Glutamic acid bacteria. In: Demain AL, Solomon NA (eds) Biology of industrial microorganisms. Benjamin/Cummings, Menlo Park, pp 115–142

Kong MC, Lee PC (2011) Metabolic engineering of menaquinone-8-pathway of *E. coli* as a microbial platform for vitamin K production. Biotechnol Bioeng 108(8):1997–2002

Kosuge T, Kamiya H (1962) Tetramethylpyrazine: a *Bacillus subtilis* growth factor. Nature 193:776

Kour A, Shawl AS, Rehman S, Sultan P, Qazi PH, Suden P, Khajuria RK, Verma V (2008) Isolation and identification of an endophytic strain of *Fusarium oxysporum* producing podophyllotoxin from *Juniperus recurva*. World J Microbiol Biotechnol 24: 1115–1121

Kruszewska D, Sahl HG, Bierbaum G, Pag G, Hynes SO, Ljungh A (2004) Mersacidin eradicates methicillin resistant *Staphylococcus aureus* (MRSA) in mouse rhinitis model. J Antimicrob Chemother 34(3):648–653

Kumaran RS, Muthumary J, Hur BK (2008) Production of taxol from *Phyllosticta spinarum*, an endophytic fungus of *Cupressus* sp. Eng Life Sci 8:438–446

Kumaran RS, Choi YK, Lee S, Jeon HJ, Jung H, Kim HJ (2012) Isolation of taxol, an anticancer drug produced by the endophytic fungus *Phoma betae*. Afr J Biotechnol 11(4):950–960

Kusari S, Zuhlke S, Spiteller M (2009) An endophytic fungus from *Camptotheca acuminata* that produces camptothecin and analogues. J Nat Prod 72:2–7

Li JY, Strobel GA, Sidhu R, Hess WM, Ford EJ (1996) Endophytic taxol-producing fungi from bald cypress, *Taxodium distichum*. Microbiology 142:2223–2226

Li JY, Sidhu RS, Ford EJ, Long DM, Hess WM, Strobel GA (1998) The induction of taxol production in the endophytic fungus-*Periconia* sp. from *Torreya grandifolia*. J Ind Microbiol Biotechnol 20:259–264

Lim SH, Choi JS, Park EY (2001) Microbial production of riboflavin using riboflavin overproducers, Ashbya gossypii, Bacillus subtilis, and Candida famate: an overview. Biotechnol Bioprocess Eng 6:75–88

Li J, Lin JC, Wang H, Peterson JW, Furie BC, Furie B, Booth SL, Volpe JJ, Rosenberg PA (2003) Novel role of vitamin K in preventing oxidative injury to oligodendrocytes and neurons. J Neurosci 23(13):5816–5826

Li W, Zhou J, Lin Z, Hu Z (2007) Study on fermentation condition for production of huperzine A from endophytic fungus 2F09P03B of *Huperzia serrata*. Chin Med Biotechnol 2:254–259

Li Y, Qui NE, Yin H, Peng H (2011) Isolation of a podophyllotoxin producing endophytic fungus from *Sinopodophyllum hexandrum* and the anti-S180 sarcoma activity of its fermentation broth in mice. Acad J Second Mil Med Univ. doi:10.3724/SP.J.1008.2011.00012

Liu JM, Ni MY, Fan JF, Tu YY, Wu ZH, Wu YL, Chou WS (1979) Structure and reaction of *Arteannuin*. Acta Chim Sin 37:129–143

Liu K, Ding X, Deng B, Chen W (2009) Isolation and characterization of endophytic taxol-producing fungi from *Taxus chinensis*. J Ind Microbiol Biotechnol 36:1171–1177

Liu K, Ding X, Deng B, Chen W (2010) 10-hydroxycamptothecin produced by new endophytic *Xylaria* sp. M 20, from *Camptotheca acuminata*. Biotechnol Lett 32(5):689–693

Lurie Y, Loebstein R, Kurnik D, Almog S, Halkin H (2010) Warfarin and vitamin K intake in the era of pharmacogenomics. Br J Clin Pharmacol 70(2):164–170

Majumdar SK, Bose SK (1958) Mycobacillin, a new antifungal antibiotic produced by Bacillus subtilis. Nature 181(4602):134–135

Mansoroi K, Khaurin P, Lohetaroenkal W, Werner RG, Gotz F, Monosroi A, Monosroi J (2010) Transdermal adsorption enhancement through rat skin of Gallidermin loaded niosomes. Int J Pharm 392(1–2):304–310

Martens, J.H., Barg H., Warren M and Jahn D. (2002) : Microbial production of vitamin B_{12} Appl Microbiol Biotechnol 58:275–285

Miao Z, Wang Y, Yu X, Guo B, Tang K (2009) A new endophytic taxane production fungus from *Taxus chinensis*. Appl Biochem Microbiol 45:81–86

Murray BE (1981) Cephalosporins. Ann Rev Med 32:559–581

Mutabingwa TK (2005) Artemisinin-based combination therapies (ACTs) best hope for malaria treatment but inaccessible to the needy! Acta Trop 95:305–315

Nadeem M, Ram M, Alam M, Ahmad MM, Qurainy A, Khan S, Abdin MZ (2012) *Fusarium solani* P1, a new endophytic podophyllotoxin producing fungus from the roots of *Podophyllum hexandrum*. Afr J Microbiol Res 6(10):2493–2499

Ogbonna JC, Tomiyamal S, Tanaka H (1998) Heterotrophic cultivation of *Euglena gracilis* Z for efficient production of α-tocopherol. J Appl Phycol 10:67–74

Ohnuma S, Suzuki M, Nishino T (1994) Archaebacterial ether linked lipid biosynthetic gene. Expression, cloning, sequencing and characterization of geranygerany-diphosphate synthase. J Biol Chem 269:14792–14797

Pandi M, Kumaran RS, Choi YK, Kim HJ, Muthumary J (2011) Isolation and detection of taxol, an anticancer drug produced by *L. theobromae*, an endophytic fungus in medicinal plant *Morinda citrifolia*. Afr J Biotechnol 10(8):1428–1435

Perkins JB, Sloma A, Hermann J, Theriault K, Zachgo E, Erdenberger T, Hannett N, Chatterjee NP, Williams V II, Rufo GA Jr, Hatch R, Pero J (1999) Genetic engineering of Bacillus subtilis for commercial production of riboflavin. J Ind Microb Biotechnol 22:8–18

Perlman D (1979) Microbial process of Riboflaviun Production In: Peppler JH, Perlman D (eds) Microbial process of riboflavin production in microbial technology, vol 1, 2nd edn. Academic, New York, pp 521–527

Piper C, Draper LA, Cotter PD, Ross RP, Hill C (2009) A comparison of the activities of Lacticin 3147 and Nisin against drug resistant *Staphylococcus aureus* and *Enterococcus* species. J Antimicrob Chem 64(3):546–551

Piper C, Carey CG, Hill C, Cotter PD, Ross RP (2012) The lantibiotic Lacticin 3147 prevents systemic spread of *Staphylococcus aureus* in Murine infection model. Int J Microbiol. doi:10.1155/2012/806230

Pollack P (2007) Fine chemicals: the industry and the business. Wiley, Hoboken. ISBN 978-0-470-05075-0

Powls R, Redfearn ER (1967) The tocopherols of blue-green algae. Biochem J 104:24C–26C

Priefert H, Rabenhorst J, Steinbüchel A (2001) Biotechnological production of vanillin. Appl Microbiol Biotechnol 56:296–314

Puri SC, Verma V, Amna T, Qazi GN, Spiteller M (2005) An endophytic fungus from *Nothapodytes foetida* that produces camptothecin. J Nat Prod 68:1717–1719

Puri SC, Nazir A, Chawla R, Arora R, Riyaz-ul-Hasan S, Amna T, Ahmed B, Verma V, Singh S, Sagar R, Sharma A, Kumar R, Sharma RK, Qazi GN (2006) The endophytic fungus *Trametes hirsuta* as a novel alternative source of podophyllotoxin and related aryl tetralin ligans. J Biotechnol 122:494–510

Qi Q, Hao M, Ng W, Slater SC, Baszis SR, Weiss JD, Valentine HE (2005) Application of the *Synechococcus* nirA promoter to establish an inducible expression system for engineering the *Synechocystis* tocopherol pathway. Appl Environ Microbiol 71:5678–5684

Rea MC, Sit CS, Clayton E, O'Conner PM, Whittal RM, Zheng J, Vederas JC, Ross RP, Hill C (2010) Thuricin CD a post translationally modified bacteriocin with narrow spectrum of activity against *Clostridium difficale*. PNAS. doi:10.1073/pnas0.913444107

Rehman S, Shawl AS, Kour A, Andrabi R, Sudan P, Sultan P, Verma V, Qazi GN (2008) An endophytic *Neurospora* sp. from *Nothapodytes foetida* producing camptothecin. Appl Biochem Microbiol 44:203–209

Rehman S, Shawl AS, Kour A, Sultan P, Ahmad K, Khajuria R, Qazi GN (2009) Comparative studies and identification of Camptothecin produced by an endophyte at shake flask and bioreactor. Nat Prod Res 23(11):1050–1057

Ro DK, Paradise EM, Ouellet M, Fisher KJ, Newman KN, Ndungu JM, Ho KA, Eachus RA, Ham TS, Kirby J, Chang MCY, Withers ST, Shiba Y, Sarpong R, Keasling JD (2006) Production of the antimalarial drug precursor artemisinic acid in engineered yeast. Nature 440:940–943. doi:10.1038/nature04640

Rommele G, Traxler P, Wehrli W (1983) Papulacandins-the relationship between chemical structure and glucan synthesis in yeast. J Antibiot 36(11):1539–1542

Roy S, Chatterjee S, Sen SK (2008) Biotechnological potential of *Phaffia rhodozyma*. J Appl Biosci 5:115–122

Sardaryan E (2002) Strain of the microorganism *Penicillium oxalicum* var. *armeniaca* and its application. US Patent 6,340,586 B1

Sasse F, Boehlendorf B, Hermann M, Kunze B, Forche E, Steinmetz H, Hoelfe G, Reinchenbach H (1999) Melithiazols, a new b-methoxyacrylate inhibitors in respiratory chain isolated from Myxobacteria. J Antibiot 52:721–729

Schatz A, Bugie E, Waksman SA (1973) Selman Abraham Waksman, Ph.D, 22 July 1888-16 August 1973. Streptomycin reported. Ann Int Med 79:678

Sharma S, Gupta C, Aggarwal S, Nagpal N (2012) Pigment extraction from fungus for textile dyeing. Indian J Fibre Text Res 37:68–73

Shweta S, Zuehlke S, Ramesha BT, Priti V, Mohana Kunar P, Ravikanth G, Spiteller M, Vasudeva R, Shaanker RU (2010) Endophytic fungal strains of *Fusarium solani*, from *Apodytes dimidiata* E. Mey. ex Arn (Icacinaceae) produce camptothecin,10-hydroxycamptothecin and 9-methoxycamptothecin. Phytochemistry 71:117–122

Silo Suh LA, Lethbridge BJ, Raffel SI, He HY, Clardy J, Handlesman J (1994) Biological activities of two fungistatic antibiotics produced by *Bacillus cereus* UW85. Appl Environ Microbiol 60:2023–2030

Smith L, Hasper H, Breukink NJ, Cerkasov J, Hillman JD, Wilson-Standford S, Orgrunty RS (2008) Elucidation of antimicrobial mechanism Mutacin 1140. Biochemistry 47:3308–3314

Solit DB, Ivy SP, Kopil C, Sikorski R, Morris MJ, Slorin SF, Kelly WK, De La Cruz A, Gerley T, Heller G, Larson S, Schwartz L, Egorin MJ, Rosen N, Scher HI (2007) Phase I clinical trial of 17- Allylamino-17-demethoxy geldanamycin in patients with advanced cancer. Clin Cancer Res 13:1775

Stierle A, Strobel G, Stierle D (1993) Taxol and taxane production by *Taxomyces andreanae*, an endophytic fungus of Pacific yew. Science 260:214–216

Strobel G, Yang XS, Sears J, Kramer R, Sidhu RS, Hess WM (1996a) Taxol from *Pestalotiopsis microspora*, an endophytic fungus of *Taxus wallachiana*. Microbiology 142:435–440

Strobel GA, Hess WM, Ford E, Sidhu RS, Yang X (1996b) Taxol from fungal endophytes and issue of biodiversity. J Ind Microbiol 17:417–423

Strobel GA, Hess WM, Li JY, Ford E, Sears J, Sidhu RS, Summerell B (1997) *Pestalotiopsis guepinii*, a taxol-producing endophyte of the wollemi pine, *Wollemia nobilis*. Aust J Bot 45:1073–1082

Sun D, Ran X, Wang J (2008) Isolation and identification of a taxol-producing endophytic fungus from *Podocrapus*. Acta Microbiol Sin 48:589–595

Sutherland R, Boon RJ, Griffin KE, Masters PJ, Slocombe B, White AR (1985) Antibacterial activity of Mupirocin (Pesudomonic acid). A new antibiotic for topical use. Antimicrob Agents Chemother 27(4):495–498

Taketomi H, Shoda K, Katsui G (1983) Results of screening test in tocopherols in the microbial realm. Vitamins (Japan) 57:133–138

Umezawa H, Maeda K, Takeuchi OY (1966) New antibiotics bleomycin A and B. J Antibiot 19(5):200–209

Unagul P, Wongsa P, Kittakoop P, Intamas S, Srikitikulchai S, Tanticharoen M (2005) Production of red pigments by insect pathogenic fungus *Cordyceps unilateralis* BCC1869. J Ind Microb Biotechnol 32(4):135–140

Valentin HE, Qi Q (2005) Biotechnological production and application of Vitamin E: current state and prospects. Appl Microbiol Biotechnol 68:436–444

Veerwal R, Wang J, Meijnen JP, Visser H, Sandmann G, van den Berg JA, van Ooyen AJJ (2007) High level production of beta-carotene in *Saccharomyces cerevisiae* by successive transformation of Carotenogenic genes from *Xanthophyllomyces dendrorhous*. Appl Environ Microbiol 73(13):4342–4350

Venkatachalam R, Subban K, Paul MJ (2008) Taxol from *Botryodiplodia theobromae* (BT 115)-an endophytic fungus of *Taxus baccata*. J Biotechnol 136:S189–S190

Vijaylaxmi P, Sarva Mangala D (2011) Fermentative production of L- Glutamic acid. Int J Adv Biotechnol Res 2(3):376–381

Vismura R, Vestri S, Kusmic C, Barsanti L, Guattieri P (2003) Natural vitamin E enrichment of *Artemia salina* fed freshwater and microalgae. J Appl Phycol 15:75–80

Waksman SA (1947) What is an antibiotic or an antibiotic substance? Mycologia 39(5):565–569

Wang J, Li G, Lu H, Zhang Z, Huang Y, Su W (2000) Taxol from Tubercularia species strain TF5 an endophytic fungus of Taxus mairei. FEMS Microbiol Lett 193(2):249–253

Wall ME, Wani MC, Cook CE, Palmer KH, McPhail AT, Sim GA (1966) Plant antitumor agents. I. The isolation and structure of camptothecin, a novel alkaloidal leukemia and tumor inhibitor from *Camptotheca acuminata*. J Am Chem Soc 88:3888–3890

Wang YT, Lo HS, Wang PH (2008) Endophytic fungi from Taxus mairei in Taiwan. First report of *Colletotrichum gloesporioides* as an endophyte in *Taxus mairei*. Bot Stud 49:39–43

Weibel EK, Hadvary P, Hochuli E, Kupfer E, Lengsfeld H (1987) Lipstatin as inhibitors of pancreatic lipase produced by *S. toxytricini*. J Antibiot 40:1081–1085

Williamson JM, Inamine E, Wilson KE, Douglas AE, Leisch JM, Schoberg GA (1985) Biosynthesis of the β-lactam antibiotic Thienamycin from *S. cattaleya*. J Biol Chem 260:4637–4647

Yang X, Zhang L, Guo B, Guo S (2004) Preliminary study of a vincristine-producing endophytic fungus isolated from leaves of *Catharanthus roseus*. Chin Tradit Herb Drugs 35:79–81

Zhang L, Guo B, Li H, Zeng S, Shao H, Gu S, Wei R (2000) Preliminary study on the isolation of endophytic fungus of *Catharanthus roseus* and its fermentation to produce products of therapeutic value. Chin Tradit Herb Drugs 31:805–807

Zhang P, Zhou P, Yu L (2009) An endophytic taxol-producing fungus from *Taxus media*, *Cladosporium cladosporioides* MD2. Curr Microbiol 59: 227–232

Zhao K, Zhao L, Jin Y, Wei H, Ping W, Zhou D (2008) Isolation of a taxol-producing endophytic fungus and inhibiting effect of the fungus metabolites on HeLa cell. Mycosystema 27:735–744

Zhao K, Ping W, Li Q, Hao S, Zhao L, Gao T, Zhou D (2009) *Aspergillus niger* var. *taxi*, a new species variant of taxol-producing fungus isolated from *Taxus cuspidata* in China. J Appl Microbiol 107:1202–1207

Zhao K, Sun L, Ma X, Li X, Wang X, Ping W, Zhou D (2011) Improved taxol production in *Nodulisporium sylviforme* derived from inactivated protoplast fusion. Afr J Biotechnol 10(20):4175–4182

Zhou S, Yang F, Lan S, Xu N, Hong Y (2009) Huperzine A producing conditions from endophytic fungus in *Huperzia serrata*. J Microbiol 29:32–36

9 Microbial Enzymes and Their Industrial Applications

9.1 Introduction

Enzymes are biological catalysts produced in living cells. They are proteinaceous in nature, the exception being catalytic RNA, which are also referred to as ribozymes. The term 'en zyme' is derived from the Greek, meaning 'in sour dough'. E. Buchner (1897) experimentally proved that cell-free extract from yeast could produce alcohol from sugars, and he referred to it as "zymase". The unique characteristics that enzymes possess are that they (1) increase the rate of reaction they catalyze, without being consumed or lost; (2) act specifically with the substrate to produce the products; and (3) remain regulated from a state of low activity to high activity and vice versa. Enzymes have been grouped into six classes based on the types of reactions they catalyze (Table 9.1). All cellular processes are controlled by a coordinated sequence of reactions that have specifically been catalyzed by a defined set of enzymes.

Enzyme preparations have been used for centuries in industrial processes such as brewing and baking. Malted barley has been used for brewing, and papaya for meat tenderization. With advances in biochemistry and analytical methods and techniques, enzymes were isolated and characterized. J.B. Sumner (1926) isolated ureases from the Jack bean and proved enzymes to be proteins that could be isolated in crystalline form. Pepsin and trypsin, the digestive enzymes, were subsequently crystallized by John Northrop and Moses Kunitz. The enzymes utilized currently are sourced from animals, plants, and microorganisms. However, the major bulk of industrially used enzymes are produced from microorganisms.

9.1.1 Advantages of Microbial Enzymes

Enzymes produced by microbes are in higher demand than are those resourced from animals and plants. This is attributed to the high growth rates of microbes on inexpensive medium compared with plant and animal cells, thus providing a higher yield of enzymes in a shorter period of time. Microorganisms exhibit a variety of catalytic activity due to their physiological, geographic, and genomic diversity. These also provide an ease of genetic manipulation by which yield improvement and undesirable product formation during their production can be controlled. Further, the fermentative process by which microbial enzymes are produced is independent of seasonal variations, and an uninterrupted standardized supply of enzyme can be obtained.

On the other hand, the quality and yield of enzymes obtained from animal or plant tissue is dependent upon variations in the season. Further, these may also contain potentially harmful materials like phenolics from plants and endogenous enzyme inhibitors in animals, thus making the isolation process very complex and cost intensive.

S. Saxena, *Applied Microbiology*,
DOI 10.1007/978-81-322-2259-0_9, © Springer India 2015

Table 9.1 Classification of enzymes based on the types of reaction catalyzed

Class	Name	Type of reaction	Examples
1	Oxido-reductases	Oxidation- reduction reactions, generally require NADP/NAD as cofactors	Lactate dehydrogenase
2	Transferases	Transfer of functional groups like phosphate, acetyl or amino groups	Hexokinase
3	Hydrolyases	Hydrolysis reactions, i.e. addition of water	α-amylase
4	Lyases	Addition or removal of groups to form double bonds	Pyruvate decarboxylase
5	Isomerases	Rearrangement of atoms within a molecule	Methionine racemase
6	Ligases	Joining of two substrates by expense of ATP hydrolysis	DNA synthetase

Hence, microbial enzymes are far more stable and efficient than their corresponding plant or animal enzymes, with more convenient and safe production methods.

9.1.2 Modest Beginnings of Enzyme Technology

Use of enzymes in the manufacture of food and alcoholic drinks has been practiced since ancient times, as is evident in old Egyptian pictures. Cheese making using enzymes dates as far back as 400 BC. Homer, in Iliad, mentioned the use of kid's stomach for making cheese. Payen and Peroz (1833) observed that extracts of germinating barley seeds exhibited hydrolysis of starch into sugar and dextrin. They further elaborated that the extracts were thermo labile, and a small amount of extract could liquefy a large amount of starch. An alcoholic precipitate of the extract was more pure and was referred to as diastase, i.e. mixture of amylases. From this developed the use of malt in breweries for hydrolysis, replacing sulfuric acid.

The early investigators were of the opinion that fermentation was the catalytic (contact) process that allowed the addition and degradation (with water) during the process. It was proposed that this process could be catalyzed by either nitrogen-containing organic substance (unorganized) or by another living body or infusorium (organized). Payen (1874) thus proposed the concept of organized and unorganized ferment. Consequently, Kühne (1878) named these unorganized ferments 'enzymes'. Christian Hansen started the first company for marketing standardized rennet in 1874, in Copenhagen (Denmark). In 1898, enzymatic synthesis of isomerase was carried out by Croft–Hill by allowing yeast extract (α-glycoside) to act on 40 % glucose solution (Sumner and Somers 1953). The twentieth century marked the beginning of plant lipases for the production of fatty acids from oils and fats (Ullmann 1914). American bakeries used US$2 million worth of malt extract for their bakery products in 1922 (Tauber 1949).

The tannery industry kept dehaired skin in the warm suspension of dung of dogs and birds. The bating action of dung was caused by enzymes such as pepsin, trypsin, and lipase present in the dung. Erodin was the first commercial bate prepared from the culture of *Bacillus erodiens* by Popp and Becker (US Patent 607549, 1895). In 1907, Rohm patented the application of a mixture of pancreatic extract and ammonium sulphate as a bating agent (Tauber 1949). He founded a company, Rohm and Haas, in 1911, which sold around 10 tons of his pancreatic extract-based product OROPON®. The enzymes were extracted from the pancreases of slaughtered animals and included proteases (trypsin and chymotrypsin), carboxypeptidases, alpha-amylases, lactases, sucrases, maltases, and lipases.

Sakaguchi and Murao (1950) reported a new enzyme, penicillin acylase from *Penicillium chrysogenum* Wisc.Q176. This enzyme splits benzylpenicillins into phenylacetic acid and 6-aminopenicillanic acid (6-APA). The 6-APA has been used as starting block for the synthesis of numerous semi-synthetic penicillins. E. Jaag (1959) developed a detergent that contained proteases isolated from *Bacillus subtilis*. Alaclase® was the first enzyme produced via fermentation, by

Novozymes, in 1965. The isolation of glucose isomerase from *Pseudomonas hydrophila*, which isomerized glucose into fructose, was reported for the first time by Marshall and Kooi (1957). A similar glucose isomerase activity that could catalyze the isomerization of D-glucose as well as D-mannose into fructose was reported from *Paracolobacterium aerogenoides*. To enable the reuse of costly enzymes, the process of 'enzyme immobilization' was initially reported by Chibata and coworkers (1967) of Tanabe Seiyaku Co., in Japan, who developed columns of immobilized *Aspergillus oryzae* aminoacylase for the resolution of synthetic DL-amino acid in correspondingly optically active enantiomers (Katzir 2005). The recognition of enzyme immobilization for commercial processes was further strengthened with the immobilization of penicillin G acylase for the production of 6-APA, and immobilization of glucose isomerase for the production of fructose syrup from glucose. Proteases were the newest class of commercially relevant enzymes until the 1980s, when amylases, lipases, and cellulases were developed and the market for them began to grow substantially. Application areas for the enzymes were slowly expanding with the discovery of new enzymes from microbial sources. The evolution of biotechnology tools and techniques has led to a rapid development in enzyme technology in the past three decades. The world value of enzymes increased rapidly with advances in the isolation and purification of enzymes. In 1960, the world market was £110 million, rising to £270 million in 1980. The market size doubled in 1985, with a gross value of £500 million and £1,000 million in 1990. The increase in market size reflects the extensive use of enzymes in starch processing, high fructose corn syrup, textile desizing, and detergent formulations, which are traded as commodity products. The majority of the enzymes find uses in industry for the development of novel products and cost-effective and environment friendly processes, and to bypass long chemical synthetic pathways.

Recombinant DNA technology has improved manufacturing processes, enabling the commercialization of enzymes that could not be produced previously. Upcoming areas of protein engineering and directed evolution have further modernized the development of industrial enzymes in terms of modifying the selectivity and specificity of the enzymes. Today, the majority of enzymes in process development, except food processing, are recombinant in nature. The enzyme industry has reached a market size of £2,100 million in 2010 and is expected to reach £2,812 million by 2015, a compound annual growth rate (CAGR) of 6 % over the 5-year forecast period.

9.2 Microbial Enzymes: Diversity and Exploitation

Microbial diversity (prokaryotes, viruses, filamentous fungi, yeasts, microalgae, and protozoans) is a great resource for the exploration of novel microorganisms for the development of products and processes. Microbes are ubiquitous, survive under inhospitable conditions across different ecosystems around the globe, and constitute approximately 60 % of the total biomass. It is estimated globally that soils and oceans consist of $4\text{–}5 \times 10^{30}$ and 3.6×10^{29} microbial cells, respectively. Hence, different approaches (e.g. random, geographic, physiological, biochemical, and molecular) can be adopted for the exploration of the vast diversity of microbes to find new enzymes. Microbes residing in extreme environments (referred as extremophiles), like those surviving in polar regions, in volcanic springs, in the sea, and under very high salt concentrations, could be used for the exploration of novel enzymes as they possess an amazing array of enzymes catalyzing biochemical reactions. Thus, these could find direct applications in industrial processes that also occur under extreme conditions. Thermophilic microbes have recently been primarily for the exploration of thermostable enzymes. Biotechnological processes at higher temperatures reduce the risk of contamination, apart from influencing the bioavailability and solubility of organic compounds.

Thus, the screening of microbes is a strategic step for the development of an enzymatic process. This could be broadly based on three criteria: (1) selection of the type of enzymatic

Table 9.2 Extremozymes isolated from organisms or from metagenome libraries from extreme environment

Enzyme	Properties	Name of the organisms/ library	Reference
β- galactosidase	Themophilic (65 °C)	*Alicyclobacillus acidocladarius* ATCC27009 (cloned in *E. coli*)	Lauro et al. (2006)
Cold active cellulase	Psychrophilic	*Pseudoalteromonas* sp. DY3	Zeng et al. (2006)
α-L- arabinofuranosidase	Hyperthermophilic (90 °C)	*Thermotoga maritima* MSB8	Miyazaki (2005)
Glucosyl -hydrolases	Different properties	Bovine gut microflora metagenomic library	Ferrer et al. (2005)
Hormone-sensitive lipase	Thermophilic/thermostable	Metagenomic library	Rhee et al. (2005)
Cellulase (Cel5A)	Halotolerant	Metagenomic library	Voget et al. (2006)
Alkaline esterase EM2L8	Psychrophilic/alkaline (10–40 °C; pH 10–11)	Deep sea sediment metagenomic library	Park et al. (2007)
Esterase	Thermophilic/alkaline (55 °C; pH-10.5)	Deep sea sediment metagenomic library	Ferrer et al. (2005)
Patatin like phospholipase, esterase (Est1)	Thermophilic/alkaline (70 °C)	Hot spring metagenomic library	Tirawangsaroj et al. (2008)
Lipase (lip G)	New lipase class	Korean tidal flat sediment metagenomic library	Lee et al. (2006)

activity for process design; (2) grouping of microorganisms and their physico-chemical features; and (3) development and design of a sensitive assay that will allow as many microorganisms as possible to be screened. It would not be out of place to mention that microorganisms cultivable under laboratory conditions only account for 1 % of the total environmental microbes. Hence, molecular tools have been of immense help in exploring cultivable microorganisms for biotechnological processes. Metagenomics as a result of DNA and RNA sequencing has been a key method in discovering new enzymes (Table 9.2). Diversa (San Diego, USA) and TerraGen Discovery (Vancouver, Canada), which is now a part of Cubist Pharmaceuticals (Massachusetts, USA), was the first to file patents on specific aspects of metagenome technology in 1996. TerraGen Discovery filed claims for the retrieval of environmental DNA sequences from soil to express industrially relevant xylanases. Diversa is a specialist biotech company that employs metagenome technologies. It has partnered with DSM, Syngenta, and BASF for the discovery and development of enzymes based on metagenome technologies. Metagenomics is being used to screen extreme environments such as deep seas, the arctic, gold ores, worm guts, and rumen to isolate approximately 36 extremozymes (Table 9.2). Using this approach, a novel esterase (0.16) has been isolated from deep sea hyper-saline anoxic basins of the Eastern Mediterranean Sea. Genomics and metagenomics, along with in vitro evolution and high throughput screening technologies, provide an unprecedented chance to bring novel biomolecules into industrial applications.

9.3 Application of Microbial Enzymes: Broad Avenues

Today, enzymes are being used for the manufacture of over 500 products involving 50 applications in different industrial sectors, which can be broadly classified into three categories: (1) technical enzymes, (2) food enzymes, and (3) feed enzymes. The term 'technical enzyme' encompasses the application of enzymes in detergent, textile, starch, pulp and paper, and personal care industries. The area of enzyme utilization is expanding with advancements in protein biochemistry, bioinformatics, molecular biology, and bioanalytical techniques. Enzymes are now

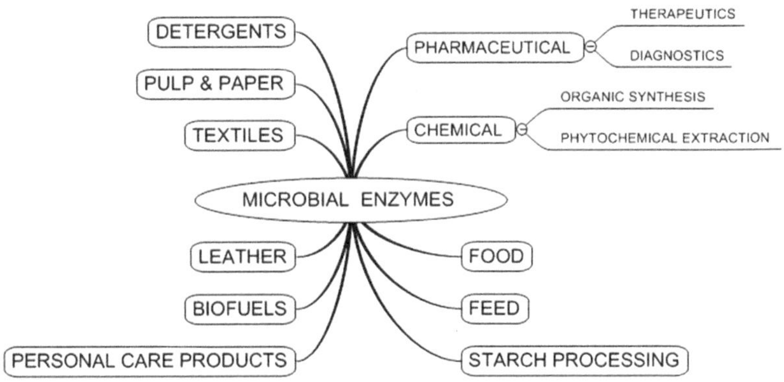

Fig. 9.1 Application of enzymes in different industrial sectors

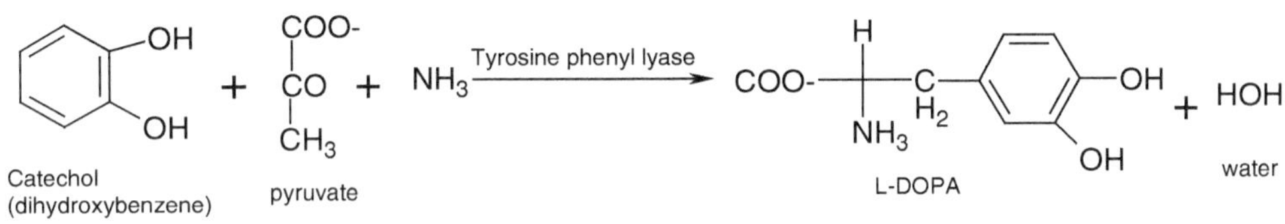

Fig. 9.2 Enzymatic conversion of catechol into l-DOPA

being used in the chemical industry for organic synthesis and for bulk extraction of phytochemicals from plants. In healthcare, they are being used in the development of clinical diagnostic methods and as therapeutic medicine in the pharmaceutical sector (Figs. 9.1 and 9.2).

9.3.1 Therapeutic Agents

Features like high target selectivity, affinity, and specificity, along with catalytic activity, distinguish enzymes as therapeutic agents when compared with classical medicines. The concept of therapeutic enzymes has evolved over around 50 years from as early as 1960 as a part of replacement therapy to overcome genetic deficiencies (de Duve 1966). The applications of microbial enzymes can be broadly classified into food disorder therapy, wound treatment, antibiotics, anticancer drugs, thrombolytic agents, and generalized therapy (Table 9.3). Microbial enzymes also find applications in clinical diagnostics.

9.3.1.1 Food Disorder Therapy by Enzymes

Parke-Davis and Company first used the medicinal properties of amylase enzyme from *A. oryzae*, which was branded as takadiastase. Takadiastase was used as a digestive aid for the treatment of dyspepsia due to incomplete digestion of starch (Takamine 1894).

Congenital sucrase-isomaltase deficiency (CSID) is a disorder where the patient cannot use sucrose. Sacrosidase (Sucraid®), a β-fructofuranoside fructohydrolyase is produced by *Saccharomyces cerevisiae* and can be administered orally. This drug hydrolyses sucrose, thereby enabling the patient to eat a normal diet, particularly young patients in whom strict compliance with a sucrose-free, low-starch diet is problematic (Treem et al. 1999).

Celiac disease or celiac sprue is an immunological disorder due to permanent intolerance of the gliadin fractions of wheat gluten and similar barley and rye proteins in susceptible subjects. Degraded gluten and similar proteins cause inappropriate T-cell-mediated immune responses, leading to intestinal inflammation and

Table 9.3 Microbial enzymes with therapeutic applications

Enzyme	Source	Therapeutic applications
Sarcosidase (β-fructofuranoside fructohydrolase)	*Saccharomyces cerevisiae*	Congenital sucrose isomaltose deficiency
Propyl endopeptidase	*Flavobacterium mengiosepticum*	Celiac disease
	Sphinogomonase capsulata	
	Myxococcus xanthus	
AN-PEP	*Aspergillus niger*	Celiac disease
Vibriolysin	*Vibrio proteolyticus*	Wound healing/debriding agent
Collagenases	*Vibrio proteolyticus ATCC53599*	Topical debriding agent
Alkaline protease	*Bacillus proteolyticus CFR3001*	Antibacterial
Lysostaphin (glycylglycine endopeptidase)	*Staphylococcus simulans*	Antibacterial
Haloperoxidase	*Curvularia verruculosa*	Antibacterial
Peptidoglycan hydrolases	*Pediococcus acidilactici ATCC 8042*	Antibacterial
L- Asparaginases	*E. coli* (*Elspar*)	Anti-cancer (acute lymphoblastic leukemia)
	Erwinia chrysanthemi	
L-Methioninase	*Clostridium sporogenes*	Anti-cancer
L-Arginine deiminase	*Mycoplasma hominis*	Anti-cancer
L-Uricase(Puricase®)	*Candida utilis*	Anti-hyperuricemic
	Arthrobacter protoformaiae	
	Bacillus fastidious	
Rasburicase	*Saccharomyces cerevisiae*	Anti-hyperuricemic
Heparinases (Heparinase I, Heparinase II and Heparinase III)	*Pedobacter heparinus*	Anti-angiogenic; Anti-cancer
Streptokinase	*Streptococcus hemolyticus*	Thrombolytic
Thrombinase	*Bacillus sphaericus*	Thrombolytic
Nattokinase	*Bacillus natto*	Thrombolytic
Verticase	*Verticillium species*	Thrombolytic
Superoxide dismutase (Orgetein/ Ontosein)	*Kluyveromyces marxianus*	Anti-inflammatory
	Aspergillus niger	Radioprotective
		Stimulate hair growth
Serratopeptidase	*Serratia marcescens* ATCC 13880	Anti-inflammatory/anti-bacterial

extra-intestinal manifestations. Prolyl oligopeptides from *Flavobacterium mengiosepticum*, *Sphingomonas capsulata*, and *Myxococcus xanthus* are capable of degrading proline-containing peptides that are otherwise resistant to degradation by proteases in the gastrointestinal tract *in vitro*. AN-PEP is a new prolyl endoprotease from *Aspergillus niger* that is resistant to gastric juice denaturation and degrades gluten and related peptides and intact proteins in the stomach prior to reaching the intestine.

9.3.1.2 Enzymes in Wound Healing

De-vitalized tissue, known as eschar, is generally found in wounds and is composed of denatured proteins like collagen, elastin, fibrin, hemoglobin, and coagulated proteins. Removal of eschar enhances faster healing and prevents infection in the wounds. Collagenases produced by *Vibrio proteolyticus* ATCC 53599 have been used as a topical debriding agent (US Patent no. 5505943). Bacterial collagenase has also been implicated in the promotion of wound healing. Clostridial collagenase has been found to improve the wound healing rate more than twofold when tested with solosite® (carboxymethyl cellulose) and regranex under in vivo conditions (Riley and Herman 2005). More recently, a recombinant vibriolysin from marine *V. proteolyticus* is under clinical phase Ib trials for application on burned skin for the removal of denatured proteins (BioMarin Inc, USA).

9.3.1.3 Antimicrobial Activity of Enzymes

The term 'enzybiotic' is a hybrid obtained by blending 'enzyme' and 'antibiotics'. Enzybiotics are used for treating bacterial and fungal infections either alone or in combination with antibiotics. They are also referred to as peptidoglycan hydrolases (PGHs) and lytic enzymes. Lysins are bacteriophage lytic enzymes that digest the bacterial cell wall for bacteriophage progeny release. It has been experimentally established that when pure recombinant lysine is added to a Gram-positive bacterium, it causes immediate lysis, leading to the death of the target bacterium. Lysins have exhibited potential control of bacterial infections on mucosa and in blood in animal models to date (Fischetti 2005).

An alkaline protease produced by *Bacillus proteolyticus* CFR3001 isolated from fish processing wastes exhibited potential antibacterial activity against pathogens like *Escherichia coli*, *Listeria monocytogenes*, *Bacillus cereus*, and *Yersinia enterocolitica* (Bhasker et al. 2007). Lysostaphin is a glycylglycine endopeptidase, an antibacterial enzyme produced by *Staphylococcus simulans* discovered by Schindler and Schuhardt in 1964. Lysostaphin effectively kills all strains of *Staphylococcus aureus*, including methicillin-resistant *S. aureus* (MRSA) and vancomycin intermediate *S. aureus* (VISA) isolates. It has been used as a topical cream for nasal decolonization of *S. aureus* and is more effective than antimicrobial activity. Hydrogen peroxide is used as a commercial sanitizer but has several side effects on the environment and the end user. Haloperoxidase (Novozymes A/S), isolated from *Curvularia verruculosa*, possesses potential antimicrobial activity against pathogenic bacteria and is 100 times more efficient in concentration than hydrogen peroxide and is expected to be less corrosive. Haloperoxidase has also been used for the inactivation of mature bacterial biofiloms. PGHs with potential antibacterial activity against *Micrococcus lysodeikticus* and *S. aureus* have been isolated from *Pediococcus acidilactici* ATCC 8042. It also exhibits lytic activity against pathogens like *B. cereus*, *L. monocytogenes*, and *Salmonella typhi* (Garcia Cano et al. 2011).

9.3.1.4 Anti-cancer Potential

Enzymatic therapy is promising in the field of anti-tumor treatment, since several enzymes have been found to interfere with the growth and proliferation of metastatic cells via differing mechanisms. Metabolite-utilizing enzymes exploit the metabolic requirement expressed by tumor cells with respect to selectively targeted tumors. Tumor cells have enhanced requirements for metabolites, i.e. they are auxotrophic, making them susceptible selective targets. Hematological cancers cannot grow and proliferate in the absence of L-asparagine, a high level of L-METHIONINE auxotrophy is observed in certain solid tumors, and melanomas and hepatocellular carcinomas have a higher requirement of L-ARGININE. Therefore, L-asparaginase, L-METHIONINASE, and L-arginine deaminase have been studied as therapeutic agents in cancer therapy.

> PEGylation is a technique whereby covalent chains of polyethylene glycol chain (PEG), a hydrophilic synthetic polymer, are attached to the protein chains. It is a well-established technology for the development of biopharmaceutical formulations used to improve the stability, solubility, bioavailability, and immunological properties of bioactive compounds.

Leukemic lymphoblasts and certain tumor cells are auxotrophic for L-asparagine since they lack the enzyme asparagine synthetase, an enzyme expressed by healthy cells. Therefore, the enzyme asparaginase can be used for the selective inhibition of tumor cells that rely on metabolic L-asparagines for their growth and proliferation. J.D. Broome (1963) first highlighted the role of L-asparaginases in guinea pig serum for their anti-lymphoma effects. Two native commercial sources of asparaginases are *E. coli* with the brand name Elspar® (marketed by Merck & Co.) and erwinia L-asparaginase isolated from *Erwinia chrysanthemi* (Ogden BioServices Pharmacy Repository, USA). Both forms of L-asparaginases do not exhibit antigenic cross reactivity, and thus allow switching from one to another. However, to overcome protein immunogenicity, PEGylation of *E. colil-asparaginase* was carried out, and PEG-L-asparaginase was developed by Enzon in 1994; this was used in combination with chemotherapy for the treatment of acute lymphoblastic leukemia (ALL). PEGylated-*L*-asparaginase is commercially sold under the brand name Oncasper® by Rhône-Poulenc Rorer. Some cancer cell lines are auxotrophic for methionine. Thus, decreases in metabolic methionine content would slow or arrest the S-G2 phase of the cell cycle. As the normal cells are resistant to exogenous L-METHIONINE restriction, this amino acid could be a target for cancer treatment. Methioninase, or methionine-α-deamino-γ-mercapto-methane-lyase or METase, is a pyridoxal-*L*-phosphate-dependent enzyme that transforms *L*-methionine into α-ketobutyrate, methanethiol, and ammonia.

The first methioninase was purified from *Clostridium sporogenes* and inhibited Walkers carcinosarcoma 256 in rats and exhibiting no toxicity. Clostridial methioninase has a Km value of 90 mM. Thus, there was a need to isolate a methioninase enzyme with a lower Km value to be used for clinical evaluation. *Pseudomonas putida* methioninase exhibited a Km value of 1 mM. The major limitation of *Pseudomonas methioninase* was the low amount of purified enzyme, which was insufficient to carry in vivo studies. To overcome this, the L-methioninase gene from the *P. putida* genome was over-expressed in *E. coli* for over-production. Clinical phase I trials have indicated low toxicity of this protein in cancer patients. Unfortunately, this protein is immunogenic, as demonstrated in studies performed in balb-C mice.

The de novo synthesis of arginine in healthy human cells utilizes citrulline in the presence of argininosuccinate synthetase (ASS) and argininosuccinate lyase (ASL). Some tumors are auxotrophic for arginine because of a deficient enzymatic pool. Some cancer cells do not express ASS and thus cannot synthesize arginine from the precursor; therefore, arginine-depleting enzymes can be useful against certain tumors. Arginine deiminase degrades arginine into citrulline and ammonia. The enzyme arginine deiminase was first isolated from *Mycoplasma hominis* infecting a cell line. Arginine deiminase is more powerful than asparaginase in killing human leukemia cells under in vitro conditions. Arginine deiminase is more specific in nature since it catalyses only one substrate compared with asparaginase, which catalyzes the conversion of asparagine and glutamine. However, the native enzyme is immunogenic and hence it is PEGylated for clinical trials.

Some enzymes are not directly involved in anti-cancer chemotherapy but are helpful in ameliorating the side effects of chemotherapy. One of the consequences of chemotherapy is hyperuricemia, which can be cured by uricase or urate oxidase enzyme. Uricase catalyzes the oxidation of uric acid to yield the more soluble allantoin, readily excreted by the kidney. The enzyme was purified from *Aspergillus flavus* and has been used in Italy and France for over two decades. A recombinant version, rasburicase, has been produced in *S. cerevisiae* and has been approved in the USA for initial management of uric acid levels in pediatric patients with leukemia, lymphoma, and solid tumor malignancies receiving anti-cancer therapy. More recently, heparinises have been implicated in the cleavage of glycosidic linkage between heparin and heparin sulphate. The process of new blood vessel formation, i.e. angiogenesis or neovascularization, is a fine balance between pro-angiogenic and anti-angiogenic growth factors and cytokines. Angiogenesis is essential for tumor growth and subsequent metastasis. Heparin and heparin sulphate are the molecules that are present on the cell surface and define the physiological characteristics of the extracellular matrix. The molecules modulate the expression and bioavailability of pro-angiogenic growth factors, e.g. vascular endothelial growth factor (VEGF) and fibroblast growth factors (FGF), and thus play a role in the neovascularization of the new tissue. Therefore, a line of treatment is under study that hypothesizes that the blocking or breakdown of heparin and heparan sulphate could modulate the process of neovascularization and inhibit tumor formation. The enzymes that use heparin/heparan sulphate as a substrate are known as heparinases and cleave their glycosidic linkage between amino sugars and uronic acids. Heparinases were initially isolated from *Pedobacter heparinus* (previously known as *Flavobacterium heparinum*). There are three classes of heparinises: heparinase I (heparinase); heparinase II (heparin lyase); and heparinase III (heparan sulphate lyase). Heparinase I exclusively acts on heparin, heparinase III on heparan sulphate, and heparinase II utilizes both heparin and heparan sulphate equally. It has been experimentally proven that heparinase I and III are the only enzymes that inhibit neovascularization in vivo, and the proliferation of capillary endothelial cells is mediated by basic FGF in vitro.

9.3.1.5 Thrombolytic Agents

The World Health Organization reports that 17 million deaths globally are due to cardiovascular diseases (CVDs). Blood clot formation or intravascular thrombosis is the major cause of CVD. Fibrin is the major protein that participates in blood clot formation via proteolysis of thrombin. Enzyme-mediated dissolution of fibrin clot is known as thrombolysis or fibrinolysis. A variety of intrinsic activators mediate the conversion of inactive plasminogen into fibrinolytic plasmin. Recombinant forms of normal human plasminogen activators, tissue plasminogen activator (tPA)/urokinase-type plasminogen activator (uPA), are used in clinical interventions but are expensive due to the high cost of production. Streptokinase is a plasminogen activator bacterial protein produced by Lancefield group C strains of f3-hemolytic streptococci (Christensen 1945). Streptokinase and uPA are the non-fibrin-specific agents approved by the US Food and Drug Administration (FDA) for clinical application. Furthermore, patients receiving streptokinase treatment can develop anti-streptococcal antibodies. Bradykinin generation due to streptokinase administration leads to hypotension (Gemmill et al. 1993). A second administration of streptokinase within 1 year may lead to allergenic reactions.

Microbial fibrinolytic enzymes and secondary metabolites have attracted attention for thrombolytic therapy due to the undesirable effects and high production costs of tPA and urokinase (Peng et al. 2005). The advantage with the microbial thrombolytic agents is that there should be no 'supply issue', since the provision of the materials could be realized through larger-scale or industrial fermentation. Fibrinolytic enzymes have been discovered from different microorganisms; streptokinase from *Streptococcus hemolyticus* has been commercially available and used for a few years now (Collen et al. 1993). Thrombinase, a fibrinolytic enzyme, has been reported from *Bacillus sphaericus* (Balaraman and Prabakaran 2001). *Bacillus natto*, producing nattokinase, was first screened from a traditional Japanese soybean-fermented food named natto (Sumi et al. 1987). Nattokinase not only directly cleaves cross-linked fibrin, but also activates the production of t-PA, resulting in the transformation of inactive plasminogen to active plasmin (Fujita et al. 1995). Furthermore, nattokinase enhances its fibrinolysis through cleavage and inactivation of plasminogen activator inhibitor (PAI)-1, which is the primary inhibitor of fibrinolysis and regulates total fibrinolytic activity by its relative ratio with tPA (Urano et al. 2001).

Fungi also offer themselves as a potential source of extracellular production of fibrinolytic enzymes (Amatayakul 1955). Fungi elaborate proteases that possess fibrinolytic activity; among these are *Aspergillus ochraceus* 513 (Batomunkueva and Egorov 2001), *Fusarium oxysporum*, *Penicillium chrysogenum* (El-Assar et al. 1990), and *Rhizopus chinensis* 12 (Liu et al. 2005). FP, a fungal protease from *Fusarium* sp. BLB, exhibited very strong fibrin degradation when compared with nattokinase in the absence of plasminogen. The plasminogen activation of FP was much higher than of nattokinase as assessed by thrombolytic activity experimentally (Sugimoto et al. 2007). In recent years, fibrinolytic enzymes produced from different mushrooms, including *Pleurotus ostreatus* (Choi and

Shin 1998), *Pleurotus sajor-caju* (Shin and Choi 1999), *Flammulina velutipes*, *Armillariella mellea* (Kim and Kim 1999), *Ganoderma lucidum* (Choi and Sa 2000), *Tricholoma saponaceum* (Kim and Kim 2001), and *Formitella fraxinea* (Lee et al. 2006a), have been successfully purified and characterized. However, reports on the production of fibrinolytic agents/enzymes biosynthesized by endophytic fungi are scarce. Very limited work exists on screening endophytic fungi possessing proteolytic as well as fibrinolytic activity. Verticase, a fibrinolytic enzyme, was isolated from endophytic *Verticillium* species and was found to be a serine protease (Li et al. 2007). An endophytic *Fusarium* CPCC 480097 produces a fibrinolytic enzyme with higher fibrinolytic activity than plasmin (Wu et al. 2009). *Bionectria* sp., from the Yungas Pedemontana forest range in Argentina has been reported to be a potential source of direct (plasminogen-independent) fibrinolytic enzymes for different therapeutic purposes (Rovati et al. 2010).

9.3.1.6 Generalized Therapy

Serratiopeptidase is a proteolytic enzyme isolated from non-pathogenic enterobacteria *Serretia* sp. E15 found in silkworms. It is used as a standard treatment for post-operative inflammation and traumatic swelling in Japan, Germany, and other European countries. Serratiopeptidases were found to possess anti-staphylococcal activity and enhanced the action of antibiotics. This enzyme has also been used in combination with paracetamol and aceclofenac (PARFLEX) for surgical pain management. It has also been used as a mucolytic expectorant to decrease the viscosity and increase the volume of sputa in broncho-pulmonary diseases.

Superoxide dismutases (SODs) catalyze the dismutation of superoxide into oxygen and hydrogen peroxide. Irwin Fridovich and Joe McCord first discovered and established SOD activity. There are three types of enzymes based on copper, manganese, and iron. SODs have a role in the prevention of oncogenesis, tumor promotion, and cardiovascular diseases. The most promising role of manganese SOD (MnSOD) in humans is the inhibition of tumorigenicity. Orgotein is a pharmaceutical dosage form of SOD that is used as a potent anti-inflammatory agent. Orgotein has been used as an aerosol (Ontosein) in the treatment of radiation-induced adverse effects in difference malignancies, notably breast, lung, bladder, prostate, cervix, and head and neck cancers. It is administered either topically or parenterally.

9.3.2 Diagnostics

Diagnostic enzymes are defined as a component of an assay system for determination of many different substances. These are generally employed for diagnosis/biomarkers of diseases and use blood, serum, and urine for analysis. Microbial enzymes are also used as indicators or biomarkers of a disease. Sialidases are bacterial enzymes that play a role in cellular interactions, bacterial nutrition, and immune response evasion, thus improving the bacterial ability to adhere to, invade, and destroy the mucosal tissue. Sialidases occur in bacteria, mycoplasma, protozoa, fungi, and viruses. Bacterial vaginosis (BV) is a disorder of the vaginal ecosystem characterized by a shift in the vaginal flora form the normally predominant *Lactobacillus* to mixed flora, including *Gardnerella vaginalis*, *Mobiluncus*, *Prevotella*, *Bacteroides*, and *Mycoplasma* species. BVBlue system (Gryphus Diagnostics, L.L.C.) is a chromogenic diagnostic test based on the presence of elevated sialidase enzymes in vaginal fluid samples (Myziuk et al. 2003). Microbial mucinases (mucin-degrading enzymes) are associated with genital tract conditions and sexually transmitted diseases (STDs).

9.4 Chemical Industry

Enzymes are vital for life, as they catalyze specific reactions as well as enhance the rate of reactions, thus forming the basis of metabolism. They also offer tremendous opportunities in industry to carry out biocatalytic conversions, thereby making the process efficient and economical. Microbial enzymes can be used as free enzymes or selectively induced intracellularly in microbial cells. Unlike many chemical processes in conventional synthetic chemistry, enzymes require non-toxic and non-corrosive conditions to carry out catalysis.

Extremozymes are enzymes produced by extremophilic microbes and are increasingly finding applications in industry, replacing the conventional catalysts as they can withstand harsh conditions. This has revolutionized the arena of synthesis of active pharmaceutical ingredients (APIs), thereby enhancing manufacturing operations. Moreover, chirally pure compounds/drugs can be synthesized (enantioselective synthesis) using microbial enzymes to avert such disasters as the thalidomide cases.

The screening of microbes that produce novel enzymes for process development is the key step. It is based on (1) selecting a process and a particular type of enzymatic activity; (2) deciding on the types of microorganisms to be selected and screened; and finally (3) developing an appropriate, sensitive, and convenient assay that would facilitate the screening. D-p-hydroxyphenylglycine and its derivatives serve as important side chain precursors for the semisynthetic penicillins and cephalosporins. Industrial production of D-amino acids involves the synthesis of hydantoin substrates. Stereospecific hydrolysis of hydantoins was catalyzed by D-hydantoinase (microbial origin) and decarbamoylation. Chemical decarbamoylation is carried out by treating the amino acid intermediate with an equimolar amount of nitrile under aerobic conditions. A novel enzyme D-carbamoylase has been found and engineered for use in the decarbamoylation process, thereby making the process cleaner and simpler. The recombinant mutant enzyme has been used for the large-scale production of D-p-hydroxyphenylglycine (~2,000 tons/year) with simultaneous use of D-hydantoinase (Ogawa and Shimizu 2000).

9.4.1 Cell-Free Biocatalysis

Biocatalysis is the use of free (extracellular) enzymes to perform chemical transformation of organic compounds.

Biotransformation is the chemical transformation of organic compounds carried out by intracellular enzymes and is often referred to as whole cell biocatalysis.

Trans-4-L-hydroxyproline is an important chiral precursor for the synthesis of antiphlogistics, carbapenem antibiotics, and angiotensin-converting enzyme (ACE) inhibitors. This requires stereoselective and region-selective hydroxylation of *L*-proline to yield hydroxy-l-proline isomers, leading to the discovery of specific proline hydroxylases: 4-hydroxylase and 3-hydroxylase in *Dactylosporangium* sp. RH1 and *Streptomyces* sp. TH1, respectively. 4-hydroxylase specifically produces trans-4-hydroxy-l-proline, whereas 3-hydroxylase produces *cis*-3-hydroxy-L-proline. These were subsequently cloned *in E. coli*. Kyowa Hakko Kogyo Co. started the production of trans-4-hydroxy-l-proline using this method.

L-DOPA (l-3,4 dihydroxyphenylalanine) is a medicine for Parkinson's disease. Tyrosine phenol lyase (TPL) is a pyridoxal phosphate (PLP) enzyme isolated from *Erwinia herbicola*, *Citrobacter freundii* that catalyzes the reversible transformation of tyrosine into phenol, pyruvate, and ammonia (Yamada and Kumagai 1975). Pyrocatechol was used as a substrate for this enzyme for the production of L-DOPA. *E. herbicola* cells accumulated 110 g/l of L-DOPA, and in 1993, Ajinomoto Co. started commercial production using this system and currently produces half of the worldwide supply of L-DOPA, which is around 250 tons/year.

Lipases have been used as biocatalysts for the synthesis of chiral compounds through kinetic

Table 9.4 Lipases in chiral synthesis of drugs

Product	Biocatalyst	Drug	Company
(S)-2-(4-isobutylphenyl) propanoic acid	*Candida rugosa* lipase	(S)- Ibuprofen	Pfizer, USA
(R)-2-(7-(4,4′-bipiperidine-1-carbonyl)-4-methyl-3-oxo-2,3,4,5-tetrahydro-1H-enzo[e][1,4]diazepin-2-yl)acetic acid	Immobilized *Candida antarctica* lipase B (CALB)	Lotrafiban S-16 antithrombotic	GlaxoSmithkline Pharmaceuticals
(S)-terc-butyl 2-carbamoyl-2,3-dihydro-1H-pyrrole-1-carboxylate	CALB	Saxagliptin (oral hypoglycemic)	Bristol-Myers Squibb (BMS)
(3R,4S)-2-oxo-4-phenylazetidin-3-yl acetate	*Pseudomonas cepacia* lipase PS30 (Amano)	Paclitaxel (anti-tumor agent)	Bristol-Myers Squibb (BMS)
	Pseudomonas sp. SC13856 lipase (BMS)		
Methyl 3-(4-ethoxyphenyl) oxirane-2-carboxylate	*Serratia marcescens* lipase	Diltiazem (calcium canal blocker)	Tanabe Pharmaceutical
(3S,3aR,6aR)-3-hydroxy-3,3a,4,6a-tetrahydro-2Hcyclopenta[b]furan-2-one	*Pseudomonas fluorescens* lipase	Carbovir (antiviral agent)	Celltech Group

resolution of racemic mixtures or enantioselective enzymatic desymmetrization of prochiral compounds. *Candida rugosa* lipase is one of the employed serine hydrolases in organic synthesis. Pfizer has prepared a procedure of (S)-ibuprofen through enantioselective hydrolysis of the corresponding methoxyethyl ester catalyzed by *C. rugosa* lipase in an immobilized bioreactor. Similarly, lipases from *Serratia* and *Pseudomonas* have been used for the enantiomeric/chiral synthesis of organic compounds (Table 9.4). Penicillin G acylase has been isolated from many microorganisms and is currently being used for the selective hydrolysis of penicillin G to afford 6-APA.

9.4.2 Whole Cell Biocatalysis

F. oxysporum has been responsible for the production of an enzyme that catalyzes the reversible hydrolysis of aldonate lactones and butyrolactones. The hydrolysis is stereospecific by recognizing the hydroxyl group configuration at the two position of carbon (Haworth projection). If the hydroxylation is below the carbon atom then hydrolysis takes place. A racemic mixture of pantolactone has been resolved into D-pantoic acid and L-pantolactone. D-pantoic acid is the chiral building block for the commercial production of the vitamin, D-pantothenic acid (vitamin B_5). The mycelium of *F. oxysporum* is entrapped in calcium alginate and incubated with the racemic mixture pantolactone for 21 h at 30 °C at pH 7; approximately 90 % of D-isomer was hydrolyzed with high optical purity between 93 and 98 %. This process has been used by Daiichi Fine Chemicals for the commercial production of D-pantolactone (~3,000 tons/year calcium D-PANTOTHENATE) (Kataoka et al. 1995).

Redox reactions play a significant role in the sustainable and productive synthesis of drugs. Oxidoreductases are employed using whole cells due to their dependence on cofactors that need to be regenerated. Whole cells of *Gluconobacter suboxydans* are responsible for the regio-selective oxidation of N-protected-1-amino-D-sorbitol. The reaction product is (3S, 4R, 5S)-1, 3, 4, 5-tetrahydroxy - 6 - (alkylamino) hexan-2-one; a sorbose derivative is a key intermediate in the synthesis of oral α-glucosidase inhibitors (Landis et al. 2002). Similarly, *Streptomyces* species has been used by Bristol-Myers Squibb to carry out biotransformation of compactin to pravastin (sodium (3R,5R)-3,5-dihydroxy-7-((1S, 2S, 6S, 8S, 8aR)-6-hydroxyl-2-methyl-8-((S)-2-methylbutanoyloxy) -1, 2, 6, 7, 8, 8a-hexahydronaphthalen-1-yl) heptanoate).

Whole cell reductions have also been carried out industrially. Whole cells of *Rhodococcus erythropolis* SC13854 reduced (1S)-[3-chloro-2-oxol-(phenylmethyl)propyl] carbamic acid, leading to the formation of tert-butyl [(2S,3R)-4-chloro-3-hydroxy-1-phenylbutan-2-yl] carbamate, an alcohol intermediate for an HIV endopeptidase inhibitor atazanavir (Patel et al. 2003).

Aryl aryl ketones have also been reduced by whole cell biocatalysis. ((S,E)-methyl 2-(2-(3-(3-(2-(7-chloroquinolin-2-yl)vinyl) phenyl)-3-hydroxypropyl)phenyl)-2-methylpropanoate) has been produced from the corresponding ketone in the synthesis of the anti-asthma drug montelukast, produced at Merck & Co. Inc., Rahway, USA, by bio-reduction catalyzed by whole cells of *Microbacterium campoquemadoensis*. The industry-relevant examples of whole cell biocatalysis are given in Table 9.5.

9.4.3 Phytochemical Extraction with Microbial Enzymes

Recently, microbial enzymes have also been used in the isolation of phytochemicals as they improve yield and protect the phytochemical from degradation. Lutein, phenolics, and lignans have been extracted using enzyme-assisted extraction methods. Lutein is a xanthophyll that has been extracted from flower petals of marigold (*Tagetes erecta*) (Tekwani and DeMello 2010). Lutein is a reactive oxygen species (ROS) scavenger and is well used in nutrition and in the cosmetic and pharmaceutical industries. Lutein is also implicated in vision improvement, protection of skin from ultraviolet (UV)-induced damage, and reduction in risk of CVDs. Conventional methods of lutein extraction from marigold flowers results in partial degradation and substantial loss of

Table 9.5 Whole cell biocatalysis in synthesis of drugs

Product	Biocatalyst	Drug	Company
(3S,4R,5S)-1,3,4,5-tetrahydroxy-6-(alkylamino) hexan-2-one	*Gluconobacter suboxydans*	α- glucosidase inhibitors	Bayer, Germany
Sodium (3R,5R)-3,5-dihydroxy-7-((1S,2S, 6S,8S,8aR)-6-hydroxy-2-methyl-8 - ((S) -2-methylbutanoyloxy)-1,2,6,7,8,8a-hexahydronaphthalen-1-yl) heptanoate)	*Streptomyces* sp. Y110	Pravastin (anti-cholesterolemic)	Bristol-Myers Squibb
Cis, Cis- Muconic acid	*Arthrobacter* sp.	Pharmaceutical and agrochemical compounds	Mitsubishi Chemical Corp.
(Tert-butyl(2S,3R)-4-chloro-3-hydroxy-1-phenylbutan-2-ylcarbamate	*Rhodococcus erythropolis* SC13854	Atazanavir (HIV endopeptidase inhibitor)	Bristol–Myers Squibb
((S)-1-(benzo[d][1,3]dioxol-5-yl) propan-2-ol	*Zygosaccharomyces rouxii*	Talampanel™	Eli-Lilly and Company
((S,E)-methyl 2-(2-(3-(3-(2-(7-chloro quinolin-2-yl)vinyl)phenyl)-3-hydroxy propyl) phenyl)-2-methylpropanoate	*Microbacterium campoquemadoensis*	Montelukast (anti-asthma)	Merck & Co.
(S)-ethyl 4-chloro-3-hydroxybutanoate	*Escherichia coli*	Atorvastatin (lateral chain)	Kaneka Co.
(S)-methyl 4-chloro-3-hydroxybutanoate	*Geotrichum candidum* SC 5469	Atorvastatin (lateral chain)	Bristol–Myers Squibb
(S)-2-amino-5-(1,3-dioxolan-2-yl)-pentanoic acid	*E. coli* cells over-expressing phenylalanine dehydrogenase from *Sporosarcina*	Omapatrilat (anti-hypertensive)	Bristol–Myers Squibb

carotenoids. Thus, enzyme treatment has been proposed as an alternate stage to solvent extraction processes to improve the yield and prevent partial degradation. A 10 % improvement in yield of lutein was observed when the extraction was carried out using cellulases and pectinases. Solvent use was also reduced, making the process more economical and greener.

Phenolic extraction from agricultural and industrial wastes has gained much attention recently as cheap and safe sources of strong antioxidants. Bioactive phytochemicals are present as soluble, suspended, or colloidal forms in complexes with the cell wall components in the plants after the cell wall is ruptured. Thus cell-wall-hydrolyzing enzymes such as cellulase, hemicellulase, and pectinase hydrolyze the plant materials, and have often been proposed as tools for extraction. Phenolics have been extracted from apple peel using cellulases apart from the recovery of sugars.

Lignans are diphenolic compounds resulting from the dimerization of two coniferyl alcohol moieties. Some lignan derivatives have been found to possess chemopreventive properties against breast and prostatic tumors after being transformed by gut flora in humans. Main flax seed (*Linum usitatissimum*) lignan secoisolariciresinol is converted after ingestion by human intestinal microbiota into enterodiol, which has been shown to reduce the development of mammary and prostatic tumors. Onuzuka RS (*T. reesei* cellulase by Merck & Co.) was found to give a better yield of secoisolariciresinol from seed hulls and whole seeds of flax as compared with β-glucosidase.

Table 9.6 Microbial enzymes commonly used in foods and feeds

Acetolactate decarboxylase	Beer maturation
Amylases	
α- amylase	Starch syrups, fermentation and animal feed
β- amylase	Brewing, maltose syrups
Glucoamylase	Manufacture of dextrose syrups and HFCS
α- galactosidase	Enhances sucrose yield
Cellulases	
β-glucanase	Animal feed, extraction and clarification of fruit and vegetable juices brewing industry
β-glucosidase	Transforms isoflavone phytoestrogens in soymilk
Dextranase	Hydrolysis of the polysaccharide dextran
Invertase	Invert syrup of cane or beet sugar
Keratinases	Commercial feather meal for poultry and fish feed
Lactase	Metabolizes/reduces lactase from dairy foods
Lipases	Oils and fats, baking, and dairy industry
Naringinases	Debittering of citrus juice and peel
Pectinase	Fruits and fruit juice processing
Phytases	Fish feeds
Tannases	Reducing undesirable effects of tannins
Transglutaminases	Texture improvement of various kinds of noodles, pasta, hams and sausages, crumb strength

9.5 Food and Feed Industry

Advanced technologies are sought by food industries to convert raw food materials to end products, and enzymes have been used as efficient tools in the food and feed industry. The application of enzymes in food manufacturing began with the cheese industry, wherein chymosin was used for cheese manufacturing. Starch hydrolysis was the first major application of enzymes in the food industry in the 1960s, wherein α-amylases and glucoamylases were used to convert 95 % of starch into glucose. In the animal feed industry, the first enzyme used commercially was β-glucanase in barley-based feed diets. Presently, a variety of enzymes are used in the processing of foods and feeds (Table 9.6).

9.5.1 Acetolactate Decarboxylase

The enzyme α-acetolactate decarboxylase is useful for the removal of a butter-like aroma from beer, present because of the presence of diacetyl, a by-product of beer fermentation. The very first organism producing α-acetolactate decarboxylase in a homogenous form was *Acetobacter aerogenes* by Loken and Stromer (1970). *Lactobacillus casei* DSM2547 also produces α-acetolactate decarboxylase with a higher specific activity, low molecular weight, and higher Km value for commercial applications.

9.5.2 Amylases

The enzymes that hydrolyze starch into such diverse products as dextrins and small polymers of glucose are referred to as amylases. They can be divided into two broad categories: endoamylases and exoamylases. Endoamylases catalyze hydrolysis in the interior of the starch molecule, while exoamylases hydrolyze the starch from the non-reducing end, thereby producing short end products.

α-amylases have been extensively used in the baking industry to provide higher volumes, improved color, and softer crumbs. Fungal amylase was permitted as a bread additive in the USA in 1955 and in the UK in 1963 after confirmation of their GRAS (Generally Recognized as Safe) status. One of the recent applications of α-amylase is preventing bread from going stale, which reduces its shelf life. Thermostable amylases are used for the conversion of insoluble starch into an aqueous solution as a result of partial hydrolysis; this is known as liquefaction. The first liquefying thermostable amylase is from *B. amyloliquefaciens* followed by *Bacillus licheniformis*. Saccharification follows liquefaction, thereby converting starch into high-fructose corn syrups, which, because of their high sweetening properties have been used as sweeteners for soft drinks and beverages. *A. oryzae*, *A. niger*, and *Rhizopus* species are prolific producers of amylases (Gupta et al. 2003).

9.5.3 α-Galactosidase

α-Galactosidase enzymes enhance the nutritional value of legume-based food by reduction in or elimination of anti-nutritive galacto-oligosaccharides (raffinose family sugars), which are responsible for flatulence. The enzyme is also used for hydrolyzing the raffinose in beet sugar syrup, thereby facilitating normal crystallization of the beet sugar.

9.5.4 Cellulases

Cellulases (endoglucanases, exoglucanases, and cellobiases) constitute the macerating enzyme complex along with pectinases (pectin lyase, pectin methylesterase, endo- and exo-polygalacturonases, pectin acetylesterase, rhamnogalacturonase, endo- and exo-arabinase), which are used for the extraction and clarification of fruit and vegetable juices to increase the yield of juices. Macerating enzymes are used to improve cloud stability and texture and to decrease viscosity of the nectars and purees from tropical fruits such as mango, peach, papaya, plum, apricot, and pear. Microbial β-glucanases have been used to reduce the viscosity of beer by hydrolyzing β-glucan. The common sources are *Penicillium emersonii*, *A. niger*, *B. subtilis*, and *Trichoderma reesei*. It has also been reported that endoglucanase II and cellobiohydrase II were responsible for maximum reductions in the degree of polymerization and wort viscosity.

Hydrolases have been predominantly used in monogastric feed to eliminate anti-nutrition factor (ANF) present in grains or vegetables by degrading them in order to improve its nutritional value. Glucanases and xylanases have been used individually or in combination to eliminate ANF and improve feed quality by eliminating non-starchy polysaccharides. Similarly for the forage diet of ruminants, enzyme preparations containing high levels of cellulase, hemicellulase, and pectinase have been used to improve the nutritive quality of forages.

9.5.5 Dextranases and Invertases

Dextrans are high-molecular-weight polysaccharides with 50 % α-1,6-linked glucose units along with α-1, 3 branch linkages and may contain other branch linkages such as α-1, 2, or α-1, 4. Dextranase is an enzyme that hydrolyses dextran, which contaminates the sugar. This happens due to the action of dextransucrase enzyme from contaminant microorganisms that home to the sugarcane sap or that attack it when its rind is damaged. These dextrans are extracted in mills along with the juices and contaminate the sugar mill flow. Dextranase (α-D-1,6-glucan-6-glucanohydrolase) is the enzyme that hydrolyses the α-1, 6 linkages mainly present in the dextran polysaccharides, breaking these bonds to form smaller oligosaccharide molecules (Jimenez 2009). The organisms producing dextranase are *Fusarium moniliforme*, *F. oxysporum*, *Chaetomium gracile*, *Lipomyces starkeyi*, *Fusarium roseum*, and *Penicillium roqueforti*. The organisms producing dextranases are listed in Table 9.7.

Currently, the three commercial dextranases used are dextranase 50 L (*Penicillium lilacinum*), dextranase plus L (*Chaetomium erraticum*) from Novozymes A/S, and dextranex L-4000 (*C. gracile*) from Genencor International.

Invert sugar contains equal amounts of glucose and fructose. Invert sugar is used as a sweetener in the baking, beverage, canning, confectionary, and dairy industries. Fructose in invert sugar provides more sweetness, satiation in products for health-conscious consumers, and is metabolically safe in patients with diabetes as it does not require insulin. Invertase enzyme is applied for the conversion of sucrose into glucose and fructose. A variety of microorganisms are found to produce invertase using sucrose as a nutrient. Commercially, invertase is biosynthesized chiefly by the yeast strains *S. cerevisiae* and *Saccharomyces carlsbergensis*.

9.5.6 Keratinases

Feather meal was initially prepared via hydrothermal treatment, which is expensive and results in the destruction of essential amino acids such as methionine, lysine, and tryptophan and is poorly digestible. Microbially produced keratinases have been used in the treatment of feathers in the

Table 9.7 Microbes producing dextranases

Class	Organism	Optimal pH	Optimal temperature
Fungi	*Penicillium lilacinum* (Novo Nordisk)	5.0–5.5	53–60 °C
	Penicillium luteum	4.0–6.0	50 °C
	Penicillium funiculosum	6.0	NR
	Penicillium aculeatum	4.5–5.6	50 °C
	Penicillium minioluteum	4.5–5.0	35 °C
	Penicillium notatum	5.0	50 °C
	Chaetomium gracile	5.5–11.0	55 and 65 °C
	Fusarium miniliforme	5.5	55 °C
	Sporothrix schenkii	5.0	NR
Bacteria	*Brevibacterium fuscum var. dextranlyticum*	7.0–7.5	NR
	Streptococcus mutans	5.5	37 °C
	Streptomyces anulatus	7.0	40 and 50 °C
	Flavobacterium sp. *M-73*	7.0	35 °C
	Thermoanaerobacter wiegelii	5.5	70 °C
	Strain of Thermoanaerobacter	4.5–5.5	80 °C
	Thermoanaerobacterium thermosaccharolyticum	5.5	65–70 °C
Yeast	*Lipomyces starkeyi*	5.0	55 °C

preparation of animal feed supplements. Microbial keratinases work at ambient temperatures, have good digestibility, and do not destroy the essential amino acids, thus making the animal feed nutritive. *B. licheniformis* PWD1 produces microbial keratinase, was transferred to BioResources Inc. (BRI), and is commercially known as Versazyme.

The Versazyme-based feather meal is also applied as a fertilizer for nitrogen supplementation. BRI has also reported the medicinal properties of Versazyme supplementation; it is capable of degrading the prions in the brain tissue of bovine spongiform encephalopathy (BSE) as well as scrapie-infected animals.

9.5.7 Lipases

Lipases are a broad class of enzymes that are used for cleaving (hydrolysis) of lipids. These are ubiquitous enzymes found in animals, plants, fungi, and bacteria. A unique feature of lipases is their ability to carry out catalysis at the interface between an aqueous and a non-aqueous phase. Microorganisms are prolific producers of extracellular lipases, which have been used commercially (Ghosh et al. 1996).

Most of the commercial lipases produced are utilized for flavor development in dairy products and processing of other foods, such as meat, vegetables, fruit, baked foods, milk products, and beer. Betapol™ 45 is the first commercial product designed using 1, 3-specific lipase, which carried out interesterification of tripalmitin with oleic, linoleic, and linolenic acid. Betapol is used as the total fat phase in infant formula, an alternative to human milk. It is a blend of vegetable fat that closely mimics the physical and chemical structure of human milk fat. In the dairy industry, lipases have been used in the hydrolysis of milk fat. A whole range of lipases from *Rhizomucor miehei*, *A. oryzae*, and *A. niger* have been used in the cheese manufacturing process. Palatase®, a 1, 3- specific lipase produced by *R. miehei* has been used for the hydrolysis of short-chain fatty acids, resulting in optimal flavor formation.

Econa Oil®, an enzymatically produced natural oil (diacylglycerol), has been produced enzymatically. It was introduced by Novozymes and Kao (Japan). The oil possesses the same energy value as triacylglycerol but is metabolized in the body and never stored as neutral fat as in the case of triglycerides. Econa Oil® can be used as cooking or frying oil and in salad dressings and mayonnaise, shortening and margarines, chocolates, ice cream fats, confectioner's fats, and baked food products. Salatrim stands for 'short- and long-chain acyl triglyceride molecules'. The commercial product is Benefat™, which is being marketed by Cultor Food Science, Canada. It is used in bakery products like cookies, pies, and cream fillings. Carpenin is yet another low-calorie fat, comprising caprylic ($C_{8:0}$), capric ($C_{10:0}$), and behenic acids ($C_{22:0}$). The usable energy value of carpenin was calculated to be 4.3 kcal/g, as against the 9.0 kcal/g of conventional fats.

Cocoa butter is a blend of palmitic acid and stearic acid, with a melting point of 37 °C, leading it to melt in the mouth and provide a cooling sensation in the majority of chocolates and cold chocolates. Cocoa butter equivalent (CBE) has been developed by interesterification of palm oil middle fraction (POMF) and stearic acid using a solvent-free system by Nova Lipase.

Recombinant DNA technology has helped in the improvement of the biochemical and catalytic features of lipase isoenzymes where the purification leads to very low yields. The expression systems used are *S. cerevisiae*, *Pichia pastoris*, and *A. oryzae*. All three expression systems can be used for the fermentative production of lipase isoenzyme, with applications in the food industry (Table 9.8).

Table 9.8 Food applications of lipases from genetically modified microorganisms

Lecitase®Novo	Degumming of vegetable oils
Lecitase®Ultra	Oil and bakery
Lipopan®	Baking
Lipozyme	Oils and fat modification
Noopazyme	Pasta/noodle
Palatase®	Dairy (cheese)
Novozymes®871	Pet food industry

9.5.8 Naringinases

Naringin (4, 5, 7,-trihydroxy flavanone-7-rhamnoglucoside), the bitter principle in citrus fruits such as grapefruit, is one of the main bitter compounds in citrus juices such as grapefruit, oranges, and kinnows. The enzyme naringinase is an α-rhamnopyranosidase type of enzyme that exhibits α-L-rhamnosidase and β-D-glucosidase activities (Ribeiro 2011). α-L-rhamnosidase activity of naringinase converts naringin to rhamnose and prunin (trihydroxyflavonone-7-glucoside). Prunin is further hydrolyzed to glucose and naringinin (4′-5, 7′-trihydronyflavonone) by the β-D-glucosidase component of naringinase (Fig. 9.3) Naringinases, alongside their anti-oxidant activity, have also been used in the citrus fruit juice industry for debittering and sweetening of the juice, thus maintaining the product stability and organoleptic characteristics. Fungi are the prolific producers of naringinases, the most prominent being *Aspergillus* species, *Penicillium* species, *Cochliobolus miyabeanus*, *Phanopsis citri*, and *Rhizopus nigricans*.

9.5.9 Pectinases

Pectinases are a heterogeneous group of enzymes that hydrolyze pectic substances. Protopectinases, polygalactouronases, lyases, and pectic esterases have been extensively studied for their applications in the food and feed industries. Pectinases constitute about 25 % of the global food enzyme sales and are predominantly produced by fungi (Pedrolli et al. 2009).

Pectinases play a very prominent role in fruit juice extraction, as the presence of pectins reduces the viscosity and turbidity of juice. Treatment of fruit pulp also helps in enhancing fruit juice volume during extraction. In apples, pears, and grapes, pectinases are used during the pressing and straining stages, whereas they are used for removal of cloudiness in mango, guava, pineapple, and papaya.

Pectinases are the most important enzymes in the wine-making process to support the extraction process, maximize juice yield, facilitate filtration, and improve flavor and aroma. The addition of pectic enzymes during extraction or

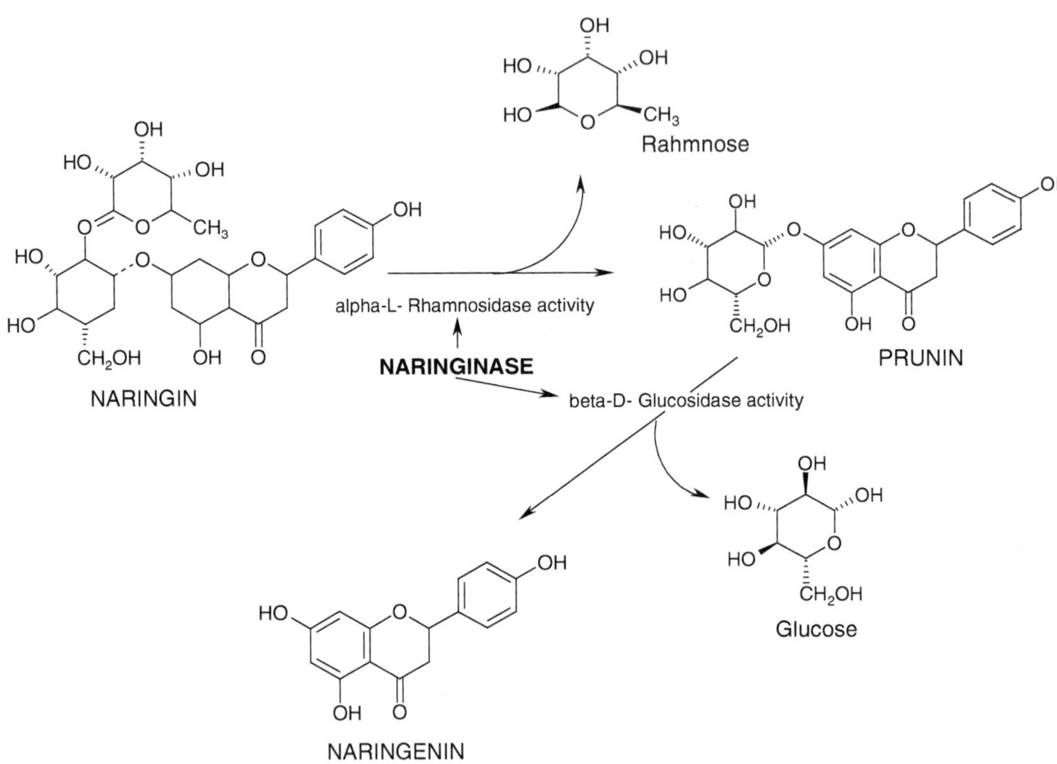

Fig. 9.3 Enzymatic action of naringinase on naringin

fermentation of the must results in an increase in aromatic precursors. Pectinases are used in cocktails, with amylases and cellulases in animal feed. Despite reports of bacteria-producing pectinases, the commercial preparations of pectinases are produced from fungal strains, the predominant producer being *A. niger*.

9.5.10 Phytases

Phytases (*myo*-inositol (1, 2, 3, 4, 5, 6) hexakisphosphate phosphohydrolases) describe a class of phosphatases possessing the in vitro capability of releasing at least one phosphate from phytate (myo-inositol-1,2,3,4,5,6-hexakisphosphates) (Haefner et al. 2005). Phytate accounts for 60–80 % of phosphorus found in plant-derived feedstuffs and serves as an ANF since the nutrients bound to them cannot be released and absorbed in the digestive tract until acted upon by phytases. Phytate also combines protein and vitamins, rendering them unavailable. Thus, phosphate becomes unavailable for monogastric/agastric animals, livestock and aquatic organisms like fish and shrimps as they lack intestinal phytases.

Natuphos® (BASF), the first phytase product derived from *A. niger*, was found to release phytate-bound phosphate. The addition of phytase promotes phosphate utilization. Studies on microbial phytases have been conducted on those originating from fungi like *Aspergillus ficuum*, *Mucor piriformis*, and *Cladosporium* species (Table 9.9). Ronozyme® P phytase (DSM Nutritional Products, Switzerland) is used as a feed additive for poultry, pigs, and sows at a minimum dose of 250, 300, 500 FYT/kg, respectively. Phytases are also used in the production of phytate-free soyabean milk. Phytase has also been used for bread improvement. Reduction in fermentation time, increase in bread volume, and improvement in crumb texture were observed with the use of phytases.

9.5.11 Tannases

Tannases or tannin acyl hydrolases are enzymes that cleave the ester linkages in tannins and gallic acids. Tannins are bitter plant polyphenolic compounds that are astringent in nature and bind to proteins. Tannase catalyzes the breakdown of tannins (tannic acid, methyl gallate, ethyl gallate, n-propyl gallate, and isoamyl gallate). The most important application of tannases is in the hydrolysis of tea tannins for the manufacture of instantaneous tea. In the food and beverage industry, it is used to remove the undesirable effects of tannins. Enzymatic treatment of fruit juices also helps in the reduction of tannin haze and sedimentation as well as the bitterness found in some new juices like pomegranate, cranberry, raspberry, and cold tea. Fungi, especially *A. niger*, has been the major producer of tannase for commercial product development. Tannase has been recently commercialized by Biocon (India), Kikkoman (Japan), ASA Specilaenzyme GmbH (Germany), and FC GmbH (Germany).

9.5.12 Transglutaminase

Transglutaminase (TGase; protein-glutamine γ-glutamyltransferase) enzyme catalyses an acyl-transfer reaction between the γ-carboxyamide group of peptide-bound glutamine residues (acyl donors) and a variety of primary amines (acyl

Table 9.9 Phytases used commercially in the food and feed industry

Product	Organism	Industry
Natuphos®	*Aspergillus niger* var. *ficuum*	BASF
RONOZYME®	*Peniophora lyci* expressed in *Aspergillus niger*	Novozymes- DSM nutritional products alliance
Phyzyme XP®	Escherichia coli	DuPont Genencor
Finase S 40	*Aspergillus niger*	DuPont Genencor

acceptors), including the ε-amino group of lysine residues in certain proteins. TGase can modify proteins by means of amine incorporation, cross-linking, and deamidation. Transglutaminases possess the ability to improve the functional characteristics of protein such as texture, flavor, and shelf life. TGase initially attracted interest because of its capacity to reconstitute small pieces of meat into a steak. It can adhere to the bonding surfaces of food such as meat, fish, eggs, and vegetables as a thin layer, and it exhibits strong adhesion in small amounts. The cross-links formed by transglutaminase strengthen the protein network structure in prepared meat products like hams and sausages and improves their elasticity and firmness. These properties make sausages resistant to high temperatures (retort treatment) and freezing due to the stability of ε-(γ -Glu) Lys cross-links, thus improving the manufacturing of retorted meat products.

The addition of transglutaminase also improves the texture of Chinese noodles, udon noodles (Japanese noodle made from wheat flour), soba noodles (made from buckwheat), and pasta products. It has been found to increase the breaking energy or firmness compared with untreated pasta. The industrial production of transglutaminase primarily uses a variant of *Streptoverticillium mobaraense* (namely MTGase).

9.6 Detergent

The most important application area for enzymes in terms of volume is detergents. Enzymes are applied to remove difficult stains and soil at low washing temperatures. Commercially available enzymes for detergents are mainly proteases, lipases, amylases, and cellulases.

9.6.1 Proteases

The first enzyme used in the detergent industry was introduced by Rohm and Haas when they introduced crude trypsin in their detergent Burnus® based on the German patent issued to Otto Rohm. Proteases are hydrolases that catalyse the hydrolysis of amide bonds within the proteinaceous substance such as blood stains, milk, egg, grass, spinach, and keratin and is therefore of interest for laundry applications. Proteases produced by microbes are generally denoted as acidic, neutral, or alkaline based on the optimal pH for their proteolytic activity. Proteases can also be distinguished on the basis of their side chain specificity and on the functional group present at the active site into serine proteases; cysteine proteases, aspartic proteases, and metalloprotease. Subtilisins like serine proteases largely find application in the detergent industry for laundering and as automatic dishwashing solutions. There are three important subtilisins: subtilisin Carlsberg isolated from *B. licheniformis*, discovered by Linderstrom, Lang, and Otteson at the Carlsberg laboratory; bacterial protease nagase (BPN) substilisin from *B. subtilis* and *B. amyloliqufaciens*; and subtilisin Novo produced from *B. subtilis*. The majority of subtilisin-type serine proteases have a molecular weight between 20 and 30 kDa and they display high activity at a pH of detergent-containing wash water. The parameter best performance of a protease in a detergent is based on its isoelectric point (pI). If the pI and pH of the detergent coincide, it is the most suitable protease for use in the detergent. They possess a broad substrate specificity. The performance of proteases is greatly influenced by the presence of non-ionic and anionic surfactants, bleaching, and sequestering agents used in the detergent formulations. Savinase® and Esperase® are two commercial preparations (Table 9.10) that are active at a high pI of 11.

Proteases are used as granular or stabilized liquid formulations in detergents and dishwashing solutions. Granulated enzymes have fewer restrictions than liquid formulations and tablet forms. In tablet forms, there is a need for highly stable enzymes as they are exposed to bleaching agents for a longer duration. New enzymes with novel properties have always been in demand to further enhance the wash properties of current enzyme-based detergents. *Bacillus* sp. have been the predominant organisms used for producing alkaline serine proteases for detergent applications; however, newer organisms are

Table 9.10 Commercial microbial alkaline proteases (subtilisins) used in the detergent industry

Industrial manufacturer	Product trade name	Enzyme properties (pH; temperature)	Microbial source
Novozymes (previously Novo Nordisk), Denmark	Alcalase	8–9; 60 °C	*Bacillus licheniformis*
	Savinase	9–11; 55 °C	*Bacillus clausii*
	Esperase	9–11; 60 °C	*Bacillus halodurans*
	Durazym[a]	10–10.5; 55 °C	*Bacillus* sp.
	Polarzym[a]	9–11, 20–40 °C	*Bacillus* sp.
Genencor International, USA (a part of DuPont)	Purafact[a]	10; 40–65 °C	*Bacillus lentus*
Gist Brocades (now a part of DSM), Netherlands	Subtilisin	n.s	*Bacillus acidophilus*
	Maxacal	11; 60 °C	*Bacillus* sp.
	Maxatase	9.5–10; 60 °C	*Bacillus* sp.
Solvay Enzymes, Germany (part of Genencor Inc, USA)	Opticlean	10–11, 50–60 °C	*Bacillus alcalophilus*
	Optimase	9–10; 60–65 °C	*Bacillus lichenformis*
	Maxapem[a]	11–12; 60 °C	Genetically modified *Bacillus* sp.
Godo Shusei, Japan	Godo-Bap	n.s	*Bacillus licheniformis*
Wuxi Synder Bioproducts, China	Wuxi	10–11; 40–50 °C	*Bacillus* sp.
Advanced Biochemicals, India	Protosol	10; 50 °C	*Bacillus* sp.

[a]Indicates that the enzyme has been developed by protein engineering/recombinant DNA technology
n.s. not specified

Table 9.11 New proteases (subtilisins) from different microbes with potential for use in the detergent industry

Microbial source	Enzyme properties (pH; temperature)	References
Virgibacillus pantotheticus (MTCC 6729)	10; 40–50 °C	Gupta et al. (2008)
Bacillus licheniformis RP1	10–11; 65–70 °C	Kamoun et al. (2008)
Brevibacillus sp. strain AS-S10-II	12.5; 45 °C	Rai and Mukherjee (2011)
Pseudomonas aeruginosa MCMB 327	8; 35 °C	Zambare et al. (2011)
Termitomyces albuminosus	10.6; 60 °C	Zheng et al. (2011)
Streptomyces fungicidus MML1614	11; 60 °C	Ramesh et al. (2009)
Vibrio metschinikovii	11; 60 °C	Jellouli et al. (2009)
Stentrophomonas maltophilia	10; 20 °C	Kuddus and Ramteke(2009)
Pseudoaltermonas sp. NJ 276	8; 30 °C	Wang et al. (2008)
Alkaliphilus transvaalensis	12.6; 40 °C	Kobayashi et al. (2006)

being explored with new properties like activity at low temperatures and enhanced stability in liquid formulations for bleach and oxidising agents and for exploration of novel protease backbones.

Some new microorganisms have recently been explored, producing enzymes with better stability to bleach agents, pH, and temperature stability for application as additives in detergent under different environmental conditions (Table 9.11). An oxidant and SDS-stable alkaline protease from *Bacillus clausii* has been used as a laundry detergent additive (Joo et al. 2003). Furthermore, a gene expressing a detergent-stable alkaline protease from the fungus *Aspergillus clavatus* ES1 has been cloned and expressed in *E. coli* for possible commercial use (Hajji et al. 2010). Hence, proteases with different characteristics find extensive usage in the detergent industry for the manufacture of laundry and diswashing products.

9.6.2 Use of Microbial Lipase as Detergent Additive

Lipases are enzymes hydrolysing triglycerides. The ability of lipases to catalyse the hydrolytic

Table 9.12 Commercially available microbial lipases for the detergent industry

Product name	Source	Class	Industrial manufacturer/supplier
Lipolase™	*Thermomyces lanuginosus*	Fungal	Novo Nordisk (Now Novozymes); 1994
Lumafast™	*Pseudomonas mendocina*	Bacteria	Genencor Intl.ltd; AU-KBC Research Centre; 1995
Lipomax™	*Pseudomonas alcaligens*	Bacteria	Genencor Intl.ltd; AU-KBC Research Centre; 1995

reaction is dependent upon their origin and their biochemical properties. Lipases are used for the removal of oil and grease from fabrics and from utensils as an additive for detergents and dishwashing formulations. The triglycerides are hydrolyzed to monoglycerols, di-glycerols, and free fatty acids that are more soluble in nature than the original fat. Novo Nordisk (now Novozymes) introduced the first lipase, Lipolase®, in detergent. It originated from the fungus *Thermomyces lanuginosus* and was expressed in *A. oryzae* in 1994. The production was observed as early as 1901 from *Bacillus prodigiosus*, *Bacillus pyocyaneus*, and *Bacillus fluorescens*. Lipases for detergents are specifically selected on the basis of low substrate specificity, stability under alkaline conditions (pH 10–11, 30–60 °C) and in the presence of surfactants (linear alkyl benzene sulfonates) and proteolytic enzymes, the constituents of many detergent formulations. Normally, fat stains are not easy to remove at low temperatures using conventional detergents, therefore lipases are required that are active at lower temperatures and can be used in detergent formulations. The use of lipases active at low temperatures in detergent formulations reduces energy consumption and the wear and tear of textile fibers (Feller and Gerday 2003; Chaplin 2004) and maintains the texture and quality of fabrics (Bjorkling et al. 1991). The addition of cold-active lipases in detergent becomes biodegradable, leaves no harmful residues, has no negative impact on sewage treatment processes, and poses no risk to aquatic life (Joseph et al. 2007).

Lipases produced by *Acinetobacter radioresistens* was found to be optimally active at pH10 and showed stability in the range of pH 6–10; therefore, it has a greater potential to be used in the detergent industry (Chen et al. 1998). Hasan et al. (2007) reported 100 % stability of lipase produced by *Bacillus* sp. FH5, at pH 10. This enzyme showed promising results when used in combination with different commercially available conventional detergents (Javed 2007). Commercial detergent formulations with high-temperature optima have been produced from *Pseudomonas mendocina* (Lumafast) and *Pseudomonas glumae* (Jaeger et al. 1994). In 1995, Lumafast and Lipomax, lipases from *P. mendocina* and *Pseudomonas alcaligenes*, respectively, were produced by Genencor International, AU-KBC Research Center, Life Sciences, Anna University (Table 9.12). More recently, cold-active lipases have been of great significance due to their catalytic activity at low temperature, low thermostability, and unusual specifications. They also find applications as additives in the detergent industry for cold washing.

9.6.3 Amylases as Detergent Additive

Amylases hydrolyze starch molecules to give diverse products, including dextrins and progressively smaller polymers composed of glucose units. Kirchhoff discovered amylases in 1811. Ohlsson suggested the classification of starch digestive enzymes as α- and β-amylases according to the anomeric type of sugars produced by the enzyme reaction. α-amylases have been in use in powder laundry detergents since 1975. Amylase-producing microorganisms generally include *Bacillus* sp., actinomycetes, and some fungi. *B. subtilis*, *Bacillus stearothermophilus*, *B. licheniformis*, and *B. amyloliquefaciens* have been considered as good producers of amylase (Table 9.13).

Presently, 90 % of all detergents contain amylases. The limitation of use of amylase in

Table 9.13 Commercially available amylases used in the detergent industry

Product name	Source	pH	Temp (°C)	Industrial manufacturer/supplier
BAN®	*Bacillus amyloliquefaciens*	6–7	70–90	Novo Nordisk (now Novozymes)
Termamyl®	*Bacillus licheniformis*	6–9	70–90	Novo Nordisk (now Novozymes)
Maxamyl®	*Alkalophilic Bacillus* species	6–8	100	Gist Brocades (now DSM)
Solvay amylase	*Thermostable Bacillus licheniformis*	5–8	75–90	Solvay, Germany
Purafect®OxAm	[a]*Engineered Bacillus licheniformis*	6–9	75–90	Genencor (now DuPont Genencor)
Duramyl	*Engineered Termamyl*	6–5	65–85	Novo Nordisk (now Novozymes)

[a]Refers to protein engineering

detergents is due to their sensitivity to calcium; stability is severely compromised in a low-calcium environment. Amylases are also sensitive to oxidants and hence require bleach stability. Protein engineering has helped in the improvement of enzymatic characteristics like bleach stability and calcium sensitivity.

Exploration of cold-active amylase-producing microbes is currently underway for detergent applications, as they offer economic benefits by saving the energy required for heating for optimal enzymatic activity.

9.6.4 Other Enzymes Used in Detergent Formulations

The application of cellulases as additives in laundry detergents began in the 1980s for cotton-based fabrics. The very first alkaline cellulase was used in Attack® by Kao Inc. The idea of incorporating cellulase was to restore the feel of the fabric, which was lost during use due to the formation of small balls of fuzz on the fabric surface (pilling). Cellulase treatment helped in removing these fibers without damaging the major fibers and thereby restoring the fabric condition. Cellulases are of two broad classes: exo-cellulases and endo-cellulases, based on the cleavage of the β-1, 4-glycosidic bonds in cellulose. Both bacteria and fungi produce cellulases, with a wide range of applications. Novozymes uses *Humicola insolens* for the production of commercial cellulase (Celluzyme®) used in the detergent industry. Carezyme® (Novozymes) is used for fuzz removal. Celluzyme is a 18.969 ptmulti-component enzyme mixture of seven cellulases, while Carezyme is a mono-component cellulase. However, damage can occur to the fiber of fabrics if 'extremely high doses' of color clarification cellulases are used in repeat launderings.

After celluloses, hemicelluloses are the second most abundant heteropolysaccharide polymer present in nature. The representative hemicelluloses are hetero-1, 4-β-D-xylans and hetero-1, 4-β- D-mannans. Mannans are generally present in ice-creams, sauces, shampoo conditions, and toothpastes as thickening agents. As stains from mannan-containing products are readily adsorbed on the cellulosic fibers, they are difficult to remove. Proctor and Gamble has collaborated with Novozymes to use alkaline microbial mannases (Mannaway®, Novozymes) in their detergent formulation to remove mannan-based stains.

9.7 Textile Industry

Historically, the use of enzymes in the textile industry began in 1912 with the use of barley for starch sizing from woven fabrics. The first microbial enzyme, 'amylase', was used in 1950 for the same starch desizing process that is nowadays a routine procedure. Today, microbial enzymes are important tools in the textile industry, as they reduce pollution and improve the economics of textile production via low resource consumption. Presently, the textile industry is employing an array of approximately 75 enzymes that could be broadly grouped as oxidoreductases and hydrolases. The different processes in which enzymes find application are textile desizing, enzymatic scouring, denim finishing, biopolishing, degumming of silk, and fiber processing.

Cotton or blended fabrics use warp threads coated with adhesive known as 'size'. Size helps in lubricating, thus protecting the yarn from abrasion and preventing the threads from breaking during weaving. Starch and its derivatives are most commonly used for this process as they have excellent film-forming capacity, are easily available, and are relatively cheaper. After weaving, the sizing agent and other non-cellulosic material present on the cotton fiber must be removed for the process of dyeing and finishing. α-amylases are used as desizing agents, due to their high specificity, and thereby remove the size. Amylases conventionally used for desizing are from *B. licheniformis*, *B. amyloliquefaciens*, or *B. stearothermophilus*. Fungal amylases can also be used for the desizing process. The preferred organism is *Aspergillus* species.

Untreated cotton or greige contains varieties of cellulosic impurities like waxes, pectins, and hemicelluloses, which give hydrophobic properties to the fiber and thereby interfere in the process of dyeing and finishing. Thus, scouring is carried out to remove these impurities and enhance wettability of the fiber. This was previously conducted with chemical methods; however, nowadays bioscouring is carried out using microbial enzymes like pectinases, proteases, lipases, and cutinases – individually or in combination. Pectinases is a general term applied for pectin esterases, polygalacturonases, and pectin lyases.

Pectinases are used as agents in bioscouring of cotton as well as for the biopreparation of bast fibers such as ramie, flax, and jute. Alkaline as well as acidic pectinases as commercial formulations are being produced by different industries (Table 9.14). Cuticle is a thin layer of cutin that is cross-linked to the primary cell wall by esterified pectic substances, thereby hindering the pectinase action on the pectin backbone. Cutin is polyester composed of epoxy and hydroxyl fatty acids. Cutinases are hydrolytic enzymes that degrade the cutin. Degani et al. (2002) were the first to report on the potential of cutinase from a bacterial source, *P. mendocina*, for wax degradation in cotton scouring. Laccases have also been used in the bioscouring of linen fabrics.

Table 9.14 Pectinases used in the textile industry

Acidic pectinases	
Forylase KL	Cognis, Germany
Viscozyme 120 L	Novozymes, Denmark
Pectinase P9179	Sigma Chemical Co, USA
Pectinase p3026	
Pectinase 62L	Biocatalysts
Multifect Pectinase	Genencor International
Alkaline pectinases	
Bioprep 3000L	Novozymes, Denmark
Pulpzyme HC	
Scourzyme L	
BayalaseEVO	Bayer, Germany
Unizim PEC	Color Centre, SA

Denims are basically cotton cloth that is conventionally dyed with indigo, having a characteristic blue color. The stonewash or worn look is popular in denim finishing and was initially achieved by laundering the denim with abrasive pumice stones. Cellulases were found to possess a similar action by loosening the indigo dye from the denim fibers, giving a faded abraded look similar to that provided by the stones. A number of cellulases are available, each with their own special properties (Table 9.15). Neutral cellulases also play an important role in this process by preventing back staining. *Melanocarpus albomyces* produces three novel cellulases for the treatment of textiles at neutral pH: 20 and 50 kDa endoglucanases and 50 kDa cellobiohydrases. Twenty kilodalton endoglucanase delivers good biostoning performance and, when combined with 50 kDa endoglucanase or 50 kDa cellobiohydrolase, it decreases back staining.

Biopolishing is the process of enzymatic treatment for improvement of cotton and natural and manmade cellulosic fibers. This prevents pilling of the fibers. A ball of fuzz in textile is referred to as a pill and they give a knotty and unattractive appearance to the fabric prepared from such fibers. Cellulases hydrolyze these pills and they break off the fiber, giving a smoother yarn

Table 9.15 Cellulases used in textile industry

Name	Microbial source	Manufacturer
Cellusoft®	*Trichoderma* sp.	Novozymes, Denmark
Celluclast®	*Trichoderma reesei*	
Denimax®Acid	*Trichoderma reesei*	
Denimax®Ultra	*Humicola* sp.	
Indiage™	*Streptomyces* sp.	Genencor Intl. USA
Primafast™		
Ecostone™		Primalco Ltd., Finland
Powerstone™		Iogen, Canada

surface. The other advantages of biopolishing are a softer texture and smoother and superior color brightness. Biopolishing usually takes place in the wet manufacturing processes of knitted and woven fabrics, which comprises the steps of desizing, scouring, bleaching, washing, dying/printing, and finishing. Biopolishing is important for the fiber lyocell, developed in 1991 from wood pulp. Treatment with cellulases enhances the silky appearance and avoids fibrillation.

The most important quality-enhancing steps in the wool finishing process to maintain the feel of softness in the texture is referred to as handle and prevents shrinkage. Proteases have been used for improvements in wool properties like the handle and shrink resistance; however, there is a slight reduction in tensile/bursting strength properties. In the processing of woolen textiles, transglutaminase helps reduce the propensity of wool fabric to shrink and maintains or increases fiber strength. There was a 25 % enhancement in tensile strength when *Streptomyces mobaraense* TGase was used alone or followed by a protease treatment.

Silk fibers are composed of approximately 75 % fibroin and 25 % sericin. Degumming is a process of removing sericin, giving a typical shiny aspect, soft handle, and elegant drape to the silk fiber that is highly valued by consumers. Uniform removal of sericin with retention of tensile properties and improvement in silk surface smoothness, handle, and luster was observed with alkaline proteases as compared with neutral and acidic proteases. A combination of lipase and protease further resulted in effective de-waxing and degumming, with positive effects on the wettability of silk.

9.8 Leather Industry

Enzymes have been used in tanneries and the leather industry for centuries as they were efficient in degrading the protein and lipid components of the hides/skin. In early days, these enzymes were derived from animal excreta and later the pancreas of cattle.

Leather-processing industries involve many sequential steps from raw hides to processed leather. The stages of leather processing are preservation, soaking, liming, dehairing, flashing, splitting, reliming, deliming, bating, degreasing, frizzing, beaching, pickling, depickling, and tanning. Raw hides undergo many of these treatments in a cascade manner before it is converted into finished leather. The pre-tanning operations require the use of harsh chemicals in large amounts, hence the leather industry is one of the worst environmental pollution offenders.

During leather manufacturing, the non-collagenous constituents of raw hides are completely removed during pre-tanning operations, of which dehairing is one of the major processes. Dehairing is the single largest process in the leather production process and requires a huge number of enzymes like proteases, amylases, and lipases to make the process environmentally friendly.

Today, microbes serve as resources of highly specific enzymes that are fast in their action and are therefore used in tanneries for soaking and dehairing processes. Other advantages of using these enzymes include the replacement of harmful chemicals that pose a threat to the environment and reductions in processing times and production costs (Table 9.16).

Table 9.16 Role of enzymes in different leather processing stages

Leather-processing stage	Enzyme involved	Function
Curing	Non-enzymatic	Preservation of hides and skin
Soaking	Alkaline/pancreatic proteases	Removal of non-fibrillar protein
Dehairing	Alkaline/neutral proteases	To improve the waste water quality
Degreasing	Lipases and proteases	To remove fats
Bating	Trypsin and alkaline protease	To make soft, supple, and pliable
Tanning	Indirect involvement of enzymes	To influence the quality of tanning

Adapted from Choudhary et al. (2004)

Curing of the hide is done to preserve them so they do not spoil before they are subject to further processing for leather development. Hides are steeped in a brine bath and dried in the sun; salt is added to the flesh side. Soaking is the first tanning operation that involves the treatment of hides with water. Soaking is used to rehydrate the skin. The degree of rehydration affects the quality of leather; the better the rehydration, the superior the leather. Soaking solutions generally comprise surfactants and antimicrobial compounds. It has been found that brine and cured hides, when soaked in protease in the presence of surfactants, reduced the soaking time by 45 %, and sulphides reduced by 40 %.

Liming is employed on improperly soaked skin. These are re-soaked in milk of lime so that the desired swelling of collagen takes place and opens up the fiber bundles. The objective of this process is to remove hairs, nails, hooves, and other keratinous materials. Alkaline proteases and alkaline lipases are used in the process of soaking and liming. The protease will open the membranes around the fat cells, making the fat accessible to lipase, resulting in the breakdown of fats. Furthermore, the breakdown products shall emulsify intact, which shall distribute throughout the felt omitting the use of degreasing surfactants. NovoLime® is a protease/lipase blend for enzyme-assisted liming of hides and skins.

Dehairing involves the removing of hairs from the hides without damaging the hide, which depends on the phenomenon of hair loosening. Specific proteases are required that can remove the hair without damaging the fibrous collagen. Clarizyme® is an alkaline serine protease from *A. flavus* by the Central Leather Research Institute (CLRI), Chennai, India, for the dehairing of skin and hides. Proteases from *Bacillus* and *Streptomyces* have been used for enzymatic dehairing processes.

Bating is a process of beating the leather cruelly with a heavy stroke using wooden logs and metal rods. This process is carried out to loosen and peptize non-collagenous skin structures via removal of interfibrillary proteins, epidermis, and scuds. For the production of soft pliable leather, used for making purses and gloves, a strong beating process must occur. This process is generally carried out in the presence of proteolytic enzymes of bacterial or pancreatic origin. Effective bating occurs under alkaline conditions conducted at 95–100 °F (30–38 °C) between a pH of 7.5 and 8.5 or else the enzyme efficiency.

Grease removal in tannery is carried out via liming. However, the grease content in some hides/skins is higher and causes fatty acid spues and uneven dyeing and finishing. A variety of microbial lipases have been used for degreasing (Table 9.17). The best stage of degreasing is pickling, as all the fat deposits are available to the surfactants. Since pickling is an aqueous phase, a combined action of disruption of fat cells and triglyceride splitting occurs, thereby improving the degreasing process. Tanning is the last stage of leather manufacture wherein collagen is cross-linked to the active group of the tanning agent, thereby irreversibly stabilizing the skin that is prone to putrefaction. This makes the collagen resistant to bacterial, enzymatic, and acid attack.

Table 9.17 Enzymes used in different processes of leather manufacture

Process	Enzyme used	Company
Soaking	Basozym S20 (protease mixture)	BASF, Germany
	Merpizyme 8008	CARPTEX, GmBH
	Adunil HR	CURTIN
	Adunil ZP	
	Debazym Prowet new	Debag Kimya, Istanbul, Turkey
	NovaCor S (protease)	Novozymes, Denmark
	Forezyme SK	La Forestal Tanica, Spain
	Trupowet SA (mix. of proteolytic enzymes)	Trumpler, Germany
Liming	Pelvit SPH	TFL (together for leather)
	Truponat HL (mix. of proteolytic enzymes)	Trumpler, Germany
	Mystozyme ECO-S (alkaline protease-based liming auxiliary)	Catomance Technologies, UK
	Basozym L10 (mixture of proteases)	BASF, Germany
	Forezym LM (bacterial proteases and lipases for dehairing)	La Forestal Tanica, Spain
	Biodart de-hairing enzyme (alkaline protease)	Southern Petrochemical Industries Corporation (SPIC), Chennai, India
	NUE (Novo UNHAIRING enzyme)	Novozymes, Denmark
Anti-wrinkling	Humectol ES-20	Cromogenia Units, Spain
	Microdep C	Tex Biosciences (P) Ltd.
Bating	Biodart alkali/acid	SPIC, Chennai, India
	Forezyme WB/PQ (acid protease)	La Forestal Tanica, Spain
	NovoCor AB	Novozymes, Denmark
	Microbate AB	Tex Biosciences (P) Ltd.
	Microbate R	
	Derobate-DK 1180 (microbial protease)	Debag Kimya, Istanbul, Turkey
Degreasing	Debazym LP1 (lipolytic proteases)	Debag Kimya, Istanbul, Turkey
	Forezym DG (bacterial lipase)	La Forestal Tanica, Spain
	Forezym WG-L (bacterial acid lipase)	
	NovoCor AD (acid lipase)	Novozymes, Denmark
	Greasex (alkaline lipase)	
Post tanning	NovaCor AX (protease mix)	Novozymes, Denmark
	NovaBate WB (neutral protease)	

Modified and adapted from Thanikaivelan et al. (2004)

9.9 Pulp and Paper Processing

Paper is composed of natural biopolymers: cellulose, hemi-cellulose, and lignin. The process of papermaking involves the use of extreme conditions and harsh chemicals to recover the cellulosic fibers. Hence, the paper industry started exploring the possibilities of using enzymes to overcome the use of harsh chemicals and make the process cost effective by reducing the energy requirements and time. Cellulases were initially introduced in the mechanical pulping process to defibrillate cellulose and enhance the fiber–fiber content, thereby enhancing the strength of fibers. The cellulase used for this process was isolated from *A. niger* and *Trametes suaveolens*. The major challenge in this experiment was to protect the change in viscosity that could result due to degradation of the cellulosic fibers.

The presence of lignins makes the wood and the pulp difficult to degrade. Lignins are removed

Table 9.18 Milestones in the application of enzymes in the pulp and paper industry

Year	Enzyme used	Process
1959	Cellulase	Pulp fibrillation
1984	Xylanases	Enzymatic beating and hemicelluloses removal
1986	Xylanases	Pre-bleaching
1989	Lipase	Pulp depitching
1993	Laccases	Pulp de-lignification
1996	Manganese peroxidase	Bleaching

from the wood during the process of pulping followed by bleaching. Lignin-degrading enzymes have spurred research on their use in the development of environmentally benign and efficient paper-making processes since 1986. In the past couple of years, a number of enzymes have been discovered for possible use in the pulp and paper industry. The major enzymes used presently are cellulases, xylanases, lipases, and laccases (Table 9.18) (Bajpai 1999). Cellulases, as a mixture of endoglucanase I, endoglucanase II, and hemicellulase, have been used for the modification of fiber properties with the aim of improving drainage and beatability. Cellulases also enhance the beachability of softwood kraft pulp, producing a brightness score comparable to that obtained with xylanase treatment.

Cellulose and hemicelluloses are the major components of lignocelluloses. Hemicellulose is a branched chain heteropolymer of a pentose and hexose sugar with xylose the most abundant. Hemicellulases are a spectrum of enzymes that cause complete hydrolysis of hemicellulose. These are endo-xylanase (endo-1, 4-β-xylanase), β-xylosidase (Xylan-1, 4- β- xylosidase), α-glucuronidase, α-arabinofuranosidase, and acetylxylan esterase. Endo-1, 4-β-xylanase, and xylan-1, 4- β- xylosidase are collectively known as xylanases and they hydrolyze the major component of hemicelluloses, the xylan (Bajpai 2011).

Xylanases for pulp treatment preferably should not possess cellulolytic activity as it will adversely affect the quality of the paper pulp. Some actinomycetes possess cellulase-free xylanases, e.g. *Chiania* sp. NCL 82-5-1, *Streptomyces roseiscleroticus*, and *Saccharomonospora viridis*.

Table 9.19 Currently used xylanases in the pulp and paper industry

Commercial xylanases	Industrial manufacturer
Pulpenzyme HA	Novozymes (earlier Novo Nordisk, Denmark)
Pulpenzyme HB	
Pulpenzyme HC	
Bleachenzyme F	Biocon, India
Cartazyme HS10	Clarient, UK
Cartazyme HT	
Cartazyme SR10	
Cartazyme PS10	
Ecopulp X100	Rohn Enzyme Oy, Finland
Ecopulp X200	
Irgazyme 40-4x	Genencor, Finland
Albazyme 40-4x	

Pulp enzyme HA® (Novozymes) was the first commercially available xylana isolated from *T. reesei* used for biobleaching of pulps (Table 9.19). The organisms producing xylanases are *Thermomyces lanuginosus*, *Aureobasidium pullulans*, *T. reesei*, *B. subtilis*, and *Streptomyces lividans*.

Xylanases of bacterial origin have an effective pH range between 6 and 9, while those of fungi have a pH range between 4 and 6. Pitch is a term used for the presence of high contents of hydrophobic components like wood resin, resin acids, triglycerides, and waxes. These pose severe problems in pulp and paper manufacture, affecting the machine runnability and reducing paper quality, thereby increasing the manufacturing cost.

Lipases or triacylglycerol acylhydrolase are hydrolases that act on carboxylic ester bonds. These hydrolyze triglycerides into monoglycerides, fatty acids, and glycerol. Lipases can be helpful in overcoming the pitch-related problems by lowering the triglyceride content in the wood pulp. The common lipase-producing microorganisms are *Rhizopus oryzae*, *Rhizopus arrhizus*, *A. niger*, *Pseudomonas alcaligenes*, and *Candida cylindrica*. EnzOx®PC is a non-selective lipase that is used as a common pitch-control agent. EnzOx®SEL (Enzymatic Deinking Technologies, LLC, USA) is a special pitch-control formulation containing one or more 1, 3-selective lipases. Resinase®A (*C. rugosa*) is a lipase and Resinase®A2X is a phospholipase manufactured by Novozymes A/B, Denmark. Alkaline lipases also find use in pitching. Lipolase100, Palatase®A, Palatase®M, and Nipozyme are formulations of alkaline lipases developed by Novozymes.

Lignin removal from chemical pulp is referred to as bleaching. It improves the properties of the paper apart from aesthetic reasons. Ligninolytic enzymes generally attack lignin directly. White rot fungi are the major producers of ligninolytic enzymes. Laccases, lignin peroxidases, and manganese peroxidase are the major lignin-degrading enzymes.

Laccases (p-diphenol: dioxygen oxidoreductase) are a blue copper family of oxidases. They are predominantly used for the de-lignification of woody fibers during the bleaching process. Laccase uses artificial mediators for pulp delignification like ABTS (2, 2′- Azino-bis (3-ethylbenzothiazoline-6-sulphonic acid)) and HBT (*N*-hydroxybenzotriazole). Laccases are employed in both mechanical and chemical pulps. Laccases are produced by both fungi and bacteria (Table 9.20). Fungal laccases have been heterologously expressed in *S. cerevisiae*, *T. reesei*, *A. oryzae*, *P. pastoris*, and *Aspergillus nidulans*. Bacterial laccases have been expressed in *E. coli* isolated from *B. subtilis*, *Thermus thermophilus*, and *Streptomyces lavendulae*.

Table 9.20 Microorganisms producing laccase

Fungi	Bacteria
Trametes versicolor	*Azospirillum lipoferum*
Trametes villosa	*Bacillus subtilis*
Lentinus edodes	*Streptomyces lavendulae*
Botrytis cineria	*Micromonas mediterranea*
Coriolus versicolor	
Penicillium sanguineus	

9.10 Biofuels

Biofuels are defined as fuels derived from biomass conversion such as biodiesel, bioethanol, biohydrogen, and biogas. The advantage of biofuels is that they produce reduced levels of particulates, carbon dioxides, and sulphur dioxide emissions compared with fossil fuels. Biodiesel (mono alkyl ester) of long-chain of fatty acid has immense potential as an alternative fuel. The feasible methods of biodiesel synthesis are pyrolysis and the use of microemulsions and transesterification. Transesterification is the best method for biodiesel production, as pyrolysis leads to more biogasoline production while microemulsion creates performance problems in the engine.

Conventional transesterification is a three-step consecutive reaction process in which diglycerides and monoglycerides are found as intermediate compounds. Every one mole of triacylglyceride gives three moles of biodiesel and one mole of glycerol (Figs. 9.4 and 9.5). The bottleneck in this process is the recovery of the catalyst, recovery of glycerol, waste water generation, and excessive energy requirements.

Lipase as a biocatalyst requires milder conditions with easy recovery of the by-product glycerol. The raw material for enzymatic

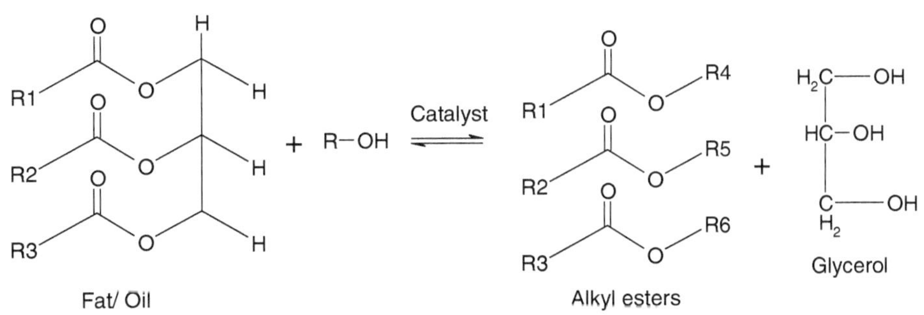

Fig. 9.4 Transesterification process

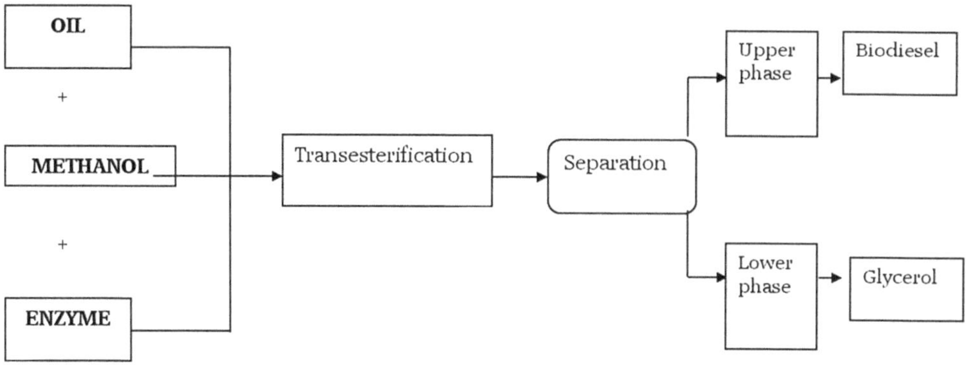

Fig. 9.5 Enzymatic transesterification process outline

Table 9.21 Microbial lipase in biodiesel production

Lipase source	Raw material	Acyl acceptor	Solvent	Time	Yield
Candida antarctica B	Waste cooking; palm oil	Methanol	Tert-butanol	4 h	79.1 %
Candida antarctica	Cotton seed oil	Methanol	Tert-butanol	24 h	97 %
Candida antarctica B	Soybean oil	Methyl acetate	Solvent-free	24 h	90–97 %
Candida antarctica	Jatropha seed oil; Karanj oil Sunflower oil	Ethyl acetate	Solvent-free	12 h	>90 %
Candida antarctica	Rapeseed oil	Methanol	Solvent-free	24 h	91.1 %
Candida antarctica	Cotton seed oil	Methanol, propanol	Solvent-free	7 h	91.5 %
Rhizomucor miehei	Soybean oil	Methanol	Solvent-free	12 h	68–95 %
Thermomyces lanuginosa	Sunflower oil; waste cooking oil	Methanol	Solvent-free	24 h	90–97%
Candida rugosa	Jatropha seed oil	Ethanol	Solvent-free	8 h	98 %
Thermomyces lanuginosa	Rapeseed oil	Methanol	Tert-butanol	12 h	95 %

Adopted from: Luković et al. (2011): Alternative fuel

transesterification for biodiesel production are materials with high free fatty acid content such as non-edible oils, waste cooking oil, and industrial waste oil. The free fatty acid content would undergo complete conversion, thereby increasing the biodiesel yield (Parawira 2009). Lipases are highly specific as chemo-, region-, and enantioselective catalysts. Lipases from different microbial sources have been evaluated for their transesterification for biodiesel production (Table 9.21). However, the most efficacious biocatalyst is *Candida antarctica* lipase B (CALB).

There is also a growing demand for renewable liquid fuels like motor grade fuel ethanol, butanol, and hydrocarbons from biomass sugars. The majority of the processes are designed for the production of ethanol, as the technology for glucose to ethanol production is very robust. Cellulases are being explored for their use in conversion of lignocellulosic materials and wastes, along with other enzymes for the preparation of mash for fermentative production of bioethanol or biobutanol.

9.11 Personal Care Products

Microbial enzymes have also been incorporated into oral care products and cleansing compositions. Oral care compositions comprise pullulanase and dextranase for mutan hydrolysis, removal of dental plaque, and prevention of dental plaque formation. Glucose oxidase and glucose oxidase-carbohydrases, proteases, amylases, and laccases have been used in the development of dentifrices and toothpastes to gain better cleansing effects and prevent bacterial growth in the oral cavity. Enzymatic cleansers have also been developed with oxidases and peroxidases for removing protein debris in contact lenses.

Hair-weaving preparations have lipases in them to facilitate penetration in the skin. Hyaluronidases, thiomucases, and lipases have been used in cosmetic/pharmaceutical preparations apart from lipases for use in skin inflammation. Topical lipase creams have also come into fashion for weight loss.

Modified virgin coconut oil (MVCO) is 1, 3-position-specific lipase-modified virgin coconut oil. MVCOs have been developed and tested for their potential antimicrobial activity against food and non-food systemic bacterial pathogens. A potent MVCO is able to kill *S. aureus* and *Candida albicans* after an incubation period of 10 min. MVCOs are readily absorbable in skin, safe for long-term application, and, most importantly, are cheap.

9.12 Summary

Microbial enzymes produced biotechnologically have great potential in industrial applications. They are comparatively more stable and efficient than their corresponding enzymes from plant or animal sources. Seasonal variations do not affect consistency and yield of microbial enzymes, unlike plant or animal sources.

The more than five decades of global development in microbial enzyme technology have resulted in patented production of a variety of enzymes used in diverse industrial applications like detergents, pulp and paper, textiles, pharmaceuticals, chemicals, leather, food, feed, personal care products, and biofuels. Market demand is huge and ever-increasing for biotechnologically produced microbial enzymes for novel products, since they make the processes cost effective and environmentally friendly.

Microbial enzymes as therapeutic agents possess higher target specificity than classical medicines in nutritional disorders, wound healing, anti-infective, anti-cancer, and thrombolytic applications. Besides this, microbial enzymes are also used in generalized therapy and in diagnostics. Extremozymes from extremophilic microbes can withstand extreme environmental conditions and therefore have revolutionized the chemical and pharmaceutical synthesis processes, leading to chirally pure drugs with less environmental damage and operational risks.

Demand for microbial enzymes in the food industries is significant, since several industrial processes are catalyzed by enzymes, the major processes being brewing, beverage production, food, livestock feed, and bakeries. Microbial enzymes have wide applications in the paper and leather industries and help in reducing environmental pollution, particularly that caused by chemical effluents. Other prominent areas of microbial enzyme application are detergents and the textile and paper industries for the development of high-end finished products through enzymatic benign processes.

Hitherto, only a fraction of microbial enzymes have been used for the development of industrial processes and products. A vast microbial biodiversity still remains to be explored for direct use of novel enzymes or evolving novel enzymes for the development of new products and processes.

Selected Reading

Amatayakul T (1955) The synthesis of fibrinolysin by fungi. Ohio J Sci 55(6):343–353

Bajpai P (1999) Applications of enzyme in pulp and paper industry. Biotechnol Prog 15:147–157

Bajpai PK (2011) Emerging applications of enzymes for energy saving in pulp and paper industry. IPPTA J 23(1):181–186

Balaraman K, Prabakaran G (2001) Production and purification of fibrinolytic enzyme from *Bacillus sphaericus*. Ind J Med Res 126:459–464

Batomunkueva BP, Egorov NS (2001) Isolation, purification and resolution of the extracellular proteinase complex of *Aspergillus ochraceus* 513 with fibrinolytic and anticoagulant activities. Microbiology 70(5):519–522

Bhasker N, Sudeepa ES, Rashmi HN, Tamil SA (2007) Partial purification and characterization of protease of *Bacillus proteolyticus* CFR3001 isolated from fish processing waste and its antibacterial activities. Bioresour Technol 98(14):2758–2764

Bjorkling F, Godtfredsen SE, Kirk O (1991) The future impact of industrial lipases. Trend Biotechnol 9:360–363

Broome JD (1963) Evidence that the L- Asparaginase of Guinea pig serum is responsible for its antilymphoma effects. J Exp Med 118(1):121–148

Chaplin M (2004) The use of enzymes in detergents. London South Bank University. http://www.lsbu.ac.uk/biology/enztech/detergent.html

Chen SJ, Cheng CY, Chen TL (1998) Production of an alkaline lipase by *Acinetobacter radioresistens*. J Ferment Bioeng 86:308–312

Choi HS, Shin PH (1998) Purification and partial characterization of a fibrinolytic protease in *Pleurotus ostreatus*. Mycologia 90:674–679

Choi HS, Sa YS (2000) Fibrinolytic and antithrombotic protease from *Ganoderma lucidum*. Myclogia 92(3):545–552

Choudhary RB, Jana AK, Jha MK (2004) Enzyme technology applications in leather processing. Indian J Chem Technol 11:659–671

Christensen LR (1945) Streptococcal fibrinolysis: a proteolytic reaction due to serum activated by streptococcal fibrinolysin. J Gen Physiol 28(4):363–383

Collen D, DeMol M, Demarsin E, Decock F, Stassen JN (1993) Isolation and conditioning of recombinant staphylokinase on man. Fibrinolysis 7:242–247

de Duve C (1966) From cytases to lysosomes. Fed Proc 23:1045–1049

Degani O, Gepstein S, Dosoretz CG (2002) Potential use of cutinase in enzymatic scouring of cotton fiber cuticle. Appl Biochem Biotechnol 102–103(1–6): 277–289

El-Assar SA, El-Badry HM, Abdel-Fattah AF (1990) The biosynthesis of proteases with fibrinolytic activity in immobilized cultures of *Penicillium chrysogenum* H9. Appl Microbiol Biotechnol 33:26–30

Feller G, Gerday C (2003) Psychrophilic enzymes: hot topics in cold adaption. Nat Rev Microbiol 1: 200–208

Ferrer M, Golyshina OV, Chernikova TN, Khachane AN, Reyes-Duarte D, Martins Dos Santos VAP, Strompl C, Elborough K, Jarvis G, Neef A, Yakimov MM, Timmis KN, Golyshin PN (2005) Novel hydrolase diversity retrieved from a metagenome library of bovine rumen microflora. Environ Microbiol 7:1996–2010

Fischetti VA (2005) Bacteriophage lytic enzymes: novel anti-infectives. Trends Microbiol 13(10):491–496

Fortney DZ, Durham DR (1996) Compositions containing protease produced by vibrio and method of use in debridement and wound healing. US Patent no. 5505943

Fujita M, Hong K, Nishimuro S (1995) Characterization of nattokinase degraded products from human fibrin or cross linked fibrin. Fibrinolysis 9:157–164

Garcia-Cano I, Velasco-Perez L, Rodriguez-Sanoja R, Sanchez S, Mendoza-Hernandez G, Llorente-Bousquets A, Farres A (2011) Detection, cellular localization and antibacterial activity of two lytic enzymes of *Pediococcus acidilactici* ATCC 8042. J Appl Microbiol 111(3):607–615

Gemmill JD, Hogg KJ, Douglas JT, Dunn FG, Lowe GDO, Pae AP, Hillis WS (1993) The incidence and mechanism of hypotension following thrombolytic therapy for acute myocardial infarction with streptokinase containing agents — lack of relationship to pretreatment streptokinase resistance. Eur Heart J 14(6):819–825

Ghosh PK, Saxena RK, Gupta R, Yadav RP, Davidson S (1996) Microbial lipases: production and applications. Sci Prog 79(2):119–157

Gupta R, Gigras P, Mohapatra H, Goswami VK, Chauhan B (2003) Microbial α-amylases: a biotechnological perspective. Process Biochem 38:1599–1617

Gupta A, Joseph B, Mani A, Thomas V (2008) Biosynthesis and properties of an extracellular thermostable serine alkaline protease from *Virgibacillus pantothenticus*. World J Microbiol Biotechnol 24:237–243

Haefner S, Knietsch A, Braun ESJ, Lohscheidt M, Zelder O (2005) Biotechnological production and applications of phytases. Appl Microbiol Biotechnol 68:588–597

Hajji M, Jellouli K, Hmidet N, Balti R, Sellami-Kamoun A, Nasri M (2010) A highly thermostable antimicrobial peptide from *Aspergillus clavatus* ES1: biochemical and molecular characterization. J Ind Microbiol Biotechnol 37(8):805–813

Hasan F, Shah AA, Hameed A (2007) Purification and characterization of a mesophilic lipase from *Bacillus subtilis* FH5 stable at high temperature and pH. Acta Biol Hung 58:115–132

Javed S (2007) Studies on the use of bacterial lipase as an additive in detergents. M. Phil thesis. Quaid-i-Azam University, Islamabad, Pakistan

Jaeger KE, Ransac S, Dijkstra BW, Colson C, van Heuvel M, Misset O (1994) Bacterial lipases. FEMS Microbiol Rev 15:29–63

Jellouli K, Bougatef A, Manni L, Agrebi R, Siala R, Younes I, Nasri M (2009) Molecular and biochemical characterization of an extracellular serine-protease from Vibrio metschnikovii J1. J Ind Microbiol Biotechnol 36:939–948

Jimenez ER (2009) Dextranse in sugar industry: a review. Sugar Tech 11(2):124–134

Joo HS, Kumar CG, Park GC, Paik SR, Chang CS (2003) Oxidant and SDS-stable alkaline protease from *Bacillus clausii* I-52: production and some properties. J Appl Microbiol 95(2):267–272

Joseph B, Ramteke PW, Thomas G, Shrivastava N (2007) Standard review cold-active microbial lipases: a versatile tool for industrial applications. Biotechnol Mol Biol Rev 2:39–48

Kademi A, Leblanc D, Houde A (2004) Microbial Enzymes: Production and Application; Lipase In:

Pandey A (ed) Concise encyclopedia of bioresource technology. Haworth Press, Binghamton, pp 552–560

Kamoun AS, Haddar A, Ali N, Ghorbel-Frikha B, Kanoun S, Nasri M (2008) Stability of thermostable alkaline protease from *Bacillus licheniformis* RP1 in commercial solid laundry detergent formulations. Microbiol Res 163(3):299–306

Kataoka M, Shimizu K, Sakamoto K, Yamada H, Shimizu S (1995) Lactonohydrolase-catalyzed optical resolution of pantoyl lactone: selection of a potent enzyme producer and optimization of culture and reaction conditions for practical resolution. Appl Microbiol Biotechnol 44:333–338

Katchalski-Katzir E (2005) My contributions to science and society. J Biol Chem 280(17):16529–16541

Kim JH, Kim YS (2001) Characterization of a metalloenzyme from a wild mushroom *Tricholoma saponaceum*. Biosci Biotechnol Biochem 65:356–362

Kim JH, Kim YS (1999) A fibrinolytic metalloprotease from the fruiting bodies of an edible mushroom *Armillariella mellea*. Biosci Biotechnol Biochem 63(12):2130–2136

Kobayashi T, Lu J, Li Z, Hung VS, Kurata A, Hatada Y, Takai K, Ito S, Horkoshi K (2006) Extremely high alkaline protease from a deep-subsurface bacterium, *Alkaliphilus transvaalensis*. Appl Microbiol Biotechnol 75(1):71–80

Kuddus M, Ramteke PW (2009) Production optimization of an extracellular cold-active alkaline protease from *Stenotrophomonas maltophilia* MTCC 7528 and its application in detergent industry. Afr J Microbiol Res 5(7):809–816

Landis BH, McLaughlin JK, Heeren R, Grabner RW, Wang PT (2002) Bioconversion of N-butylglucamine to 6-deoxy-6-butylamino sorbose by *Gluconobacter oxydans*. Org Process Res Dev 6(4):547–552

Lauro BD, Rossi M, Moracci M (2006) Characterization of a β-glycosidase from the thermoacidophilic bacterium *Alicyclobacillus acidocaldarius*. Extremophiles 10(4):301–310

Lee JS, Back HS, Park SS (2006a) Purification and characterization of two novel fibrinolytic proteases from mushroom *Fomitella fraxinea*. J Ind Microbiol Biotechnol 16:264–271

Lee MH, Lee CH, Oh TK, Song JK, Yoon JH (2006b) Isolation and characterization of a novel lipase from a metagenomic library of tidal flat sediments: evidence for a new family of bacterial lipases. Appl Environ Microbiol 72(11):7406–7409

Li Y, Shuang JL, Yuan WW, Huang WY, Tan RX (2007) Verticase: a fibrinolytic enzyme produced by *Verticillium* sp. Tj33, an endophyte of *Trachelospermum jasminoides*. J Integr Plant Biol 49(11):1548–1554

Liang J, Mundorff E, Voladri R, Jenne S, Gilson L, Conway A, Krebber A, Wong J, Huisman G, Truesdell S, Lalonde J (2010) Highly enantioselective reduction of a small heterocyclic ketone: biocatalytic reduction of tetrahydrothiophene-3-one to the corresponding (R)-Alcohol. Org Process Res Dev 14(1):188–192

Liu XL, Du LX, Lu FP, Zheng XQ, Xio J (2005) Purification and characterization of a novel fibrinolytic enzyme from *Rhizopus chinensis*. Appl Microbiol Biotechnol 67:209–214

Locken JP, Stromer FC (1970) Acetolactate decarboxylase from *Aerobacter aerogenes*. Purification and properties. Eur J Biochem 14:133–137

Luković N, Knežević-Jugović Z, Bezbradica D (2011) Biodiesel fuel production by enzymatic transesterification of oils: recent trends, challenges and future perspectives, alternative fuel. In: Manzanera M (ed), ISBN: 978-953-307-372-9. InTech. Available from: http://www.intechopen.com/books/alternative-fuel/biodiesel-fuel-production-by-enzymatic-transesterification-of-oils-recent-trends-challenges-and-futu

Marshall RO, Kooi ER (1957) Enzymatic conversion of D-glucose to D-fructose. Science 125:648–649

McCord JM, Fridovich I (1988) Superoxide dismutase: the first twenty years (1968–1988). Free Radic Biol Med 5(5–6):363–369

Miyazaki K (2005) Hyperthermophilic α-L -arabinofuranosidase from *Thermotoga maritima* MSB8: molecular cloning, gene expression, and characterization of the recombinant protein. Extremophiles 9(5):399–406

Myziuk L, Romanowski B, Johnson SC (2003) BVBlue test for diagnosis of bacterial vaginosis. J Clin Microbiol 41(5):1925–1928

Ogawa J, Shimizu S (2000) Stereoselective synthesis using hydantoinases and carbamoylases. In: Patel RN (ed) Stereoselective biocatalysis. Marcel Dekker, Inc, New York, pp 1–21

Parawira W (2009) Biotechnological production of biodiesel fuel using biocatalysed transesterification: a review. Crit Rev Biotechnol 29(2):82–93. ISSN 0738-8551

Park HJ, Jeon JH, Kang SG, Lee JH, Lee SA, Kim HK (2007) Functional expression and refolding of new alkaline esterase, EM2L8 from deep-sea sediment metagenome. Protein Expr Purif 52(2):340–347

Patel RN, Chu L, Mueller R (2003) Diastereoselective microbial reduction of (S)-[3-chloro-2-oxo-1-(phenylmethyl)propyl]carbamic acid, 1,1-dimethylethyl ester. Tetrahedron-Asymmetry 14(20):3105–3109

Payen A (1874) Handbuch der technischen Chemie. In: Stohmann F, Engler C (eds), vol 2. E. Schweizerbartsche Verlags-buchhandlung, Stuttgart, p 127

Payen A, Persoz JF (1833) Memoir on diastase, the principal products of its reactions and their applications to the industrial arts. Annales de Chimie et de Physique, 2nd Series 53:73–92

Pedrolli DB, Monteiro AC, Gomes E, Carmona EC (2009) Pectin and pectinases: production, characterization and industrial application of microbial pectinolytic enzymes. Open Biotechnol J 3:9–18

Peng Y, Yang XJ, Zhang YZ (2005) Microbial fibrinolytic enzymes: an overview of source, production, properties and thrombolytic activity in vivo. Appl Microbiol Biotechnol 69:126–132

Popp G, Becker H (1895) A process for preparing hides for tanning. US Patent 607549 (1898)

Rai SK, Mukherjee AK (2011) Optimization of production of an anti-oxidant and alkaline stable β-keratinase from *Brevibacillus* sp. AS-S10-11: applications of enzyme in laundry detergent formulations and in leather industry. Biochem Eng J 54:47–56

Ramesh S, Rajesh M, Mathivanan N (2009) Characterization of a thermostable alkaline protease produced by marine *Streptomyces fungicidicus* MML1614. Bioprocess Biosyst Eng 32(6):791–800

Rhee JK, Ahn DG, Kim YG, Oh JW (2005) New thermophilic and thermostable esterase with sequence similarity to the hormone-sensitive lipase family, cloned from a metagenomic library. Appl Environ Microbiol 71(2):817–825

Ribeiro MH (2011) Naringinases: occurrence, characteristics and applications. Appl Microbiol Biotechnol 90:1883–1895

Riley KN, Herman I (2005) Collagenase promotes the cellular response to injury and wound healing in vitro. J Burn Wound 4:e8

Rovati JI, Osvaldo D, Lucıa I, Figueroa C, Farina JI (2010) A novel source of fibrinolytic activity: *Bionectria* sp., an unconventional enzyme-producing fungus isolated from Las Yungas rainforest (Tucuma'n, Argentina). World J Microbiol Biotechnol 26:55–62

Sakaguchi K, Murao S (1950) A preliminary report on a new enzyme, "penicillin-amidase". J Agric Chem Soc Jpn 23:411

Sellami-Kamoun A, Haddar A, El-Hadj A, Ali N-H, Ghorbel-Frikha B, Kanoun S, Nasri M (2008) Stability of thermostable alkaline protease from *Bacillus licheniformis* RP1 in commercial solid laundry detergent formulations. Microbiol Res 163:299–306

Shin HH, Choi HS (1999) Purification and characterization of metalloproteases from *Pleurotus sajor-caju*. J Microbiol Biotechnol 9:675–678

Sugimoto S, Fujii T, Morimiya T, Johdo O, Nakamura T (2007) The fibrinolytic activity of a novel protease derived from a tempeh producing fungus, *Fusarium* sp. BLB. Biosci Biotechnol Biochem 71(9): 2184–2189

Sumi H, Hamada H, Tsushima H, Mihara H, Miwaki H (1987) A novel fibrinolytic enzyme (nattokinase) in vegetable cheese Natto: a typical and popular soyabean food in Japanese diet. Experientia 43:1110–1111

Sumner JB, Somers GF (1953) Chemistry and methods of enzymes, 3rd edn. Academic Press Inc, New York, p 462

Takamine J (1894) Process of making diastatic enzyme. US Patent 525,823

Tauber H (1949) The chemistry and technology of enzymes. Wiley, New York

Tekwani S, De Mello PM (2010) Enzyme assisted extraction of lutein from marigold flowers and its evaluation by HPLC. Int J Adv Pharmaceut Sci 2:381–386

Thanikaivalen P, Rao JR, Nair BU, Ramasami T (2004) Progress and recent trends in biotechnological methods for leather processing. Trends Biotechnol 22(4):181–188

Tirawongsaroj P, Sriprang R, Harnpicharnchai P, Thongaram T, Champreda V, Tanapongpipat S, Pootanakit K, Eurwilaichitr L (2008) Novel thermophilic and thermostable lipolytic enzymes from a Thailand hot spring metagenomic library. J Biotechnol 133(1):42–49

Tosa T, Mori T, Fuse N, Chibata I (1966) Studies on continuous enzyme reactions. I. Screening of carriers for preparation of water-insoluble aminoacylase. Enzymologia 31:214–224

Treem WR, McAdams L, Stanford L, Kastoff G, Justinich C, Hyams J (1999) Sacrosidase therapy for congenital sucrase-isomaltase deficiency. J Pediatr Gastroenterol Nutr 28:137–142

Ullmann F (1914) Enzyklopädie der technischen Chemie, vol 5. Urban und Schwarzenberg, Berlin, p 445

Urano T, Ihara H, Umemura K, Suzuki Y, Oke M, Akite S, Tsukamoto Y, Suzuki I, Takeda A (2001) The profibrinolytic enzyme subtilisin NAT purified from *Bacillus subtilis* cleaves and inactivates plasminogen activator inhibitor type I. J Biol Chem 276(27): 24690–24696

Voget S, Steele HL, Streit WR (2006) Characterization of metagenome derived halotolerant cellulase. J Biotechnol 126(1):26–36

Wang Q-F, Hou Y-H, Xu Z, Miao J-L, Li G-Y (2008) Purification and properties of an extracellular cold-active protease from the psychrophilic bacterium *Pseudoalteromonas* sp. NJ276. Biochem Eng J 38: 362–368

Wu B, Wu L, Ruan L, Ge M, Choi D (2009) Screening endophytic fungi with antithrombotic activity and identification from endophytic fungal strain CPCC 480097. Curr Microbiol 58:522–527

Yamada H, Kumagai H (1975) Synthesis of L-tyrosine-related amino acids by L -tyrosinase. Adv Appl Microbiol 19:249–288

Zambare V, Nilegaonkar S, Kanekar P (2011) A novel extracellular protease from *Pseudomonas aeruginosa* MCM B-327: enzyme production and its partial characterization. New Biotechnol 28(2):173–181

Zeng R, Xiong P, Jianjun W (2006) Characterization and gene cloning of a cold-active cellulase from a deep-sea psychotrophic bacterium *Pseudoalteromonas* sp. DY3. Extremophiles 10(1):79–82

Zheng S, Wang H, Zhang G (2011) A novel alkaline protease from wild edible mushroom *Termitomyces albuminosus*. Acta Biochemica Polonica 58(2): 269–273

10 Strategies of Strain Improvement of Industrial Microbes

Classical and Recombinant DNA Technology in Improving the Characteristics of Industrially Relevant Microbes

10.1 Introduction

Microbes produce a variety of products in very low concentrations which have been used as antibiotics, drugs, vitamins, enzymes, bulk organic compounds, polymers, amino acids, biofuels, etc. Prerequisite for efficient biotechnological processes at industrial scale requires the use of microbial strains which produce high titre of the desired product. However, this is not an inherent property of the selected microorganism(s); hence, modifications in their genetic material could possibly help in overcoming this limitation. Thus, industrially relevant microbes are subjected to a variety of treatments using physical, chemical or genetic tools to overproduce the desired metabolite and make the process cost efficient. This process of enhancing the biosynthetic capabilities of microbes to produce desired product in higher quantities is defined as microbial strain improvement.

The term microbial strain improvement encompasses development of strains which possess enhanced capacity to: (1) utilise complex raw material and efficiently assimilate making the process inexpensive, (2) reduce or eliminate undesired by-products of the microbial process, (3) enhance extracellular release of the by-product, (4) reduce the toxic threshold of the end product as to facilitate high accumulation with minimal cell death, (5) reduce the fermentation period and (6) overproduce native or foreign products after genetic recombination.

The widespread use of penicillin during the Second World War necessitated the production of penicillin as global necessity. Hence, early studies on microbial genetics were directed to induce a change in the genetic makeup of *Penicillium chrysogenum* by using physical (UV rays, X-rays, etc.) and chemical agents (EMS, NNTG) generally called mutagens. Cells which survived the treatment underwent a 'permanent' heritable change in the characteristics of microorganism due to alteration in the genomic organisation which is referred to as mutation, and the altered organism is called as a mutant. Mutation has been the major factor involved in the 100- to 1,000-fold increases obtained in the production of microbial metabolites. High penicillin yielding strains were developed using classical or random mutagenesis which comprised the use of physical and chemical mutagens singly as well as in combination.

10.2 Spontaneous Mutations

Mutations that occur naturally in the cells are referred to as spontaneous mutations. The various mechanisms which are responsible for spontaneous mutations are: (1) mis-pairing errors during replication, (2) de-purination, (3) deletions, (4) insertion sequences and (5) error-prone DNA repair mechanism.

Spontaneous mutations generally occur at a very low frequency of 10^{-10} to 10^{-6} per generation

S. Saxena, *Applied Microbiology*,
DOI 10.1007/978-81-322-2259-0_10, © Springer India 2015

per gene. However, selection pressure could be adopted as a method of screening and isolating spontaneous mutants from populations which possess elevated mutation rates. These could be further subjected to mutagenesis by using physical or chemical methods in the development of an industrial strain. Spontaneous mutants of the wild strain as well as of mutants have been used in the industrial strain improvement programme of *Penicillium chrysogenum* (Kardos and Demain 2011).

Spontaneous mutations are generally linked to normal chemical changes in the organism that can alter the structure of the sequence of genes. All four bases of DNA have unusual tautomeric forms, i.e. enol forms, which cause the change in the hydrogen bonding characteristics of bases. When a purine is substituted by another purine and pyrimidine by a pyrimidine, it is referred to as transitions, and when a purine is substituted by pyrimidine or vice versa, it is referred to as transversions. Both transitions and transversions are point mutations. Spontaneous mutations also result due to methylation of purine or pyrimidines which often leads to the loss of those bases spontaneously or through base cleavage.

10.3 Classical Mutagenesis

The simplicity of the mutation techniques had tremendous appeal to microbiologists. Classical mutagenesis generally involves the use of physical and chemical agents (Table 10.1) which can bring about random changes in the genetic structure of the microbe to improve the desired characteristics. The general procedure of mutagenesis is divided in three broad steps: (a) exposing the parent strain to a mutagen to induce genetic variability, (b) random selection and screening of the surviving population using a small-scale model fermentation and (c) assay of fermentation broth/agar for improved products and statistical scoring of the improved strains. Each time, the improved strain obtained after mutation is used as a parent strain for the next cycle of mutation. The whole process is repeated until a strain confirmed to be statistically superior in performance is carried head to head using the clone cell of parent (non-mutated) and improved (mutated) strain.

Physical mutagens like the ionising radiations are responsible for bringing a change in structure of the genetic material due to fragmentation leading to deletions, whereas ultraviolet light brings about dimerisation of adjacent thymine bases thereby interrupting the process of translation, i.e. protein synthesis. M. Demerec first used X-rays to isolate a mutant strain of *Penicillium chrysogenum*, X-1612 which was three times as much productive when compared to wild strain 1951-B25. *Penicillium* was also the first organism to be treated with UV rays, and a mutant Q-176 was isolated which exhibited three-time higher activity when compared to the mutant X-1612. UV irradiation at 250 nm induces dimerisation of pyrimidine bases especially thymine, and this leads to deletion mutation when a modified strand

Table 10.1 Mutagens and mutational changes for microbial strain improvement

Mutagen class	Mutagen type	Type of mutation	Impact of DNA
Physical	X-rays; γ-rays	Deletions; structural changes	Single-stranded or double-stranded breakage of DNA
	UV rays	Transversion, deletion, frameshift, transitions GC→AT	Pyrimidine dimerisation and cross links in the DNA
Chemical	5- Bromouracil	Transitions, AT→GC, GC→AT	Faulty pairing
	5- Chlorouracil	Transitions, AT→GC, GC→AT	Faulty pairing
	Hydroxylamine (NH_2OH)	GC→AT transition	Deamination of cytosine
	Nitrous acid (HNO_2)	Bidirectional translation, deletion, AT→GC, GC→AT	Deamination of A, C and G
	N-methyl-*N*′-nitro-N-nitrosoguanidine	GC→AT transition	Methylation, high pH, alkylation of C and A
	Ethyl methane sulphonate	GC→AT transition	Alkylation of bases C and A

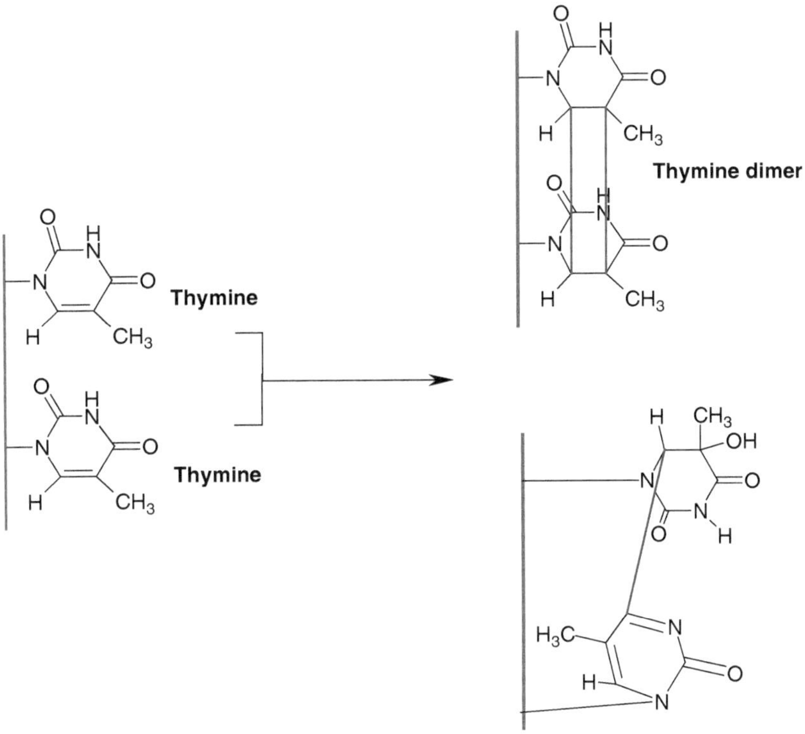

Fig. 10.1 Mutations or aberrations introduced in the pairing of nitrogenous bases in DNA due to UV radiation

is copied. Another aberration caused by UV is the formation of a photoproduct (6-4) known as a lesion wherein carbon atoms number 6 and 4 of adjacent pyrimidines are linked together covalently. Both dimers and lesion interfere in the accurate replication of the DNA leading to erroneous transcription (Fig. 10.1).

UV mutagenesis has been extensively used in the strain improvement programme of *Acremonium chrysogenum* for production of cephalosporin. M8650 was a mutant strain developed from the original strain isolated by Brotzu. Eli Lily and Co. developed a mutant CW19 which produced three times more cephalosporin than Brotzu's strain. Further, optimisation of fermentation conditions produced 15 times more cephalosporin C than the progenitor wild type (Elander and Espenshade 1976) (Fig. 10.2).

Chemical mutagens comprise of natural and synthetic organic and inorganic chemicals which can react with the genetic material, i.e. DNA, and alter its properties. The errors introduced by chemicals include faulty base pairing, deamination of cytosine, transitions transversions and frameshift mutations. Classical mutagenesis is a random process wherein it is not possible to predict which type of mutation would yield an improvement in particular strain. Charlotte Auerbach and J.M. Robson first demonstrated that mustard gas was the cause of mutations in fruit flies, and subsequently, a large number of chemical mutagens have been discovered.

Alkylating agents are the most commonly used mutagens which have been used in the classical mutagenesis programme. Ethyl methane sulphonate (EMS), methyl methane sulphonate (MMS), diethyl sulphate (DES) and nitrosoguanidine (NTG) are the common alkylating agents. Alkylating agents add the alkyl group to the nitrogen at the seventh position of the purine resulting in a labile *N*-glycosidic bond and that hydrolyses to have a depurinated site. This depurinated site is filled by other base leading to unmatched bases, thus restricting

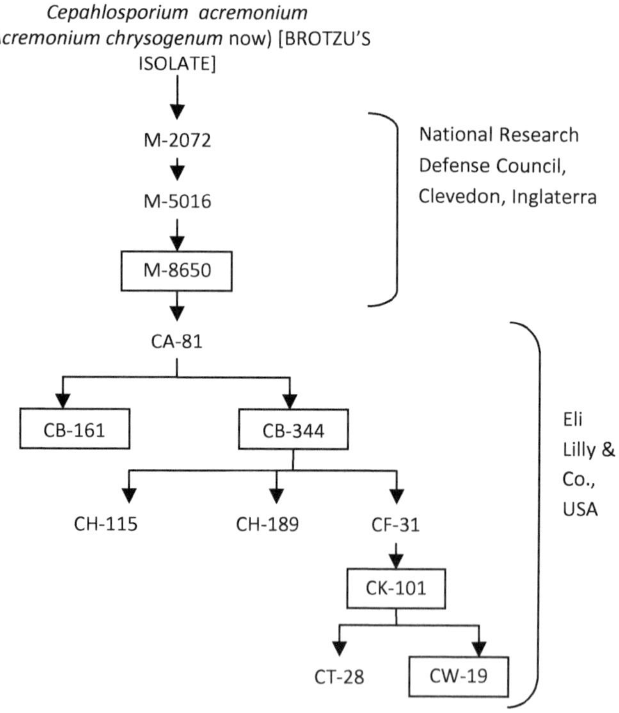

Fig. 10.2 Strain improvement of *Acremonium chrysogenum* by UV mutagenesis programme for cephalosporin production

DNA replication. NNTG (*N*-methyl-*N*′-nitro-N-nitrosoguanidine) primarily acts at the replication fork.

Nitrous acid (HNO_2) is also used as a chemical mutagen as it causes oxidative deamination of nitrogenous bases. Adenine is converted into hypoxanthine and cytosine to uracil, and guanine is converted into xanthine leading to miscoding type of mutations directly. The condition of the selection of a mutagen depends upon the ease of use and safety. Ethyl methane sulphonate (EMS) is more safe and easier to handle than NTG since it is in a liquid state. Similarly, UV exposure is much easier as it can induce a variety of mutations apart from easy application as compared to X-rays or γ-rays.

The best example is mutagenesis of *Penicillium chrysogenum* NRRL-1951. In order to make penicillin production economically viable, mutagenesis was carried out using radiations and chemical mutagens by four research groups, viz., USDA Peoria laboratory, Carnegie Institute of Washington and University of Minnesota, University of Wisconsin and Eli Lilly Industries. The mutagens used were nitrogen mustard (NM), ultraviolet radiation (UV) and X-radiation (X), and the advantage of spontaneous (S) mutation was also taken in the *Penicillium chrysogenum* NRRL-1951 strain improvement programme. Around 21 rounds of mutation and improvement by were taken by different laboratories to increase the yield of Penicillin by a factor of 55. Eli Lilly eventually had a modified *P. chrysogenum* E15.1 which produced 7 g/l penicillin in shake flask conditions when compared to Florey's yield by *P. chrysogenum* NRRL-1951 (Elander 1999). Further optimisation of fermentation conditions in a bioreactor under submerged conditions yielded 20 g/l of penicillin. With this development, penicillin became the very first commercial antibiotic (Kardos and Demain 2011) (Fig. 10.3).

10.4 Mutant Selection in Classical Mutagenesis

Classical mutagenesis is a random procedure to generate aberrations in the genetic material, and hence, the selection of the mutant is largely based on change in morphology or colour, nutritional

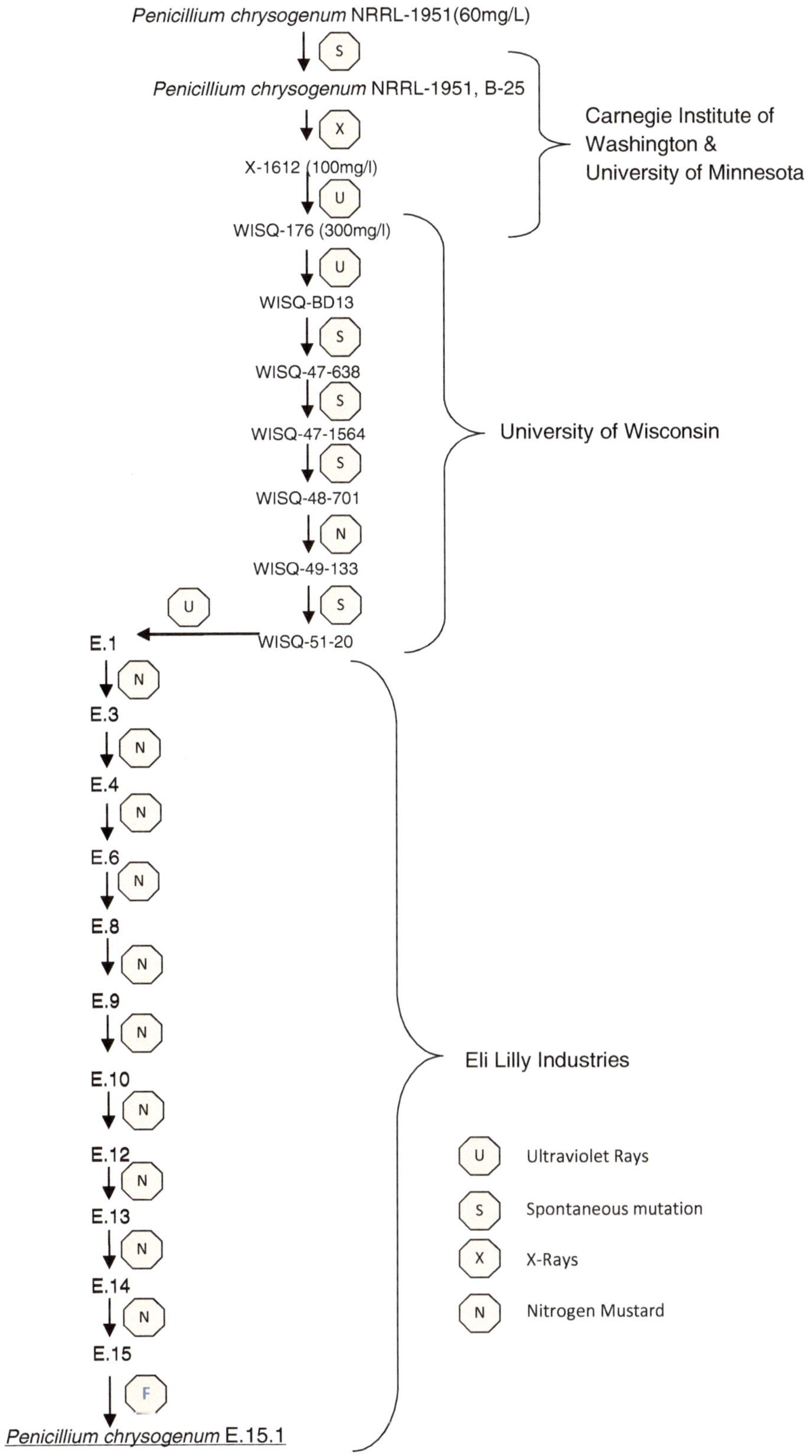

Fig. 10.3 Strain improvement of *Penicillium chrysogenum* by classical mutagenesis for penicillin production

requirements (biochemical mutants) and screening of enhanced production of the end product.

10.4.1 Morphological Mutants

In morphological mutants, the production ability of the end product is improved; however, the exact mechanism of this improvement remains uncertain. Gibberellic acid production by *Gibberella fujikuroi* is commonly carried out by aerobic fermentation. A variety of morphological mutants possessing different mycelial and soluble pigmentation have been generated from *Gibberella fujikuroi* upon exposure to UV radiation. It was observed that the unpigmented morphological mutant of *Gibberella* resulted in the lowering of the viscosity of the fermentation broth as well as resulted in an increase in the production of gibberellic acid (Lale et al. 2006).

Streptomyces species have been designated as the most prolific producers of antibiotics in actinomycetes which find use in agriculture, medicine and veterinary purpose. Bald colonies have been exhibited by *Streptomyces coeruleorubidus* which produces higher titre of daunorubicin (Blumauerova et al. 1978). Bald mutants in *Streptomyces coelicolor* are found to be defective in carbon utilisation, morphogenesis and cell signalling.

In *Aspergillus nidulans*, mutagenesis leads to the formation of mutants which lack conidia apart from being coloured differently. Wet, white-coloured mutant of *Aspergillus nidulans* was obtained by random mutagenesis using nitrous acid. In these mutants, the conidia are colourless and become brown and accumulate water droplets. Abacus is an aconidial mutant of *A. nidulans* generally produced by the treatment of nitrous acid, UV radiation and NNTG. It bears rod-like structures with swellings at intervals in place of conidia (Clutterbuck 1969). *Actinoplanes teichomyceticus* is a teicoplanin-producing strain that when mutated exhibits pink instead of brown mycelia and leads to 25-fold higher production of teicoplanin than the parent strain (Lee et al. 2003).

10.4.2 Auxotrophic Mutants

These microorganisms possess specific nutritional requirements as a result of mutation. Mutant auxotrophs of *Corynebacterium glutamicum* have been developed which possess high rates of amino acid production (Nakayama 1985). The development of these auxotrophs has been done through the iterative process of mutagenesis and selection. Optimisation of the mutant strains has been further carried out by chemical mutagenesis and UV irradiation to release enzymes which help in the regulation of metabolite production by feedback inhibition (Shiio et al. 1990).

Streptomyces has been studied using various aerial mycelium negative (amy$^-$) mutants or by biochemical analyses during their exploitation in the industry (Hopwood 1978). During the production of bicozamycin (previously bicyclomycin) by *Streptomyces griseofulvin*, a mutant which was an arginine auxotroph could grow with citrulline and ornithine and could produce maximum aerial mycelium with citrulline (Godfrey 1973).

α-Ketoglutarate has applications in agrochemical, food industry, pharmaceuticals and cosmetics. A *Yarrowia lipolytica* isolate producing a very high titre was isolated when production of single-cell protein was carried out using petroleum as the raw material (Finogenova et al. 2005). *Torulopsis glabrata* and *Y. lipolytica* are the only thiamine-auxotrophic species which have been studied as α- ketoglutarate producers (Liu et al. 2007; Zhang et al. 2009). Thiamine limitation essentially deactivates pyruvate decarboxylase and α-ketoglutarate dehydrogenase complex leading to overflow of α- ketoglutarate (Morgunov et al. 2004; Nosaka et al. 2008).

10.4.3 Mutants Exhibiting Resistance to Antimetabolites

Compounds that resemble structurally to natural metabolites are the metabolic analogues and interfere with the normal function of the cells, thus referred as antimetabolites. The simplest example of antimetabolite is sulfa drugs. *N*-flourophenylalanine is an antimetabolite of tyrosine, and it was observed that *E. coli* mutants resistant to *p*-fluorophenylalanine secreted high amount of tyrosine (Adelberg 1958). Canavanine-resistant mutants of *E. coli* and *S. cerevisiae* and arginine hydroxamate-resistant mutants of *Bacillus subtilis* have been recognised to excrete or accumulate *L*-arginine (Kubota et al. 1973).

Phosphate has been a key regulator of biosynthesis of peptide, polyene and tetracycline class of antibiotics. Candicidin synthetases in *Streptomyces griseus* are regulated by phosphate, and therefore, mutants which can resist a phosphate concentration higher than 10 mM are of great industrial interest, suggesting the loss of regulatory control by phosphate in candicidin biosynthesis (Martin et al. 1979). Wild strains of *Penicillium chrysogenum* produce low titres of penicillin V in medium containing excess glucose; however, GRI (glucose repression insensitive) mutants have been reported to produce high penicillin V than the wild-type strain consisting of lactose as the main carbon source in the medium (Chang et al. 1990). Thus, GRI selection mutants can be used as selective markers for higher penicillin-producing strains.

High astaxanthin-producing strains of *Phaffia rhodozyma* were screened using diphenylamine, and the mutants did not exhibit variation in size or numbers but appeared as salmon-coloured colonies against pink salmon of wild type. Diphenylamine-resistant mutant exhibited a twofold increase in astaxanthin production over wild type (Chumpolkulwong et al. 1997). *Ashbya gossypii*, a hemiascomycete fungus, is characterised as a natural riboflavin producer and has been utilised for industrial production of riboflavin. Isocitrate lyase was identified as the key enzyme responsible for riboflavin production using soybean oil as sole carbon source. Isocitrate lyase gets strongly inhibited by oxalate or itaconate. An oxalate-resistant *Ashbya gossypii* has been recently reported to produce threefold higher riboflavin production when compared to the wild strain (Sugimoto et al. 2010). Roseoflavin has been used successfully for screening mutants of *Bacillus subtilis* for high riboflavin production.

Coenzyme Q_{10} is an isoprenylated benzoquinone which is a well-known component of electron transport chain in eukaryotes and is used as an oral nutritional supplement in several disorders like cardiomyopathy, diabetes and neurodegenerative diseases associated with CoQ_{10} deficiency. *Sporidiobolus johnsonii* is the only heterobasidiomycete yeast strain known to produce CoQ10 under submerged fermentation conditions. UV and EMS mutagenesis has been attempted for strain improvement of *S. johnsonii* which has been screened using atorvastatin resistance marker. EMS induced the atorvastatin-resistant mutant strain EA22 generated from *S. johnsonii* ATCC 20490 which exhibited a twofold increase in the CoQ_{10} titre (Ranadive et al. 2011).

10.4.4 Enhanced Production of the End Product: Agar Zone Mutants

Agar zone assay is a screening method of mutants based on their fermentation performance and high yield of the desired end product. Lovastatin is an inhibitor of HMG-CoA reductase thereby limiting cholesterol biosynthesis for the treatment of hypercholesterolemia. Mutant *Aspergillus terreus* colonies were screened by preparing individual agar plug of each colony and then extracted with ethyl acetate for 15 min at 50 °C and subsequently the supernatant removed and tested by transferring on a 6-mm paper disc and placed on a 90-mm *Candida albicans* plate. The positive control comprised of a known concentration of lovastatin loaded on a 6-mm paper disc. Subsequently, the inhibition zones of the mutants were compared with positive control for selection of mutants with high titre of lovastatin (Ferron et al. 2005). Starch iodine assay has been used to screen and select high glucoamylase-producing deregulated mutants of *Thermomucor indicae-seudaticae* after sequential exposure to nitrous acid and γ-radiation (Kumar and Satyanarayana 2009). Zone diameter assay has also been used in screening and selection of UV and NTG mutants of *Trichoderma reesei* MTCC 3929 for higher alkaline protease production (Zambare 2010).

Methylene blue agar zone assay has been used for screening and selection of mutants of *Streptomyces diastatochromogenes* with enhanced ε-poly-*L*-lysine (Wang et al. 2012).

Streptoverticillium rimofaciens produces mildiomycin nucleoside antibiotic which is potentially effective against powdery mildew disease on different crops. Enhanced production by an improved strain can be determined by analysing extracts from agar plugs by HPLC/TLC and diffusion assays. A bioassay method has been developed to report improved strains by assessing the mildiomycin content in complex broth using a strain of *Rhodotorula rubra* AS 2.166 as an indicator organism. Mildiomycin gives a clear inhibition zone with sharp edges when the mutant colonies were punched into individual agar plugs and then placed on the test agar medium seeded with the test microorganisms. The zone formation was directly related to higher production of mildiomycin by the mutant *S. rimofaciens* (Xie et al. 2005).

10.5 Recombination

Mutagenesis and screening was a dominant strategy for microbial strain improvement as compared to genetic recombination. This mainly was associated with the low frequency of genetic recombination in industrial microorganisms; the major microorganisms which were initially exploited from commercial aspect were *Cephalosporium* and *Penicillium* species involved in the production of cephalosporin and penicillin, respectively, which had only parasexual mode of reproduction. Recombination would have been more successful if it would have been exploited as a complementary technique to mutagenesis rather than alternative. Recombination generally involves the fusion of two different ancestors or genotypes or strain so that a superior hybrid is generated which does not resemble either of the parents. Protoplast fusion technology has a great potential in the development of superior fusants as compared to contributing parents which has been exploited for the development of industrial microorganisms.

Recombination is also helpful in removal of neutral as well as deleterious mutations which accumulate during the random mutagenesis. Clavulanic acid produced by *Streptomyces clavuligerus* NRRL3585 was randomly mutated to produce a strain which exhibited a tenfold enhancement in production of clavulanic acid. However, when arginine and cysteine auxotrophs of *S. clavuligerus* were fused, the fusant produced 30-folds more clavulanic acid than the wild type. Another important example of recombination via protoplast fusion technology is enhancement in the production of cephalosporin C. Two isolates of *Acremonium chrysogenum* from the commercial strain improvement programme were selected. A low titre rapidly growing spore-forming strain having optimal methionine requirement for cephalosporin C production was fused with a high titre slow-growing asporogeneous strain which used inorganic phosphate, and the fusant grew rapidly, sporulated and produced cephalosporin C from sulphate and 40 % more antibiotic than the higher titre parent. Similarly, higher cephamycin C-producing *Nocardia* strain has been developed by recombination of two improved cephamycin C-producing *Nocardia* strains.

DNA shuffling was introduced in 1994 by Stemmer which was the first homologous recombination method. This technique involves the digestion of genes by DNAase I into random fragments and reassembling the fragments into full length gene by primerless PCR. By this technique, Stemmer established that genomic recombination within a population of bacteria can efficiently generate combinatorial libraries of new strains of which some may possess marked improvement in the phenotype as well as production potential. Protoplast fusion is a method which leads to genome shuffling. Genome shuffling, on the contrary, is the recombination between multiple parents of each generation in which several rounds of genome fusion are carried out. The major advantage of this technology is that the new strain developed is not considered a genetically modified microorganism (GMMO) which restricts its applicability in the food industry. Genome shuffling exploits the diversity that exists among population in organisms and eliminates the nonessential/deleterious genes documented during random mutagenesis by back crossing of progeny with the parents.

High production of polyketide antibiotic tylosin via protoplast fusion was achieved by creation of a new strain of *Streptomyces fradiae* by genome shuffling. Intrastrain fusion has been carried out in *Trichoderma reesei* strain Ptr2 for the development of fusants SFTr2/SFTr3 possessing high CMCase activity exhibited on CMCA (carboxymethylcellulose agar) (Prabavathy et al. 2006). The production of riboflavin has been enhanced by genome shuffling in *Bacillus subtilis* (Chen et al. 2004). 65.3 % improvement in teicoplanin was recorded by carrying out genome shuffling in *Actinoplanes teichomyceticus* (Xu et al. 2006). Lipase production was increased by 317 % over the starting strain *Penicillium expansum* FS8486 by two rounds of genome shuffling (Lin et al. 2007). Similarly, the production of epithiolone, the anticancer drug produced by the bacterium *Sorangium cellulosum* strain So157-2, was improved in fusants and increased about 130 times after two rounds of genome shuffling (Gong et al. 2009). Thus, a lot of time is saved which was initially invested in classical mutagenesis by the process of genome shuffling. The technique of genome shuffling was demonstrated as an efficient method for the evolution of the environmental microorganism; *Sphingobium chlorophenolicum* can degrade pentachlorophenol (PCP), which is a highly toxic anthropogenic pesticide (Dai and Copley 2004). Thus, genome shuffling is a global technique for engineering phenotypes of industrial microorganisms at whole genome level which has been used for improvement in substrate uptake, higher tolerance to end products and higher production yields (Table 10.2) and can be integrated with metabolic engineering to promote evolution of a complex genotype.

At times, protoplast fusion technology has been used in discovery of new antibiotics. For example, protoplast fusion between nonantibiotic producing mutants of *Streptomyces griseus* and *S. tenjimariensis* led to the production of a new antibiotic indolizomycin (Gomi et al. 1984).

10.6 Recombinant DNA Technology

Recombinant DNA technology has played an enormously potential role in microbial strain improvement of industrial microorganisms. Today, advances in molecular biological techniques have led to the development of basic cloning systems in bacteria, yeasts as well as in filamentous fungi. Initially, the recombinant DNA technology was used in improving the yields of primary metabolites like amino acids and extracellular enzymes (Gouka et al. 1997). Using genetic engineering techniques, removal of homoserine dehydrogenase in wild-type producing *Corynebac-terium* is converted into lysine overproducing mutant which cannot grow unless methionine and threonine are added to the medium (Eggeling et al. 1998).

Recombinant microbes have been developed to produce biotin by cloning the biotin operon (bioABFCD) on a multicopy plasmid which allowed *E. coli* to produce 10,000 times more biotin than the wild type thereby giving economic advantage over traditional chemical synthesis (Levy-Schil et al. 1993). A recombinant *E. coli* strain was developed by introducing a proline-4-hydroxylase gene from *Dactylosporangium* species thereby producing hydroxyproline at 25 g/l when proline was added; then, the production reached 41 g/l (Shibasaki et al. 1999). Cloning of extra copies of threonine export genes in *E. coli* led to the increased threonine production (Kruse et al. 2002). Similarly, riboflavin synthesis genes were cloned and overexpressed in its producing organism *Corynebacterium ammoniagenes* which resulted in the production of 15.3 g/l of riboflavin in 3 days. Many enzymes resourced from microbes existing in diverse environment find applications in industrial process chemistry and tremendous applications in the pharmaceutical industries.

In antibiotic production, the benzyl penicillin acylase gene of *E. coli* was cloned on multicopy plasmid which resulted in 45-fold increase in production of this enzyme as compared to the

Table 10.2 Recombination methods for microbial strain improvement

Category of improvement	Microorganisms	Results	Reference
Substrate uptake	*Lactobacillus delbrueckii* and *Bacillus amyloliquefaciens*	Non-fastidious strain obtained after three rounds of genome shuffling for direct conversion of starch to lactic acid	John et al. (2008)
	Sphingobium chlorophenolicum	Higher tolerance level of pentachlorophenol for degradation at a faster rate	Dai and Copley (2004)
Strain tolerance to end product	*L. rhamnosus*	More accumulation of lactic acid at 3.6 pH compared to wild type	Wang et al. (2007)
	Streptomyces pristinaespiralis	Four rounds of protoplast fusion generated 100-μg/ml pristinamycin resistant recombinant	Xu et al. (2008)
	Saccharomyces cerevisiae	Improved strain maintaining cell viability till 55 °C and 25 % ethanol stress after three rounds of genome shuffling	Shi et al. (2009)
Product yield enhancement	*Streptomyces fradiae*	Two rounds of genome shuffling led to six times higher titre of tylosin	Zhang et al. (2002)
	Streptomyces gilvosporeus	High natamycin-producing strain – approximately 153 % higher than the parent strains and 1.17 times more than present strain	Zhu et al. (2006)
	Phaffia rhodozyma	Two cycles of recursive protoplast fusion, a shuffled strain was selected and 1.43 times higher yield of astaxanthin was obtained	Zheng and Zhao (2008)
	Sorangium cellulosum	The epothiolone production of the fusant was increased about 130 times compared to the starting strain by 3 rounds of genome shuffling	Gong et al. (2009)
	Clostridium diolis DSM 15410	Improvement of 1,3-propanediol production by 4 rounds of genome shuffling by 80 %	Otte et al. (2009)
	Propionibacterium shermanii	Enhanced vitamin B_{12} production after genome shuffling	Zhang et al. (2010)

uninduced wild type. Penicillin G is converted into 6-aminopenicillanic acid by an enzyme penicillin V amidase which is then used as an intermediate for synthesis of a variety of semisynthetic penicillins. Therefore, the production of penicillin V amidase was enhanced by cloning the penicillin V amidase gene in wild-type *Fusarium oxysporum* XF 4, a soil isolate increased enzyme titre by 130-fold. *F. oxysporum* f.sp. *lycopersici* (ATCC 16322) served as a heterologous host and was used for DNA transformation. Thus, both these have industrial relevance in the production of antibiotic production using core structure of penicillin.

α-Amylase gene from *Bacillus amyloliquefaciens* was cloned in a mulitcopy plasmid pUB110 in *Bacillus subtilis* which led to enhancement of α- amylase by 2,500-fold (Palva 1982). Genes of captopril esterase from *Pseudomonas putida* were cloned in *E. coli* where its yield was improved by 38-folds. It has been commercially used for the production of chirally pure side chain of captopril (Elander 1995). Genencor and DuPont have carried out collaborative project for the development of an economical process of conversion of glucose into 1, 3- propanediol using recombinant *E. coli* which consists of two metabolic pathways for conversion of glucose into propanediol (Tong et al. 1991; Laffend et al. 1996). Propanediol is a precursor of a new biodegradable polymer 'Sorona'.

Heterologous expression is generally adapted when difficulties are encountered with the functional genetic system or detectable production conditions for a particular secondary metabolite. *E. coli* has been engineered to produce 6-deoxyerythrronoide B as the same level as the host *Streptomyces coelicolor* by homologous recombination methods. Daptomycin (Cubicin®) gene cluster has been characterised and cloned from *Streptomyces roseosporus*. A bacterial artificial chromosome (BAC) clone containing the entire 12 gene cluster on 128 Kb DNA has been introduced in *Streptomyces lividans* for heterologous expression resulting in 18 mg/l of Daptomycin. Compactin is a potent HMG (hydroxymethylglutaryl) – CoA reductase produced by *Penicillium citrinium* and is widely used as a substrate for the production of pravastatin, widely used as an anti-hypercholestrolemic drug. It was observed that the introduction of compactin biosynthesis cluster or regulatory gene *mlc*R into high compactin-producing strains of *Penicillium citrinium* further enhanced over-production of compactin. The best compactin-producing strain exhibited 50 % higher production of compactin and consisted of five copies of *mlc*R gene (Baba et al. 2009). Productions of many antibiotics have been improved by heterologous expression (Table 10.3).

Cephalosporins like cefuroxime, cefoxitin and cefcapene pivoxil contain carbamoyl side chain at the position -3, and these require adipoyl-7-amino-3-carbamoyloxymethyl-3-cepham-4-carboxylic acid (ad7-ACCCA) as a preferred precursor in semisynthesis. Ad7-ACCCA cannot be produced from 7-ACA or 7-ACDA precursors that can be produced by fermentation, and there is a growing demand of it as the active pharmaceutical ingredient. *Penicillium chrysogenum* DS17690 strain has been reprogrammed by the introduction of *cef EF* genes from *Acremonium chrysogenum* and *cmcH* gene from *Streptomyces clavuligerus* to form ad7-ACCCA, a carbamoylated derivative of adipoyl-7-amino de-acetoxy cephalosporanic acid. The *cefT* gene of *A. chrysogenum* which encodes cephalosporin C transporter is further introduced producing *Penicillium chrysogenum* resulting in almost twofold increase in cephalosporin production with concomitant decrease in penicillin by-product formation (Nijland et al. 2008).

Plant isoprenoids have also been produced through their heterologous expression in microbes. The microorganisms used for heterologous expression are *Saccharomyces cerevisiae* or *Escherichia coli*. Artemisinin, a plant natural product used as an antimalarial drug, is an excellent therapeutic intervention produced by the plant *Artemisia annua* to treat clinical episodes of malaria caused by *Plasmodium falciparum* and *Plasmodium vivax*. However, the requirement of

Table 10.3 Use of heterologous expression systems for pharmaceutical agents

Drug/drug precursor	Host	Application	Engineering technology	Reference
Shikimic acid	*E. coli*	Antiviral	Heterologous expression of glucose facilitator protein, glucokinase both from *Zymomonas mobilis*	Chandran et al. (2003)
Taxadien-5 α- acetoxy-10β-ol (taxol precursor)	*S. cerevisiae*	Antitumor	Heterologous expression of taxol biosynthetic genes	DeJong (2006)
Lycopene	*E. coli*	Antioxidant	Heterologous expression of carotenoid genes from *Pantoea agglomerans*	Yoon et al. (2007)
Tylactone	*S. venezuelae*	Antibiotic	Heterologous expression of tylosin polyketide synthase in *S. venezuelae*	Jung et al. (2006)
Macrolide 6- deoxyerythomycin-D	*E. coli*	Antibiotic	Heterologous expression	Lee and Khosla (2007)

the number of doses to treat the afflicted population in third world countries and other parts of the globe is not fulfilled due to its dependence on plant for the production and extraction of artemisinin. Heterologous expression has provided opportunity in the production of plant drug precursors for commercial bulk production through microbial biosynthetic pathway via fermentation route. Amorphadiene was the first precursor which was expressed in *E. coli*. Unlike the modular synthesis of polyketides and non-ribosomal peptides, the isoprenoid biosynthesis requires multiple genes as a part of the overall pathway.

S. cerevisiae being eukaryotic offers promise for heterologous expression of plant genes including membrane proteins which pose difficulty in functionalising in the bacterial systems. Ideally, the yeast has to be engineered to have increased FPP (farnesyl pyrophosphate) production and decreased sterol biosynthesis. Amorphadiene synthase (ADS) gene has been transferred in *Saccharomyces* using plasmid mediated and genome-based methods, and their comparative effects have been studied for the production of artemisinin production. Yeast cells harbouring the ADS gene via episomal plasmid using a galactose-inducible promoter produced 600 μg/l amorpha-4,11-diene, while yeast cells which carried ADS gene through homologous recombination produced 100 μg/l amorpha-4,11-diene in 16-day batch cultivation (Lindahl et al. 2006). It was evident that the production of amorpha-4,11-diene is positively correlated with the gene dosage but insufficient pool of FPP. Transferring multiple genes of artemisinin biosynthetic pathway has led to the development of genetically engineered yeast strains capable of producing artemisinic acid (Ro et al. 2006). In these strains, the ADS and HMGR genes have been transferred thereby enhancing the FPP and amorpha-4,11-diene pool (Fig. 10.4). Reduction in sterol biosynthesis by repression of SS gene along with methionine repressible promoter further augmented FPP and amorpha-4,11-diene production. Three-step oxidation of amorpha-4, 11-diene to produce artemisinic acid has been enabled by the introduction of CYP71 AV1. The methionine repressible promoter downregulates the sterol biosynthetic gene ERG9, thereby enhancing amorpha-4,11-diene production. Hydroxymethylglutaryl reductase gene (HMR) further enhanced the production of amorpha-4,11-diene bringing it to approx. 149 mg/l.

All these multiple gene manipulations lead to 153 mg/l of amorpha-4,11-diene accumulation in the fermentation medium, and finally, addition of the CYP71 AV1 gene and CPR genes lead to the production of 32-mg/l of artemisinic acid in shake flask cultures. Further, production can be enhanced by optimisation of fermentation conditions using bioreactors from bench to pilot scale. Thus, genetically engineered strains of yeasts expressing plant genes could be effectively used for the production of artemisinic acid at levels comparable to the plant *Artemisia annua* on biomass basis but in much shorter time say 4–5 days as compared to several months in plants thereby reducing the production cycle and meeting the global drug demands.

Hence, various innovative strategies of recombinant DNA technology developed in the past two decades have drastically improved the residential traits of industrial microorganisms or altogether developed engineered microorganisms expressing novel traits of different organisms for industrial processes and products.

10.7 Integrated Strain Improvement: Precision Engineering Technology

Integrated strain development or precision engineering technology is a new aspect which also emphasises on the overall impact and performance of industrial microorganisms which have been improved by classical mutagenesis, recombination methods or recombinant DNA technology. Previously, the strain improvement methods overlooked the detrimental effects in the industrial microorganism, i.e. slow growth, substrate specificity, weak stress tolerance and formation of undesired products. Therefore, there is a paradigm shift towards metabolic engineering approaches which is based on rationally induced beneficial genetic modifications by understanding

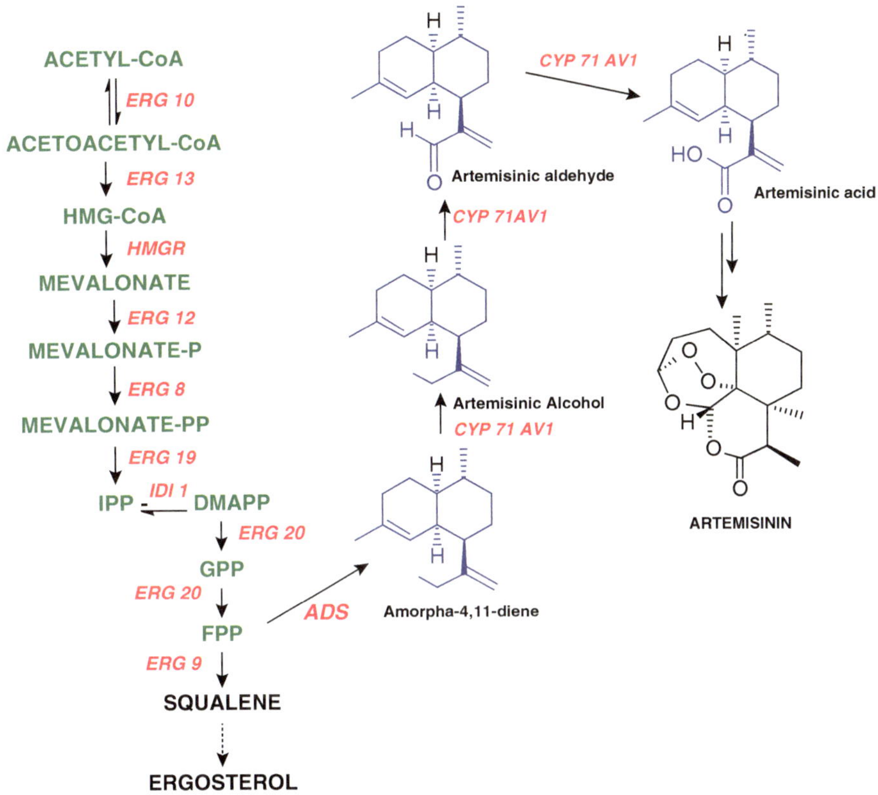

Fig. 10.4 Genetically engineered pathway of artemisinin production in *Saccharomyces cerevisiae* (Adopted from Ro et al. 2006)

of the underlying metabolic and regulatory networks on the whole system.

Precision engineering technology involves the pairing of classical metabolic engineering and conventional screening methods with profiling technologies which provide more comprehensive understanding of the genetics and physiology associated with metabolite production (Fig. 10.5).

Precision engineering technology developed by Microbia has core strength of association analysis which essentially employs computational method of profiling and identifying gene expression patterns which correlate with the desirable characteristics. This helps in the identification of novel genes and reduction of impurities. This technology also uses the concept of regulator engineering where a set of evolutionary conserved genes which function both to integrate signalling from multiple pathways and exert coordinate control through global regulons. The approach was initially developed for fungal systems, for strain improvement as well as for the identification of the most relevant pathway of a specific metabolite.

10.7.1 Production of High Lovastatin-Producing Strains Through Precision Engineering

Best industrial strains generally have product yields ($Y_{P/S}$, gram product/gram carbohydrate) approaching to 40 % of the theoretical yield in appropriately managed fermentations, levels which are rarely attained in fermentation of secondary metabolites. Microbia has used the collection of *Aspergillus terreus* strains producing lovastatin. Metabolite detection method was used to identify the amount of lovastatin and

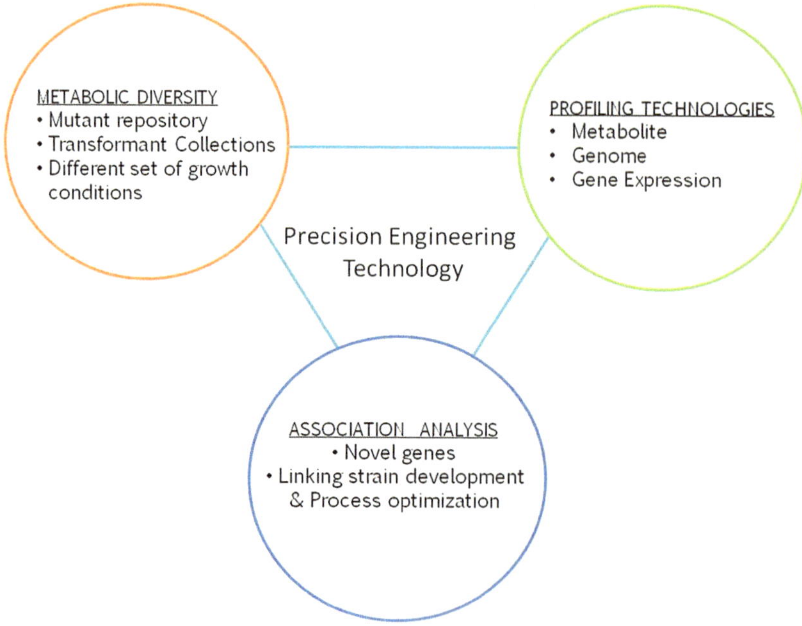

Fig. 10.5 Components of precision engineering technology for microbial strain improvement

(+)-geodin being produced by these strains in the fermentation broths. Microbia in one case used a strain of *Aspergillus terreus* exhibiting increased resistance to lovastatin developed by classical mutagenesis. Subsequently, a hyperactivated regulator gene identified by associative analysis was introduced into the *A. terreus* strain, resulting in a titre improvement of approximately 3.5-fold. This strain of *A. terreus* was further subjected to metabolic engineering, followed by mutation, and screening yielded a strain in less than 6 months which produced nearly 100-fold more lovastatin than the starting strain (*A. terreus* MF-753). These results emphasise the importance of partnering hypothesis driven (molecular biology) with open-ended (mutagenesis) approaches (Askenazi et al. 2003).

In another project, Microbia developed a strain with very high productivity but unacceptable levels of specific impurities that track the lovastatin in downstream processing methods (*A. terreus* MF-874). Thus, there was a need to improve the expression of an enzyme or gene of the existing strain to remove the chemical impurities. Thus, by this method, a new strain of *A. terreus* (MF-906) was developed which had dramatically reduced level of impurities but also had enhanced lovastatin production (Bailey 2005).

Precision engineering still requires further development before its full potential can be fully realised. Its use is limited largely by the global knowledge of the cell, the translation of the functions and characteristics of the biological systems. This technology requires further integration of different high-throughput technology for rationally designing the microbes for efficient production of commercially valuable primary as well as secondary metabolites. Thus, precision engineering is a good framework of old and new methodologies being exploited in a conclusive manner which is going to have a tremendous impact on industrial biotechnology.

10.8 Summary

This chapter comprehensively highlights the need of microbial strain improvement in industry using traditional as well as most modern methods involving spontaneous mutations, selection and use of mutagens, recombinant DNA technologies and genome shuffling to enhance the biosynthetic abilities of microorganisms for their exploitation at industrial scale. The most modern method which has been developed is a logical association between traditional mutagenesis and modern recombinant methods known as the precision engineering technology platform developed by Microbia which helps correlates the gene expression with the desirable product to identify nonobvious genes which could be used to enhance the biosynthetic capabilities. The precision engineering technology developed by Microbia has a few success stories in the Industries like Ranbaxy Laboratories (Now Sankyo Daichii), Teva Pharmaceuticals, Novus International, Tate & Lyle and DuPont Central Research & Development. Microbia now has been acquired by DSM, Netherlands.

Selected Reading

Adelberg EA (1958) Selection of bacterial mutants which excrete antagonists of anti-metabolites. J Bacteriol 76:326

Askenazi M, Driggers EM, Holtzmann DA, Norman TC, Iverson S, Zimmer DP, Boers ME, Blomquist PR, Martinez EJ, Monreal AW, Fiebelman TP, Mayorga MM, Maxon ME, Tobin SK, Cordero E, Salama SR, Trueheart J, Royer JC, Madden KT (2003) Integrating transcriptional and metabolite profiles to direct the engineering of lovastatin producing fungal strains. Nat Biotechnol 21:150–156

Baba S, Abe Y, Suzuki T, Ono C, Iwamoto K, Nihira T, Hosobuchi M (2009) Improvement of compactin (ML-236B) production by genetic engineering in compactin high-producing *Penicillium citrinum*. Appl Microbiol Biotechnol 83:697–704

Bailey RB (2005) Rewiring cellular systems to enhance biomanufacturing. Spec Chem Mag, July/August. www.specchemonline.com

Blumauerova M, Pokorny V, Stastna J, Hostalek Z, Vanek Z (1978) Developmental mutants of *Streptomyces coeruleorubidis*, a producer of anthracyclines: isolation and preliminary characterization. Folia Microbiol 23:177–182

Chandran SS, Yi J, Draths KM, von Daeniken R, Weber W, Frost JW (2003) Phosphoenol- pyruvate availability and the biosynthesis of shikimic acid. Biotechnol Prog 19:808–814

Chang LT, McGrory EL, Elander RP (1990) Penicillin production by glucose derepressed mutants of *Penicillium chrysogenum*. J Ind Microbiol 6:165–169

Chen T, Wang JY, Zhou SQ, Chen X, Ban R, Zhao XM (2004) Trait improvement of riboflavin producing *Bacillus subtilis* by genome shuffling and metabolic flux analysis. J Chem Ind Eng 55:1842–1848

Chumpolkulwong N, Kakizono T, Nagai S, Nishio N (1997) Increased astaxanthin production by *Phaffia rhodozyma* mutants isolated as resistant to diphenylamine. J Ferment Bioeng 83:429–434

Clutterbuck AJ (1969) A mutational analysis of conidial development in *Aspergillus nidulans*. Genetics 63:317–327

Dai MH, Copley SD (2004) Improved production of ethanol by novel genome shuffling in *Saccharomyces cerevisiae*. Appl Environ Microbiol 70:2391–2397

DeJong JM, Liu Y, Bollon AP, Long RM, Jennewein S, Williams D, Croteau RB (2006) Genetic engineering of taxol biosynthetic genes in *Saccharomyces cerevisiae*. Biotechnol Bioeng 93:212–224

Eggeling L, Oberle S, Sahm S (1998) Improved L- Lysine yield with *Corynebacterium glutamicum*: use of *dapA* resulting in increased flux combined with growth limitation. Appl Microbiol Biotechnol 49:24–30

Elander RP (1995) Genetic engineering applications in the development of selected industrial enzymes and therapeutic proteins. In: Sankaran R, Manja KS (eds) Microbes for better living. Defense Food Research Laboratory, Mysore, pp 619–628

Elander RP (1999) Two decades of strain development in antibiotic producing microorganisms. J Ind Microbiol Biotechnol 22:241–253

Elander RP, Espenshade MA (1976) The role of microbial genetics. In: Miller BM, Litsky W (eds) Industrial microbiology. McGraw-Hill, New York, pp 192–256

Ferron MAV, Lopez JLC, Perez JAS, Sevilla JMF, Chisti Y (2005) Rapid screening of Aspergillus *terreus* mutants for overproduction of lovastatin. World J Microbiol Biotechnol 21:123–125

Finogenova TV, Morgunov IG, Kamzolova SV, Chernyavskaya OG (2005) Organic acid production by the yeast *Yarrowia lipolytica*: a review of prospects. Appl Biochem Microbiol 41:418–425

Godfrey OW (1973) Isolation of regulatory mutants of the aspartic and pyruvic acid families and their effect on antibiotic production in *Streptomyces lipmanii*. Antimicrob Agents Chemother 4(2):73–79

Gomi S, Ikeda D, Nakamura H, Naganawa H, Yamashita F, Hotta K, Kondo S, Okami Y, Umezawa H, Iitaka Y (1984) Isolation and structure of a new antibiotic, indolizomycin, produced by a strain SK2-52 obtained by interspecies fusion treatment. J Antibiot 37:1491–1494

Gong J, Zheng H, Wu Z, Chen T, Zhao X (2009) Genome shuffling: progress and applications for phenotype improvement. Biotechnol Adv 27:996–1005

Gouka RJ, Punt PJ, van den Hondel CAMJJ (1997) Efficient production of secreted proteins by *Aspergillus*: progress, limitations and prospects. Appl Microbiol Biotechnol 47:1–11

Hopwood DA (1978) Extrachromosomally determined antibiotic production. Annu Rev Microbiol 32:373–392

John RP, Gangadharan D, Madhavan Nampoothiri K (2008) Genome shuffling of *Lactobacillus delbrueckii* mutant and *Bacillus amyloliquefaciens* through protoplasmic fusion for L-lactic acid production from starchy wastes. Bioresour Technol 99(17):8008–8015

Jung W, Lee S, Hong J, Park S, Jeong S, Han A, Sohn J, Kim B, Choi C, Sherman D, Yoon Y (2006) Heterologous expression of tylosin polyketide synthase and production of a hybrid bioactive macrolide in *Streptomyces venezuelae*. Appl Microbiol Biotechnol 72:763–769

Kardos N, Demain AL (2011) Penicillin: the medicine with the greatest impact on therapeutic outcomes. Appl Microbiol Biotechnol 92:677–687

Kruse D, Krämer R, Eggeling L, Rieping M, Pfefferle W, Tchieu JH, Chung YJ, Saier MH Jr, Burkovski A (2002) Influence of threonine exporters on threonine production in *Escherichia coli*. Appl Microbiol Biotechnol 59:205–210

Kubota K, Onoda T, Kamijo H, Yoshinaga F, Okumura S (1973) Production of L-arginine by mutants of glutamic acid-producing bacteria. J Gen Appl Microbiol 19:339–352

Kumar P, Satyanarayana T (2009) Overproduction of glucoamylase by a deregulated mutant of a thermophilic mould *Thermomucor indicae-seudaticae*. Appl Biochem Biotechnol 158:113–125

Laffend LA, Nagarajan V, Nakamura CE (1996) Bioconversion of a fermentable carbon source to 1, 3-propanediol by a single microorganism. Patent WO 96/53.796 (E. I. DuPont de Nemours and Genencor International)

Lale G, Jogdand VV, Gadre RV (2006) Morphological mutants of *Gibberella fujikuroi* for enhanced production of gibberellic acid. J Appl Microbiol 100:65–72

Lee HY, Khosla C (2007) Bioassay-guided evolution of glycosylated macrolide antibiotics in *Escherichia coli*. PLoS Biol 5:e45

Lee J-C, Park H-R, Park D-J, Son KH, Yoon K-H, Kim Y-B, Kim C-J (2003) Production of teicoplanin by a mutant of *Actinoplanes teicomyceticus*. Biotechnol Lett 25:537–540

Levy-Schil S, Debussche L, Rigault S, Soubrier F, Bacchette F, Lagneaux D, Schleuniger J, Blanche F, Crouzet J, Mayaux JF (1993) Biotin biosynthetic pathway in a recombinant strain of *Escherichia coli* over expressing bio genes: evidence for a limiting step upstream from KAPA. Appl Microbiol Biotechnol 38:755–762

Lin J, Shi BH, Shi QQ, He YX, Wang MZ (2007) Rapid improvement in lipase production of *Penicillium expansum* by genome shuffling. Chin J Biotechnol 23(4):672–676

Lindahl A-L, Olsson ME, Mercke P, Tollbom O, Schelin J, Brodelius M, Brodelius PE (2006) Production of the artemisinin precursor amorpha-4,11-diene by engineered *Saccharomyces cerevisiae*. Biotechnol Lett 28:571–580

Liu LM, Li Y, Zhu Y, Du GC, Chen J (2007) Redistribution of carbon flux in *Torulopsis glabrata* by altering vitamin and calcium level. Metab Eng 9(1):21–29

Martín JF, Naharro G, Liras P, Villanueva JR (1979) Isolation of mutants deregulated in phosphate control of candicidin biosynthesis. J Antibiot 32(6):600–606

Morgunov IG, Kamzolova SV, Perevoznikova OA, Shishkanova NV, Finogenova TV (2004) Pyruvic acid production by a thiamine auxotroph of *Yarrowia lipolytica*. Process Biochem 39:1469–1474

Nakayama K (1985) Lysine. In: Moo-Young M, Blanch HW, Drews G, Wang DIC (eds) Comprehensive biotechnology, vol 3. Pergamon Press, Oxford, pp 607–620

Nijland JG, Kovalchuk A, van den Berg MA, Bovenberg RAL, Driessen AJM (2008) Expression of the transporter encoded by the cefT gene of *Acremonium chrysogenum* increases cephalosporin production in *Penicillium chrysogenum*. Fungal Genet Biol 45:1415–1421

Nosaka K, Onozuka M, Konno H, Akaji K (2008) Thiamin-dependent transactivation activity of PDC2 in *Saccharomyces cerevisiae*. FEBS Lett 582:3991–3996

Otte B, Grunwaldt E, Mahmoud O, Jennewein S (2009) Genome shuffling in *Clostridium diolis* DSM 15410 for improved 1,3-propanediol production. Appl Environ Microbiol 75:7610–7616

Palva I (1982) Molecular cloning of a-amylase gene from *Bacillus amyloliquefaciens* and its expression in *Bacillus subtilis*. Gene 19:81–87

Prabavathy VR, Mathivanan N, Sagadevan E, Murugesan K, Lalithakumari D (2006) Intrastrain protoplast fusion enhances carboxymethyl cellulase activity in *Trichoderma reesei*. Enzym Microb Technol 38:719–723

Ranadive P, Mehta A, George G (2011) Strain improvement of *Sporidiobolus johnsonii* –ATCC 20490 for biotechnological production of coenzyme Q10. Int J Chem Eng Appl 2(3):216–220

Ro DK, Paradise EM, Ouellet M, Fisher KJ, Newman KL, Ndungu JM, Ho KA, Eachus RA, Ham TS, Kirby J, Chang MC, Withers ST, Shiba Y, Sarpong R, Keasling JD (2006) Production of the antimalarial drug precursor artemisinic acid in engineered yeast. Nature 440:940–943

Shi DJ, Wang CL, Wang KM (2009) Genome shuffling to improve thermotolerance, ethanol tolerance and ethanol productivity of *Saccharomyces cerevisiae*. J Ind Microbiol Biotechnol 36:139–147

Shibasaki T, Mori H, Chiba S, Ozaki A (1999) Microbial proline 4-hydroxylase screening and gene cloning. Appl Environ Microbiol 65(9):4028–4031

Shiio I, Yoshino H, Sugimoto S (1990) Isolation and properties of lysine-producing mutants with feedback-resistant aspartokinase derived from a *Brevibacterium flavum* strain with citrate synthase and pyruvate kinase defects and feedback resistant phosphoenol pyruvate carboxylase. Agric Biol Chem 1990(54):3275–3282

Sugimoto T, Morimoto A, Nariyama M, Kato T, Park EY (2010) Isolation of oxalate resistant *Ashbya gossypii* strain and its improved riboflavin production. J Ind Microbiol Biotechnol 37:57–64

Tong IT, Liao HH, Cameron DC (1991) 1, 3-propanediol production by *Escherichia coli* expressing genes from the *Klebsiella pneumoniae* dha regulon. Appl Environ Microbiol 57(12):3541–3546

Wang Y, Li Y, Pei X, Yu L, Feng Y (2007) Genome shuffling improved acid tolerance and L-lactic acid volumetric productivity in *Lactobacillus rhamnosus*. J Biotechnol 129(3):510–515

Wang T, Jia S, Tan Z, Dai Y, Song S, Wang G (2012) Mutagenesis and selective breeding of a high producing ε-poly-L-lysine strain. Front Chem Sci Eng 6(2):179–183

Xie ZP, Xu ZN, Shen WH, Cen PL (2005) Bioassay of mildiomycin and a rapid, cost-effective agar plug method for screening high-yielding mutants of mildiomycin. World J Microbiol Biotechnol 21:1433–1437

Xu B, Wang MR, Xia Y, Yang K, Zhang CY (2006) Improvement of the output of teicoplanin by genome shuffling. Chin J Antibiot 31:237–242

Xu B, Jin Z, Wang H, Jin Q, Jin X, Cen P (2008) Evaluation of *Streptomyces pristinaespiralis* for resistance and production of pristinamycin by genome shuffling. Appl Microbiol Biotechnol 80:261–267

Yoon SH, Kim JE, Lee SH, Park HM, Choi MS, Kim JY, Lee SH, Shin YC, Keasling JD, Kim SW (2007) Engineering the lycopene synthetic pathway in *E. coli* by comparison of the carotenoid genes of *Pantoea agglomerans* and *Pantoea ananatis*. Appl Microbiol Biotechnol 74:131–139

Zambare V (2010) Strain improvement of alkaline protease from *Trichoderma reesei* MTCC-3929 by physical and chemical mutagen. IIOAB J 1(1):25–28

Zhang YX, Perry K, Vinci VA, Powell K, Stemmer WPC, del Cardayre SB (2002) Genome shuffling leads to rapid phenotypic improvement in bacteria. Nature 415:644–646

Zhang DD, Liang N, Shi ZP, Liu LM, Chen J, Du GC (2009) Enhancement of alpha-ketoglutarate production in *Torulopsis glabrata*: redistribution of carbon flux from pyruvate to alphaketoglutarate. Biotechnol Bioproc Eng 14(2):134–139

Zhang Y, Liu JZ, Huang JS, Mao ZW (2010) Genome shuffling of *Propionibacterium shermanii* for improving vitamin B12 production and comparative proteome analysis. J Biotechnol 148(2–3):139–143

Zheng ZB, Zhao XM (2008) Astaxanthin-producing strain breeding by genome shuffling. J Biotechnol 136S:S310–S311

Zhu H, Jin ZH, Cen PL (2006) Natamycin-producing strain breeding by genome shuffling. Chin J Antibiot 31(12):739–742

Vaccines and Their Production 11

Bacterial and Viral Vaccines and Their Development

11.1 Introduction

One of the greatest contributions of modern medicine for humanity is the development of vaccine which is a cost-effective and powerful prophylactic measure to protect against deadly diseases. The discovery of vaccine and immunisation began with Edward Jenner, an English practitioner living in Berkley, England, who performed the world's first vaccination in 1796. Edward Jenner inoculated an 8-year-old boy, James Phipps, with a cowpox lesion on a milkmaid's hand and later with smallpox, and the boy was unaffected by this and subsequent exposures. Louis Pasteur in 1885 developed a rabies vaccine which basically comprised of rabies antitoxin that functioned as a postinfection antidote. Subsequently, a variety of vaccines were developed and used for the prevention of disease against a variety of bacterial, viral and parasitic organisms.

Today, vaccine is defined as a suspension of live (usually attenuated) or inactivated microorganisms, e.g. bacteria or viruses or fractions thereof, administered to induce immunity and prevent infectious diseases or its sequelae. Vaccines today can be classified as traditional and modern vaccines (Hilleman 2000). Traditional vaccines generally are referred to as those which have been developed before the advent of recombinant DNA technology.

11.2 Traditional Vaccines

The traditional vaccines have been developed against viral and bacterial pathogens and to a lesser extent parasitic diseases. However, with the advent of recombinant DNA technology and greater understanding of the molecular mechanisms of different diseases, vaccines are also being developed to treat/prevent cancer and autoimmune conditions. Traditional vaccines can be classified into four broad groups: (a) live attenuated bacteria/virus, (b) dead or inactivated bacteria/virus, (c) toxoids and (d) pathogen-derived antigens.

11.2.1 Live Attenuated Vaccines

Live attenuated vaccines generally comprise of attenuated or weakened viral and bacterial strains. Attenuation is the process of weakening or reducing the virulence of a microorganism by using empirical procedures like prolonged storage and cultivation under suboptimal conditions (passaging). An example of live attenuated vaccines is represented by Bacillus Calmette–Guerin (BCG) which basically is a strain of tubercle bacillus (*Mycobacterium bovis*) that fails to cause tuberculosis but retains the antigenicity of the pathogen. Viruses are also attenuated in appropriate animal cell culture systems for many

S. Saxena, *Applied Microbiology*,
DOI 10.1007/978-81-322-2259-0_11, © Springer India 2015

Table 11.1 Cell culture systems used for passaging viral particles for use as vaccines

Cell culture system	Viral particle/vaccine
Chick egg embryos	Yellow fever virus
Chick egg embryo cells	Measles virus, mumps virus
Monkey kidney tissue culture	Polio (Sabin's vaccine)
Human diploid fibroblasts	Hepatitis A viral vaccine

generations under a stressful environment. The viral attenuation apart from animal cell culture systems is also carried out in fertilised eggs or cultures of chick embryo tissue (Table 11.1). Live attenuated viruses have also been used for the manufacture of vaccines for measles, mumps and yellow fever. The Jeryl Lynn strain of mumps vaccine is propagated in chick embryo cell culture.

The advantages of live attenuated vaccines are their low cost of preparation and their ability to elicit the desired immunological response in a single-dose application. The disadvantages include the potential to revert and become virulent and a limited shelf life.

11.2.2 Dead, Inactivated Vaccines

The treatment of pathogenic microorganisms, viz. bacteria or virus, with chemicals or high temperature or radiation renders them inactivated or kills them, thereby making them suitable as vaccines. The inactivated or killed microorganism should retain the immunological properties as present in the active or live pathogen. The process of killing is a very crucial step as it should be 100 % effective to prevent accidental transmission of the live pathogen. The chemicals which are used for inactivation or killing of the pathogenic microorganisms are formaldehyde, phenol and acetone.

The advantages of this method are that there are no chances of reversal into a virulent strain and a relatively stable shelf life. However, there are some disadvantages of this method like (1) the higher cost of vaccine development since one has to ensure that the process used for killing or inactivation is reliable as well as 100 % efficient and (2) the possibility of reduced immune response due to treatment processes; hence, multiple booster vaccinations would be required.

Cholera vaccine is made up of sterile liquid suspensions of killed *Vibrio cholerae* selected for high antigenic efficiency. The dead strains used for the preparation of cholera vaccine are Inaba and Ogawa. A typical 1-ml dose contains less than eight billion *V. cholerae* particles and phenol (up to 0.5 %) which is generally added as a preservative. The liquid vaccine has a shelf life of 18 months, while in dried form the shelf life is 5 years. Similarly, hepatitis A vaccine is made up of a formaldehyde-inactivated preparation of the HM175 strain of hepatitis A. The other vaccines which are prepared using this method are pertussis, typhoid and influenza vaccines.

11.2.3 Toxoids

Some pathogenic bacteria produce toxins which play an important role in pathogenesis. A toxoid is derived from the treatment of active toxin produced by the bacterium with formaldehyde. The product is generally sold as a sterile aqueous preparation. Two commonly toxoid-based vaccine preparations are tetanus and diphtheria vaccines. Tetanus toxin is produced by *Clostridium tetani* which is cultured on a suitable medium; the toxin is recovered and then inactivated by treatment with formaldehyde. It is marketed as tetanus toxoid, a sterile aqueous product. *Corynebacterium diphtheriae* produces the diphtheria toxin which is also used as a toxoid for the treatment of whooping cough. The advantages and disadvantages of toxoids are similar to those for killed or inactivated vaccines.

11.2.4 Pathogen-Derived Antigens

The pathogen-derived antigens constitute the traditional antigenic vaccine preparations and generally comprise of antigenic portions of the pathogens which are generally surface derived, most commonly surface polysaccharides. It has been generally observed that carbohydrate-based substances of pathogenic bacteria are less

Table 11.2 Representative examples of traditional vaccine preparations which are being used clinically

Product	Vaccine class	Description	Application
BCG (Bacillus Calmette–Guerin) vaccine	Live attenuated vaccine	Attenuated strain of *Mycobacterium tuberculosis*	Immunisation against tuberculosis
Cytomegalovirus vaccine	Live attenuated vaccine	Attenuated strain of human cytomegalovirus	Immunisation against cytomegalovirus
Measles vaccine	Live attenuated vaccine	Attenuated strain of measles virus	Immunisation against measles
Poliomyelitis vaccine	Live attenuated vaccine	Attenuated strains of poliomyelitis virus	Immunisation against polio
Rabies vaccine	Inactivated vaccine	Inactivated rabies vaccine	Immunisation against rabies
Japanese encephalitis vaccine	Inactivated vaccine	Inactivated Japanese encephalitis virus	Immunisation against Japanese encephalitis
Pertussis vaccine	Killed vaccine	Killed strain of *Bordetella pertussis*	Immunisation against whooping cough
Hepatitis A vaccine	Inactivated hepatitis A virus	Formaldehyde-treated hepatitis A virus	Prevention against hepatitis A
Diphtheria vaccine	Toxoid	Treatment of diphtheria toxin by formaldehyde	Prevention of diphtheria by immunisation
Tetanus vaccine	Toxoid	*Clostridium tetani* toxin treated with formaldehyde	Prevention of tetanus by immunisation
Pneumococcal vaccine	Pathogen-derived antigen	Mixture of purified surface polysaccharide antigens obtained from different serotypes of *Streptococcus pneumoniae*	Immunisation against infections caused by *Streptococcus pneumoniae*
Hepatitis B vaccine	Pathogen-derived antigen	Suspension of surface antigen of hepatitis B (HBsAg)	Immunisation against hepatitis B
Meningococcal vaccine	Pathogen-derived antigen	Purified surface polysaccharide antigens of one or more strains of *Neisseria meningitidis*	Immunisation against infections caused by *Neisseria meningitidis*

immunogenic to their protein counterparts. The antigenicity of these substances can be improved by conjugating them with a protein-based antigen. *Haemophilus influenzae* capsular polysaccharide has been conjugated with diphtheria toxoid or the outer membrane protein of *Neisseria meningitides*. Similarly, for anthrax vaccine, antigen found in the sterile filtrate of *Bacillus anthracis* has been used. For the development of meningococcal vaccines, purified surface polysaccharide from *Neisseria meningitides* groups A or C has been used. Approximately 30 traditional vaccines are currently being used clinically (Table 11.2).

11.3 Modern Vaccines

The advances in molecular biology, genomics and recombinant DNA technology have tremendously propelled the discovery, development and manufacture of vaccines with least chances of virulence. Today, vaccination technology primarily targets decoupling the virulence and immunity functions so that the vaccine is safely administered. The modern vaccines can be classified as subunit, DNA-based recombinant and peptide vaccines (Jackwood et al. 2008).

11.3.1 Subunit Vaccines

Instead of using the entire bacterium or virus for vaccination, subunit vaccines comprise of a part of a bacteria or virus which can best stimulate the immune systems without rendering the person susceptible to the disease. As subunit vaccine contains essential antigens and not all molecules that make up the microbes, chances of adverse reactions are therefore very low. The subunits generally comprise of envelope proteins and antigenic surface proteins known as epitopes. There are two methods of developing a subunit vaccine, viz. (a) to grow the bacteria in the laboratory and then break apart to recover and purify the desired antigens and (b) by using recombinant DNA technology clone and express the antigenic moiety in a non-pathogen, generally referred to as recombinant subunit vaccine.

More recently, the recombinant subunit vaccine production method has become a preferred method since in this process the pathogen-derived polypeptide (antigen) is expressed in a non-recombinant host, removing all possibilities of the presence of any undetected pathogen in the end product. Secondly, since the process can be scaled up in a fermenter, there is a consistent unlimited supply.

The development of recombinant subunit vaccine requires the knowledge of the genome sequence of the pathogen by identification of the open reading frames (ORFs) that potentially encode for the antigenic surface proteins. Subsequent to identification, these ORFs are cloned to express the epitopes using self-replicating plasmids. ELISA and FACS are the two common methods to study the binding properties of these epitopes. After laboratory testing, the leading candidates of epitopes are tested in experimental animals so as to confirm their propensity to generate an immunological response, i.e. proliferation of antibodies. Those which generate a significant response are selected and optimised to become vaccine candidates with further tests prior to human clinical trials. The first recombinant subunit vaccine was HBsAg (hepatitis B surface antigen) which was approved by the US FDA in 1986. Previously, this antigen was being directly purified from the blood of people suffering with hepatitis B.

The HBsAg gene has been cloned and expressed in bacterial (*E. coli*), yeast (*S. cerevisiae*) and a number of mammalian cell lines. The yeast expression system has been found to produce the polypeptide as well as assemble it into particles which are found in the blood of hepatitis B-infected individuals. The common brands of hepatitis vaccine available are Recombivax HB (Merck), Genevac B (Serum Institute, India), Engerix-B, GSK (GlaxoSmithKline Beecham). More recently, Gardasil, a tetravalent subunit recombinant vaccine, has been developed against human papillomavirus (HPV) which is implicated in cervical cancer. Some recombinant subunit vaccines which are being used for humans are listed in Table 11.3.

11.3.1.1 Virus-Like Particles (VLPs)

Virus-like particles are basically viral structural proteins that are expressed in cells and possess native conformational epitope. However, they are a class of subunit vaccines as they do not contain the genome and therefore cannot spread infection. They have the capacity to spread both cellular and humoral responses. Both Recombivax and Engerix-B were first to use virus-like particles.

11.3.2 Conjugate Vaccines

Conjugate vaccines are a special type of subunit vaccine. These generally use polysaccharide antigens which are large molecules present in the cell wall of the pathogenic bacteria. These are not processed by antigen-presenting cells (APCs) but interact directly with B cells, inducing antibody synthesis in the absence of T cells. Despite being immunogenic, the polysaccharide antigens have a limited antibody response and are not able to induce immunologic memory. Goebel and Avery (1929) first observed the need to improve the immunogenicity of the polysaccharide antigens. They enhanced the immunogenicity of purified *S. pneumoniae* type 3 polysaccharide in rabbits by conjugating it with a protein carrier. This formed the basis of the modern development of conjugate

Table 11.3 Clinically used recombinant subunit vaccines

Product	Description	Company	Application
Recombivax	rHBsAg produced in *Saccharomyces cerevisiae*	Merck	Prevention of hepatitis B
Engerix-B	rHBsAg produced in *Saccharomyces cerevisiae*	GSK	Prevention of hepatitis B
Hepacare	rS, pre-S and pre-S2 hepatitis B surface antigens produced in murine cell line	Medeva Pharma	Hepatitis B immunisation
Twinrix	Recombinant protein of hepatitis B + hepatitis A killed virus	GSK	Hepatitis immunisation
HBVAXPRO	rHBsAg produced in *Saccharomyces cerevisiae*	Aventis Pharma	Hepatitis B immunisation
Gardasil	Recombinant VLPs assembled from the L1 proteins of HPV types 6, 11, 16 and 18	Merck	Immunisation against human papillomavirus

vaccines. The first glycoconjugate vaccine for human use was *Haemophilus influenzae* type b (Hib) conjugate, which was licensed in the USA in 1987 and shortly thereafter was introduced into the US infant immunisation schedule. The Vi polysaccharide of *Salmonella typhi* was improved by conjugation with tetanus or diphtheria toxoids, cholera toxin B subunit or recombinant mutant *Pseudomonas aeruginosa* exoprotein A (rEPA) for the development of an improved typhoid conjugate vaccine.

Prevnar13® is a pneumococcal vaccine by Pfizer that contains polysaccharides of 13 bacterial serotypes conjugated to diphtheria CRM197 carrier protein, a non-toxic variant of diphtheria toxin.

11.3.3 Recombinant Vaccines

The recombinant vaccines basically comprise of live genetically modified organisms, recombinant subunit vaccines and genetic vaccines. Subunit vaccines and recombinant subunit vaccines have been already discussed in the previous section.

11.3.3.1 Live Genetically Modified Organisms

The vaccines generally comprise of bacteria or viruses with one or more deleted or inactivated genes or else could be vaccines carrying a gene from other disease agents which are referred to as vaccine vectors. Generally, two (double knock-out) or more genes are deleted or inactivated so that the vaccine remains stable and does not revert back into a pathogenic state. This type of vaccine development requires knowledge about the gene(s) responsible for the pathogenicity of the organism and assumes that those genes are not the same genes governing viability and the ability of the modified organism to induce an immune response.

Gene-deleted vaccines have been developed for *Salmonella* infections in poultry and pseudorabies virus vaccines for pigs. Another strategy is the creation of an infectious clone of a disease agent. The infectious clone is created by isolating the entire genome of the disease agent (usually viruses) in the laboratory. This isolated or cloned genome is purposefully and specifically modified in the laboratory and then used to re-create the live genetically modified organism.

Vector-based vaccines are bacteria, viruses or plants carrying genes of another disease agent which is expressed and induces an immune response when the host is vaccinated. VectorVax FP-N (Zeon Corporation, Japan) is the first commercial vaccine vector for turkeys which consists of fowl pox vaccine virus that carries genes from Newcastle disease virus. Edible vaccines comprise of delivering the antigens of some diseases to induce immune response when delivered orally. Foreign genes have been inserted in potatoes, soybeans and corn plants and fed to the animals in order to immunise them against diseases.

11.3.3.2 Genetic or DNA Vaccines

As the name indicates, the recombinant vaccines only comprise of the DNA. These usually consist of circular pieces of DNA called the plasmids which contain a foreign gene from the disease-causing agent and a promoter which is helpful in expressing that gene as a protein in the target animal, thereby generating an immune response. The recombinant plasmids containing a foreign gene are purified from the bacteria, and the 'naked' DNA is injected directly, usually intramuscularly or intradermally (into the skin). Bird flu DNA vaccine was approved in June 2006, and subsequently a veterinary vaccine for horses to prevent the West Nile virus was approved in 2007. The use of DNA vaccines for humans is in various stages of development for targets such as HIV, cancer and multiple sclerosis (Stuve et al. 2007).

11.3.4 Peptide Vaccines

These are chemically synthesised peptides comprising of 8–24 amino acids. As compared to proteins, these are relatively small and have been referred to as peptidomimetic vaccines as they mimic the epitopes. Complex structures of cyclic components, branched chains or other configurations can be built into the peptide chain, and thus with these kind of conformations, they mimic the epitopes. An in silico vaccine design approach has been used to find potential epitopes. A critical aspect of peptide vaccines is to produce 3D structures similar to the native epitopes of the pathogen.

11.4 Summary

Vaccines are generally designed to prevent illness and are generally given to people who are not sick but are at risk of getting ill. Today, with improved technology and research methods, the time from basic research to a licensed vaccine has drastically been reduced, and more safe products are being designed to prevent humans from contracting different diseases (Josefsberg and Buckland 2012).

Selected Reading

Hilleman MR (2000) Vaccines in historic evolution and perspective: a narrative of vaccine discoveries. Vaccines 18:1436–1447

Goebel WF, Avery OT (1929) Chemo-immunological studies on conjugated carbohydrate-proteins. II Immunological specificity of synthetic sugar-proteins. J Exp Med 50:521–533

Josefsberg JO, Buckland B (2012) Vaccine process technology. Biotechnol Bioeng 109:1443–1460

Understanding vaccines: what they are? How they work? U.S. Department of Health and Human Services, National Institutes of Health, USA

Jackwood MW, Hickle L, Kapil S, Silva R (2008) Vaccine development using recombinant DNA technology, CAST Issue paper no. 38, May 2008

Stuve O, Eagar TN, Frohman EM, Cravens PD (2007) DNA plasmid vaccination for multiple sclerosis. Arch Neurol 64(10):1385–1386

12 Immobilisation and Biosensors

12.1 Introduction

Immobilisation is defined as the technique of fixing the cells, organelles or enzymes/other proteins (monoclonal antibodies) onto a solid support system, into a solid support matrix or retained by a membrane, in order to maintain stability and make possible their repeated or continued use. The immobilised cell technologies comprise of modifications of the technique developed for enzymes. However the microbial size has a significant impact on these techniques. The immobilisation of microbial cells occurs as a natural phenomenon or through artificial process. The artificially immobilised cells are allowed restricted growth.

The early attempts of whole cell immobilisation were developed from processes applied to the immobilisation of enzymes and generally involved non-viable cells, i.e. cells impaired by physical or chemical treatment to perform single step enzyme reactions. The obvious benefit derived from using whole cell is to avoid enzyme extraction/purification steps which consequently have an effect on enzyme activity, stability and cost. These techniques in due course were extended to viable cells as they were exploited in bioreactors and fermentation systems. The advantages of viable immobilised culture systems are manifold. High cell densities are expected as viable immobilised cells multiply during the substrate metabolisation process while remaining confined with the immobilised matrix, and thus high volumetric reaction rates are expected in the immobilised cell culture; further regeneration of immobilised cultures is possible even under hostile incubation conditions like low nutrients or the presence of toxic compounds. In continuous processes efficient biomass retention is ensured minimising cell washout which normally occurs at high dilution rates by whole cell immobilisation. Immobilisation of whole cells also facilitated cell/liquid separation thereby easing the downstream processing in fermentation processes using immobilised whole cells. There are three basic methods which have been used for immobilisation of microorganisms, viz. (1) attachment to a support, i.e. carrier binding, (2) entrapment and (3) self-aggregation (Fig. 12.1). The overall composition of immobilised cell systems is less chemically defined as compared to the immobilised enzyme system. In this chapter the emphasis would be on various strategies which have been developed for immobilisation of the whole cells and their exploitation in different industrial settings.

12.2 Strategies of Whole Cell Immobilisation

12.2.1 Adsorption

This process relies on the tendencies of the cell to aggregate or adhere to particular surfaces or settle in the pores of the framework. This kind of

S. Saxena, *Applied Microbiology*,
DOI 10.1007/978-81-322-2259-0_12, © Springer India 2015

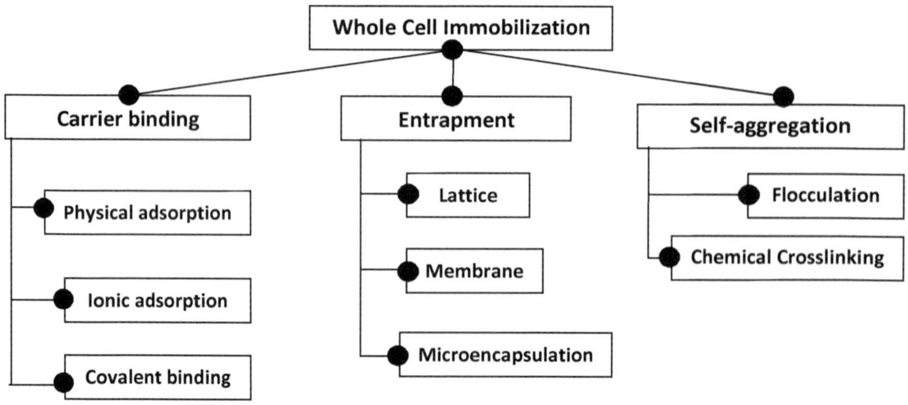

Fig. 12.1 Broad classes of whole cell immobilisation

Table 12.1 Immobilisation of whole cells by covalent bonding

Microbe	Covalent binding agent	Product
Aspergillus niger	Glycidylmethacrylate polymer + glutaraldehyde	Formation of gluconic acid from glucose
Proteus rettgeri	Carriers with epoxy and halocarbonyl groups	Conversion of penicillin G to 6-APA
Acetobacter sp.	Ti(IV) oxide	Production of vinegar
Burkholderia cepacia	Polyurethane foam	Trichloroethylene degradation
Saccharomyces cerevisiae	Cellulose + cyanuric chloride	Conversion of glucose to ethanol

cell immobilisation is usually achieved by keeping the support material and actively growing cells in physical contact for a specific duration. One of the classical examples is related to the process of vinegar production using woodchips as carriers of *Acetobacter* adsorption. A variety of substrates like zeolites, baked clay, glass beads, sponge rubber, cellulose acetate fibre and activated carbon fibre have been employed for the immobilisation of the whole cells' adhesion or adsorption.

At times pretreatment of cells either by starvation or washing of the cells or activation of the support material or cells may be beneficial for the improvement of the adsorption characteristics. Aluminium was used to neutralise the surface charge of *Saccharomyces cerevisiae*, and its absorption helped in its immobilisation on glass plates (Van Haecht et al. 1985). *Erwinia rhapontici* was immobilised on diethylaminoethyl (DEAE) cellulose by mixing 2 g of cells with 10 ml of thick DEAE slurry at pH 7 (Cheetham et al. 1985). A variety of support material has been used for the immobilisation of the whole cells for different applications (Table 12.1).

The adhesive strength of adsorption-based immobilisation is not very strong until there exists an exceptional mechanism in the microbe for the surface-anchored growth. The major advantages of adsorption are simplicity and negligible changes on the cell physiology; however, the drawbacks are limited cell loading and limited adhesion stability compared to cell entrapment.

12.2.2 Covalent Binding

Covalent binding is yet another way of attaching cells to the surface of the carrier. This method has been extensively exploited in enzyme immobilisation. It was realised that to achieve high efficiency binding, stability is to be enhanced, and this could be achieved by creating covalent linkages between the cell and the support surface. *Saccharomyces cerevisiae* has been immobilised on silanised silica beads using α-aminopropyl triethoxy silane as a coupling agent (Navarro and Durand 1977). *Pseudomonas stutzeri* has been immobilised on polyethylene surface by chlorosulphonic acid and chlorosulphonic acid combined with polyethyleneimine (Choi et al. 1999). Various coupling agents which have been used for covalent immobilisation of cells are aminosilane, carbodiimide and glutaraldehyde, which introduce specific groups on the carrier surface and subsequently can interact with reactive groups on the cell surface (Table 12.1).

12.2.3 Cell to Cell Cross-Linking

Flocculation is one of the simplest methods of achieving cell aggregation in the form of larger particles with high cell densities. However the capacity of microorganisms to flocculate naturally is limited. Hence chemical cross-linking is the most appropriate method to enhance flocculation and stabilise the cell aggregates. Commonly used cross-linking agents are glutaraldehyde, polyamines, polyethyleneimine, polystyrene sulphonates, polyvinyl alcohols, etc. Cross-linking reduces the chances of washout and improves the mechanical strength of the cell. Some common applications have been presented in Table 12.2.

12.2.4 Encapsulation

The encapsulation technique generally employs the use of polymeric beads to immobilise the whole cells. This is broadly divided as macroencapsulation and microencapsulation. Microspheres or microcapsules are usually spherical particles less than 1,000 μm in which liquid or suspension is enclosed by the dense but semipermeable polymeric film. The major limitation of this technique is the transport of nutrients across the membrane. Probiotic microbes like *Lactobacillus acidophilus*, *Lactobacillus casei* and *Bifidobacterium bifidum* have been microencapsulated in substances like gelatine, carrageenan, etc. (Kailasapathy 2002). The microencapsulation technique has applications in various fields like pharmaceuticals, agrochemicals, nutrition and therapeutics. The different methods used for the process of microencapsulation are extrusion, spray drying, emulsification and coacervation. The different methods adopted for immobilisation of microbial cells by microencapsulation are given in Table 12.3.

Microspheres are mechanically stronger then macrospheres and exhibit efficient diffusion of nutrients, oxygen and metabolites. Microencapsulation is an advantageous method in the fermentation industry since it not only carries out the process efficiently due to larger specific area for nutrient utilisation and metabolite

Table 12.2 Immobilisation of whole cells by cell to cell cross-linking

Microbe	Covalent binding agent	Product
Saccharomyces cerevisiae	1 % albumin + 0.25 % glutaraldehyde	Fructose-1,6 diphosphate production
Erwinia ariodea TSMPMV-2970	N′, N′- m- phenylene disasperimide (PDAI)	Production of L-aspartate-β-decarboxylase
Bacillus subtilis TSMPMV-259 M	N′, N′-m-phenylene disasperimide (PDAI)	Production of L-aspartate-β-decarboxylase
Aspergillus niger	Flocculation by polyelectrolytes	Production of gluconic acid from invert sugar
Lactobacillus brevis	Flocculation by chitosan	Production of fructose from glucose
Saccharomyces formanensis	Polymer of 2-hydroxyethcyrylate-and methoxypolyethylene glycerol methacrylate using γ-rays and tetraethylene glycerol dimethacrylate as cross-linking agent	Production of ethanol

Table 12.3 Exploitation of encapsulation as a method of whole cell immobilisation

Microbe	Covalent binding agent	Encapsulation/microencapsulation	Product
Pantoea agglomerans E25	Sodium alginate + calcium chloride	Modified precision particle fabrication	Prevention against harsh environmental conditions
Lactobacillus delbruckeii subsp. *bulgaricus NBRC13953*	Sodium alginate + calcium chloride	Emulsification	Prevention against harsh environmental conditions
Saccharomyces cerevisiae	Eudragit®	Extension	Production of ethanol from glucose
Pseudomonas sp. *SA01*	Sodium alginate + calcium chloride	Extrusion	Phenol degradation
Bifidobacterium lactis	Gellan/xanthan gum blend, calcium chloride	Extrusion	Protection against harsh gastrointestinal conditions
Pseudomonas sp.	Polyvinyl alcohol	Extrusion	Biodegradation of phthalic acid ester

production but also allows easy separation of the cells thereby minimising the cell washout. The technique also enhances the possibilities of the reuse of cells due to improved tolerance to undesirable processes like end product inhibition or contamination.

12.2.5 Entrapment

Physical entrapment of polymeric carriers of defined porosity is a widely used method for whole cell immobilisation. Different strategies of entrapment are gelation, precipitation, ionotropic gels and polycondensation. During the gelation process a temperature-controlled phase transition of the polymer solvent system is carried out wherein it is transformed into a homogenous system without change in the composition. Calcium alginate gels appear to be most compatible for immobilisation of living cells. Besides gelatine, agar and agarose have also been used for the process of gelation. The only concern with calcium alginate cells is the high affinity for calcium which destabilises the cell. Other matrices which have been employed are agar, alginate, k-carrageenan, cellulose and its derivatives, collagen, epoxy resin, polyacrylamide, polyester, polystyrene and polyurethane.

12.2.5.1 Precipitation of Polymers

Polymer solution is primarily used as a dispersion medium of cells. However coagulation of the polymer is achieved by changing the physicochemical parameters other than temperature, i.e. solvent and pH. The major drawback of this procedure is intensive contact of viable cells with non-physiological solvents thereby limiting its applications (Table 12.4).

12.2.5.2 Ionotropic Gelation of Polymers

Ionotropic gelation of polymers is based on the ability of a polyelectrolyte to cross-link in the presence of counterions to form hydrogels. Ionotropic gelation is generally used to have highly water-swollen structures with controlled morphology. A well-known example is calcium alginate cells. Chitosan is a polycation which also finds application in cell immobilisation. This process is generally very simple and non-toxic; various cells could be immobilised by complete preservation of viability (Table 12.5).

Table 12.4 Encapsulation by precipitation of polymers in whole cell immobilisation

Microbe	Encapsulation/microencapsulation	Product
Escherichia coli	Eudragit®	Conversion of L-serine and indole into tryptophan
Actinoplanes missouriensis	Cellulose	Conversion of glucose to fructose
Candida tropicalis	Polystyrene	Degradation of phenol

Table 12.5 Encapsulation ionotropic gelation of polymers

Microbe	Encapsulation/microencapsulation	Product
Aspergillus niger	Carrageenan–polyethylenimine	Production of citric acid
Candida tropicalis	Al copolystyrene maleic acid	Degradation of phenol
Saccharomyces cerevisiae	Chitosan alginate	Production of ethanol
Escherichia coli	Chitosan polyphosphate	Production of tryptophan from serine and indole
Lactobacillus rhamnosus	Modified alginate and chitosan	Safety of the probiotic culture in gastric environment

12.3 Alginate Method of Whole Cell Immobilisation

The use of the alginate method for whole cell immobilisation was first reported in 1975. Alginate is a heteropolymer carboxylic acid coupled with α 1→4 glycosidic bonds of β-d-mannuronic acid and α-L-guluronic acid. A broad range of particles can be prepared by using calcium alginate. The process of cell immobilisation in calcium alginate gels involves the mixing of viable cells like yeast. This process generally involves dissolving of 9 g of sodium alginate in 300 ml of growth medium stirring slowly to avoid formation of clumps (i.e. 3 % sodium alginate solution by weight). Subsequently 250 g of wet cells are suspended slowly in the sodium alginate solution by continuous stirring to achieve homogenisation.

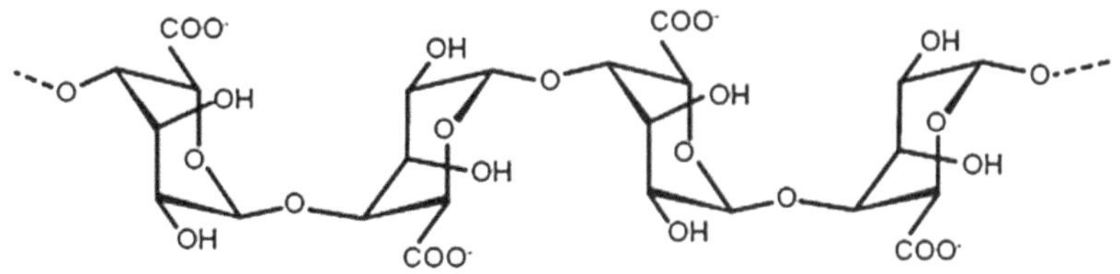

A cross-linking solution is prepared by addition of 0.005 M $CaCl_2$ in the growth medium which is chilled at 4 °C. Subsequently a yeast alginate mixture is dropped from a height of 20 cm into 1 L of cross-linking solution, using a syringe and needle. A diameter of 0.5–2 mm alginate beads can be prepared by this method and possess minimal mass transfer resistance. The beads are hardened within 1–2 h. These can then be stored between 10 and 20 °C after washing the beads with fresh cross-linking solvents. Some typical applications of alginates in whole cell immobilisation of microbes have been presented in Table 12.6.

Table 12.6 Use of alginate in immobilisation of whole cells

Microbe	Product	Alginate type
Chrysobacterium	Transformation of rifamycin B to rifamycin S	Sodium alginate
Saccharomyces cerevisiae	Ethanol production	Calcium alginate
Acetobacter sp. *CCTCC M209061*	Asymmetric reduction of ketones	
E. coli (recombinant with B. licheniformis genes)	L-ribulose production	Sodium alginate
Bacillus megaterium	Hydroxamic acid synthesis	Sodium alginate
Bacillus subtilis DJ6	Hyperthermostable extracellular β-amylase	Calcium alginate Barium alginate
Aspergillus awamori	Lactic acid	Calcium alginate
Gibberella fujikuroi	Gibberellic acid	Calcium alginate
Aspergillus sydowii	Xylanase	Calcium alginate

Table 12.7 Amperometric microbial biosensors

Microorganisms	Target	Limit of detection	Reference
T. cutaneum and *B. subtilis*	BOD	0.5 mg/L	Jia et al. (2003)
K. oxytoca AS1	BOD	<44 mg/L	Ohki et al. (1994)
P. putida SG10	BOD	1 mg/L	Chee et al. (2005)
G. oxydans or *P. methanolica*	Ethanol	0.05 mM	Reshetilov et al. (2001)
G. suboxydans	Ethanol	0–25 mg/L	Kitagawa et al. (1987)
S. ellipsoideus	Ethanol	69 μM	Rotariu and Bala (2003)
S. cerevisiae	Sucrose	6–100 mM	Rotariu et al. (2002)
E. coli K12	Mono-and/disaccharides	0–4 mM for disaccharides 0–2.5 mM for monosaccharide	Held et al. (2002)
R. erythropolis	2,4-Dinitrophenol	2–20 μM	Emelyanova and Reshetilov (2002)
T. ferrooxidans	Cyanide	0.5 μM	Okochi et al. (2004)
A. peroxydans	Hydrogen peroxide	0.1–9.5 μM	Rajasekar et al. (2000)
Recombinant *S. cerevisiae*	Cu2+	0.5–2 mM	Lehmann et al. (2000)

The immobilisation technology had a significant impact on microbial fermentation technology for improving the yield of fermentation-based products like organic acids, antibiotics, enzymes and alcohols as well as carrying out biotransformations (Table 12.7). An important area of research is bioreactor design and development using immobilised cells and long-term reactor operation. Further immobilisation technology has been extensively exploited in the development of biosensors exploiting whole cells as well as pure enzymes.

12.4 Microbes as Biosensors

Currently whole cell-based biosensors are being considered more advantageous to enzyme-based biosensors since they offer low cost and long-term stability. A biosensor is an analytical device which comprises of biological molecules as the recognition element with physical transducer and provides quantitative and semi-quantitative analytical data corresponding to the target concentration. Numerous biosensors with high sensitivity and accuracy have been developed by taking advantage of the high specificity and tight interaction between biomolecules and target compounds. Generally enzymes have been extensively exploited in developing biosensors, but the processes of isolation and purification of enzymes make them highly expensive which in turn enhance the cost of the sensor. Whole cells on the other hand appear to be good alternatives to enzymes as they are less expensive and have more stability. The basis of a microbial biosensor is the close contact between the microorganisms and the transducer which is generally based on immobilisation on the transducer. Hence immobilisation technology is very crucial and selection of appropriate immobilisation method is very critical for the development of a biosensor. Both physical and chemical methods of immobilisation are exploited for the development of a microbial biosensor. Based on transducers the microbial biosensors can be classified as electrochemical, optical and others.

12.4.1 Microbial Electrochemical Biosensors

Microbial electrochemical biosensors are of three types: amperometric, potentiometric and conductometric. Amperometric microbial biosensors work at a fixed potential with respect to the reference electrode and involves the detection of current generated by the oxidation or reduction of the species at the surface of the electrode. Amperometric microbial biosensors have been developed for the estimation of biochemical oxygen demand (BOD) for the measurement of biodegradable organic matter in aqueous samples. Some microorganisms which have been exploited for the development of amperometric biosensor are *Torulopsis candida*, *Klebsiella oxytoca* AS1, *Bacillus subtilis*, *Serratia marcescens*, *Hansenula anomala*, etc. (Lei et al. 2006). More than one microbial strain has been used for the development of biosensors to broaden the substrate and hence analyte spectrum with a stable performance.

The first commercial BOD biosensor was produced by Nisshin Denki (Electric) in 1983. Since then several BOD biosensors have been developed and commercialised. Some important manufacturers of these BOD biosensors are Autoteam FmbH, Germany; Dr. Lange GmbH & Co., Germany; Kelma, Belgium; Bioscience Inc., USA; and US Filter, USA. After BOD, amperometric biosensors have been developed for the detection of ethanol. The microorganisms which metabolise ethanol have been generally exploited for the fabrication of an ethanol biosensor by immobilising them on an oxygen electrode; however, the selectivity of this electrode is poor. The microorganisms which have been exploited for the development of ethanol include *Trichosporon brassicae*, *Gluconobacter suboxydans*, *Acetobacter aceti*, *Candida vini*, *Aspergillus niger*, *Saccharomyces ellipsoideus* and *Pichia methanolica*. Replacement of oxygen with a ferricyanide electron acceptor has been found to improve the selectivity of ethanol in *G. oxydans-immobilised* ethanol biosensors (Tkac et al. 2002). Amperometric microbial sensors have also been used for the detection of other analytes like sugars, phenols and substituted phenols. Some examples of amperometric microbial biosensors and their uses have been summarised in Table 12.7.

12.4.1.1 Microbial Potentiometric Biosensors

The microbial potentiometric biosensor is developed using a conventional ion-selective electrode (pH, ammonium chloride, etc.) or a gas sensing electrode (p_{CO2} and p_{NH3}) coated with an immobilised microbe layer. The microorganisms consume the analyte thereby bringing a change in the potential resulting from ion accumulation or depletion. The potentiometric transducers compare the change between the working electrode and the reference electrode, and the signal is correlated to the concentration of the analyte. The simplest potentiometric microbial sensor is based on an ion-selective electrode. Microbial sensors for the detection of the antibiotic penicillin

Table 12.8 Potentiometric microbial biosensors

Microorganisms	Target	Limit of detection	Transducer	Reference
Pseudomonas aeruginosa	Cephalosporins	11 mM	pH electrode	Kumar et al. (2008)
Recombinant *E. coli*	Penicillins	5–30 mM	Flat pH electrode	Galindo et al. (1990)
Flavobacterium species	Organophosphates	0.025–0.4 mM	pH electrode	Gaberlein et al. (2000)
Recombinant *E. coli*	Organophosphates	3 μm	pH electrode	Mulchandani et al. (1998)
E. coli WP2	Tryptophan	0–12 μm	LAPS	Seki et al. (2003)
Bacillus sp.	Urea	0.55–550 μm	NH_4^+ ion-selective electrode	Verma and Singh (2003)
S. ellipsoideus	Ethanol	0.02–50 μm	Oxygen	Rotariu et al. (2004)
S. cerevisiae	Sucrose	3.2 μm	Oxygen	Rotariu et al. (2002)

have been developed using recombinant *E. coli* harbouring plasmids encoding for β-lactamase, and penicillinase synthesis has been immobilised on pH electrode using gluten and acetylcellulose membrane entrapment, respectively (Galindo et al. 1990).

Bacillus species were utilised for developing a disposable biosensor for monitoring contamination of urea in the milk (Verma and Singh 2003). Some potentiometric microbial biosensors which have been developed for different targets such as chloride ion, ethanol, sucrose, tryptophan, etc. have been listed in Table 12.8.

12.4.1.2 Microbial Conductometric Biosensors

Conductometry is a technique which measures conductivity changes in the solution due to the production or consumption of ionic species. A majority of microbe-catalysed reactions involved a change in the ionic species. *C. vulgaris*, a microalgae that acts as a bioreceptor, was constructed to detect heavy metal ions and pesticides in a water sample. A platinum interdigitated electrode was used to immobilise *C. vulgaris* using a bovine serum albumin membrane (Chouteau et al. 2005). *Brevibacterium ammoniagenes* was used in the development of a conductometric biosensor by immobilising the lyophilised culture in pH-sensitive polyaniline on a Pt twin wire electrode to detect urea (Jha et al. 2009).

12.4.2 Optical Microbial Biosensors

Optical microbial biosensors are generally based on the changes in the optical properties, viz. absorption (UV-Vis), bioluminescence, chemiluminescence, reflectance and fluorescence, brought by the interaction of the biocatalyst with the target analyte. The advantages of optical biosensors are compactness, flexibility, resistance to electrical noise and small probe size.

12.4.2.1 Fluorescence Biosensor

Fluorescence spectroscopy is a sensitive analytical technique which can detect very low concentrations of the analyte. There is a direct correlation between the fluorescence emission intensity and concentration at low analyte concentrations. The fluorescent microbial biosensors have been divided into two broad categories based on the detection mode: in vivo and in vitro. The in vivo fluorescent microbial sensor generally incorporates the use of a genetically engineered microorganism which is able to express a reporter gene encoding a protein. Green fluorescent protein (GFP) encoded by the gene *gfp* is the most popular tool due to its stability and sensitivity. The fluorescence emitted by GFP can be easily and conveniently detected by modern optical instruments with little or no damage to the host system (Pickup et al. 2005). The mechanism used for the development of biosensors is quantitative induction of the promoter gene by the target analyte;

Table 12.9 Fluorescence microbial biosensors

Microorganisms	Target	Limit of detection	Transducer	Reference
E. coli ΔlysA mini-Tn5-Km-gfpmut3	Lysine	3 μg/ml	Fluorescent	Chalova et al. (2008)
E. coli DH5α pTOLGFP	Toluene (Bioavailable BTEX)	25 μM	Fluorescent	Li et al. (2008)
Caulobacter crescentus NJH371	Uranium	0.5 μM	Fluorescent	Hillson et al. (2007)
Bacillus megaterium pSD202	Zinc	1 μm	Fluorescent	Date et al. (2007)
E. coli MG1655	Mitomycin C	2.5 ng/g soil	Fluorescent	Norman et al. (2006)
Recombinant *Pseudomonas syringae*	Bioavailable iron	10^{-7} M	Fluorescent	Joyner and Lindow (2000)
Pseudomonas fluorescence A506 (pTolLHB)	Bioavailable toluene	0.02 μM	Fluorescent	Casavant et al. (2003)

thus, concentration of the analyte could be correlated with fluorescence intensity. Using this mechanism a whole cell biosensor using *Pseudomonas putida* was engineered to determine the bioavailability of naphthalene (Kohlmeier et al. 2008). A microbial biosensor using genetically engineered *E. coli* was developed for measuring aromatic aldehydes in the aqueous systems using the same principle (Fiorentino et al. 2009). Another example is a uranium biosensor which was developed by using a bioengineered strain of *Caulobacter crescentus*. This organism became fluorescently green in the presence of toxic levels of uranium. The promoter *urcA* was strongly up-regulated upon exposure to uranium and expressed GFPuv which could be detected in situ using a hand-held UV lamp when the concentration of the uranyl cation was higher than 0.5 μm (Hillson et al. 2007). Similarly *Sinorhizobium meliloti* has been genetically engineered by fusing the *gfp* gene with the *melA* promoter which is induced on the exposure to galactose and galactosides (Bringhurst et al. 2001). Some examples of fluorescence-based microbial biosensors have been listed in Table 12.9.

12.4.2.2 Bioluminescent Microbial Biosensor

The bioluminescent microbial biosensor measures the changes in luminosity of the living microorganisms. The *lux* gene which encodes for luciferase is widely applied as a reporter either in inducible or constitutive manner to observe the change in luminosity by responding to the target analyte in a dose-dependent manner.

In constitutive manner the lux gene exists constitutively in the sensing microbe, and the luminescence of the microbial system would change directly on the addition/interaction of the target analyte. Thus it could be a case where the light intensity produced by a microbe would decrease on interaction with a toxic compound. A *Vibrio fischeri*-based microbial biosensor was developed on the same principle for the determination of toxicity of some common environmental pollutants in a continuous flow system (Stolper et al. 2008). *A bioluminescent biosensor using Pseudomonas fluorescens* HK44 pUTK21 recognition element has been designed to detect the available fraction of naphthalene in soil since there was a linear relationship between the luminescence from the microbe and naphthalene concentration (Paton et al. 2009).

In inducible manner the reporter *lux* gene is generally fused with a promoter gene which is regulated by the existence and concentration of the target analyte. It has been found that the inducible manner of *lux* gene exploitation in biosensor development is more sensitive and specific in nature. *Ralstonia eutropha* AE2515 was constructed by transcriptionally fusing the *cnr*YXH regulatory gene to the bioluminescent *lux*CDABE report system to fabricate a whole cell biosensor to detect the bioavailable concentration

Table 12.10 Bioluminescence microbial biosensors

Microorganisms	Target	Limit of detection	Transducer	Reference
Ralstonia eutropha AE2515	Ni^{2+} and Co^{2+}	0.1 μM Ni^{2+}, 9 μM Co^{2+}	Luminescence	Tibazarwa et al. (2001)
E. coli HMS174 with *mer-lux* plasmid pR27 of pRB28	Hg^{2+}	0.2 ng/g	Luminescence	Rassmussen et al. (2000)
Recombinant E. coli (DL-2-haloacid dehalogenase and lux operon)	Halogenated organic acids	>100 mg/L	Luminescence	Tauber et al. (2001)
E. coli K12 pTetLux1	Tetracycline	25 ng/g tissue	Luminescence	Virolainen et al. (2008)
E. coli K12 pTetLux1	Doxycycline	5 ng/g tissue	Luminescence	Virolainen et al. (2008)
E. coli K12 pTetLux1	Chlortetracycline	7.5 ng/g tissue	Luminescence	Virolainen et al. (2008)
V. fischeri	2,3,5,6,-tetrachlorophenol	4–30 mg/L	Luminescence	Stolper et al. (2008)
E. coli EMS500	Ofloxacin	0.05 μg/ml	Luminescence	Shapiro and Baneyx (2007)
E. coli DH5α p TOLLUX	Toluene (Bioavailable BTEX)	7.5 μM	Luminescence	Li et al. (2008)

of Ni^{2+} and Co^{2+} in soil (Tibazarwa et al. 2001). A tetracycline luminescent biosensor was developed by using *Photorhabdus luminescens* bacterial luciferase operon *lux*CDABE which was inserted as an EcoRI fragment under the control of tetracycline-inducible *tet A* promoter and has been used for the rapid and specific tetracycline residue assay in poultry muscle tissue (Virolianen et al. 2008). A BTEX (benzene, toluene, ethylbenzene and xylene) biosensor was developed using *Pseudomonas putida* containing a lux reporter fused with a BTEX-specific *tod* promoter to assess the biodegradation of BTEX (Dawson et al. 2008). Table 12.10 summarises different bioluminescence-based biosensors.

Over the decades microbial biosensors have been under extensive investigation with some fruitful applications developed in the areas of environmental monitoring, food, military and biomedical devices. When compared to enzymatic biosensors, the development of highly sensitive microbial biosensor is yet to come as there are inherent drawbacks such as long response time, low sensitivity and poor selectivity which are to be overcome. With advances in technology such as nanotechnology and nanostructured materials, these inherent drawbacks could be potentially rectified thereby potentially improving the sensitivity as well as specificity of the microbial biosensors.

12.5 Summary

The use and application of immobilisation of microorganisms have greatly influenced the fermentation industry for the production of a variety of antibiotics, enzymes, chemicals and fine chemicals and for carrying out industrial biotransformations. It has further influenced the development of microbial biosensors. Advancements in nanotechnology and nanomaterial structures would definitely influence the process of immobilisation of microbial cells on the transducers for improving the selectivity and sensitivity of the microbial biosensors for industrial as well as research applications.

Selected Reading

Bringhurst RM, Cardos ZG, Gage DJ (2001) Galactosides in the rhizosphere: utilization of *Sinorhizobium meliloti* and development of a biosensor. Proc Natl Acad Sci U S A 98:4540

Casavant NG, Thompson D, Beattle GA, Phillips GJ, Halverson LJ (2003) Use of site specific recombination based on biosensor for detecting bioavailable toluene and related compounds on roots. Environ Microbiol 5:238

Chalova VI, Zabala-Diaz IB, Woodward CL, Ricke SC (2008) Development of a whole cell green fluorescent sensor for lysine quantification. World J Microbiol Biotechnol 24:353–359

Chee GJ, Nomura Y, Ikebukuro K, Karube I (2005) Development of a photocatalytic biosensor for evaluation of biochemical oxygen demand. Biosens Bioelectron 21:67–73

Cheetham PSJ, Garrett C, Clark J (1985) Isomaltulose production using immobilized cells. Biotechnol Bioeng 27:471–481

Choi HS, Shin MS, Kim JA (1999) Enhancement of microbial adhesion on chemically modified polyethylene surface. Environ Eng Res 4:127–133

Chouteau C, Dzyadevych S, Durrien C, Chovelon JM (2005) A bi-enzymatic whole cell conductometric biosensor for heavy metal ions and pesticide detection in water samples. Biosens Bioelectron 21:273–281

Date A, Pasini P, Daunert S (2007) Construction of spores for portable bacterial whole cell biosensing systems. Anal Chem 79:9391–9397

Dawson JJC, Iroegbu CO, Maciel H, Paton GI (2008) Application of luminescent biosensors for the monitoring the degradation and toxicity of BTEX compounds in soils. J Appl Microbiol 104:141–151

Emelyanova EV, Reshetilov AN (2002) *Rhodococcus erythropolis* as the receptor of cell based sensor for 2, 4-dinitrophenol detection: effect of co-oxidation. Process Biochem 37:683–692

Fiorentino G, Ronca R, Bartolucci S (2009) A novel *E. coli* biosensor for selecting aromatic aldehydes based on inducible archaeal promoter fused with green fluorescent protein. Appl Microbiol Biotechnol 82:67–77

Gaberlein S, Spener F, Zaborosch C (2000) Microbial and cytoplasmic membrane based potentiometric biosensors for direct determination of organophosphorus insecticides. Appl Microbiol Biotechnol 54:652

Galindo E, Bautista D, Garcia JL, Quintero R (1990) Microbial sensors for penicillin's using a recombinant strain of *E. coli*. Enzym Microb Technol 12:642

Held M, Schuhmann W, Jahreis K, Schmidt HL (2002) Microbial biosensor array with transport mutants of *Escherichia coli* K12 for simultaneous determination of mono and disaccharides. Biosens Bioelectron 17:1089–1094

Hillson NJ, Andersen GL, Shapiro L (2007) *Caulobacter crescentus* as a whole cell uranium biosensor. Appl Environ Microbiol 73:7615–7621

Jha SK, Kanungo M, Math A, D' Souza SF (2009) Entrapment of live microbial cells in electropolymerized polyaniline and their use as a urea biosensor. Biosens Bioelectron 24:2637–2642

Jia J, Tang M, Chen X, Qi L, Dong S (2003) Co-immobilized microbial biosensor for BOD estimation based on sol-gel derived composite material. Biosens Bioelectron 18:1023–1029

Joyner DC, Lindow SE (2000) Heterogeneity of iron bioavailability in plants assessed with whole cell green fluorescent protein based bacterial biosensor. Microbiology 146:2435–2445

Kailasapathy K (2002) Microencapsulation of probiotic bacteria: technology and potential applications. Curr Issues Intest Microbiol 3:39–48

Kitagawa Y, Ameyama M, Nakashima K, Tamaiya E, Karube I (1987) Amperometric alcohol sensor based on immobilized bacteria cell membrane. Analyst 112:1747–1749

Kohlmeier S, Mancuso M, Deepthike U, Tecon R, van der Meer JR, Harms H, Wells M (2008) Comparison of naphthalene bioavailability determined by whole cell bioassay and availability determined by extraction of Tenax. Environ Pollut 156:803–808

Kumar S, Kundu S, Pakshirajan K, Dasu VV (2008) Cephalosporin determination with a novel microbial biosensor based on permeabilized Pseudomonas areugionosa whole cells. Appl Biochem Biotechnol 151:653–664

Lehmann M, Riedel K, Adler K, Kunze G (2000) Amperometric measurement of copper ions with a deputy substrate using a novel *Saccharomyces cerevisiae*. Biosens Bioelectron 15:211–219

Lei Y, Chen W, Mulchandani A (2006) Microbial biosensors. Anal Chim Acta 568:200–210

Li FX, Li FY, Ho CL, Liao VHC (2008) Construction and comparison of fluorescence and bioluminescence bacterial biosensor for the detection of bioavailable toluene and related compounds. Environ Pollut 152:123–129

Mulchandani A, Mulchandani P, Kaneva I, Chen W (1998) Biosensor for direct determination of organophosphate nerve agents using recombinant *Escherichia coli* with surface expressed organophosphorus hydrolase 1. Potentiometric microbial electrode. Anal Chem 70:4140–4145

Navarro JM, Durand GC (1977) Modification of yeast by immobilization onto a porous glass. Eur J Appl Microbiol 4:243–254

Norman A, Hansen LH, Sorensen SJ (2006) A flow cytometry optimized assay using an SOS-green fluorescent protein whole cell biosensor for the detection of genotoxins in complex environments. Mutat Res Genet Toxicol Environ Mutagen 603:164–172

Okhi A, Shinohara K, Ito O, Naka K, Maeda S, Sato T, Akano H, Kato N, Kawamura Y (1994) A BoD sensor using *Klebsiella oxytoca* AS1. Int J Environ Anal Chem 56:261–269

Okochi M, Mima k, Miyata M, Shinozaki Y, Haraguchi S, Fujisawa M, Kaneka M, Masukata T, Matsunaga T (2004) Development of an automated water toxicity biosensor using *Thiobacillus ferrooxidans* for monitoring cyanides in natural water for a water filtering plant. Biotechnol Bioeng 87:905–911

Paton GI, Reid BJ, Sempled KT (2009) Application of a luminescence based biosensor for assessing naphthalene biodegradation in soils from a gas manufacturing plant. Environ Pollut 157:1643–1648

Pickup JC, Hussain F, Evans ND, Rolinski OJ, Birch DJS (2005) Fluorescence based glucose sensors. Biosens Bioelectron 20:2555–2565

Rajasekar C, Rajasekar R, Narasimhan KC (2000) *Acetobacter peroxydans* based electrochemical biosensor for hydrogen peroxide. Bull Electrochem 16:25–28

Rasmussen LD, Sørensen SJ, Turner RR, Barkay T (2000) Application of a mer-lux biosensor for estimating bioavailable mercury in soil. Soil Biol Biochem 32:639–646

Reshetilov AN, Trotsenko JA, Morozova NO, Iliasov PU, Ashin VV (2001) Characteristics of *Gluconobacter oxydans* B-1280 and *Pichia methanolica* MN4 cell based biosensor for the detection of ethanol. Process Biochem 36:1015–1020

Rotariu L, Bala C (2003) New type of ethanol microbial biosensor based on a highly sensitive amperometric oxygen electrode and yeast cells. Anal Lett 36:2459–2471

Rotariu L, Bala C, Magearu V (2000) Use of yeast cells for selective determination of glucose. Rev Roum Chem 45:21–26

Rotariu L, Bala C, Magearu V (2002) Yeast cell sucrose biosensor based on a potentiometric oxygen electrode. Anal Chim Acta 458:215–222

Rotariu L, Bala C, Magearu V (2004) New potentiometric microbial biosensor for ethanol determination in alcoholic beverages. Anal Chim Acta 513:119–123

Seki A, Kawakubo K, Iga M, Nomura S (2003) Microbial assay for tryptophan using silicon based transducer. Sensors Actuators B 94:253–256

Shapiro E, Baneyx F (2007) Stress-activated bioluminescent Escherichia coli sensors for antimicrobial agents detection. J Biotechnol 132:487–493

Stolper P, Faber S, Weller MG, Knopp D, Niessner R (2008) Whole cell luminescence based flow through biodetector for toxicity testing. Anal Bioanal Chem 390:1181–1187

Tauber M, Rosen R, Belkin S (2001) Whole-cell biodetection of halogenated organic acids. Talanta 55:959–964

Tibazarwa C, Corbisier P, Mench M, Bossus A, Solda P, Mergeay M, Wyns L, van der Lelie D (2001) A microbial biosensor to predict bioavailable nickel in soil and its transfer to plants. Environ Pollut 113:19–26

Tkac J, Vostiar I, Gemanier P, Sturdik E (2002) Monitoring ethanol during fermentation using a microbial biosensor with enhanced selectivity. Bioelectrochemistry 56:127–129

Van Haecht JL, Bolipombo M, Rouxhet PG (1985) Immobilization of *Saccharomyces cerevisiae* by adhesion treatment of cells by Al ions. Biotechnol Bioeng 27:217–224

Verma N, Singh M (2003) A disposable microbial based sensor for quality control in milk. Biosens Bioelectron 18:1219–1224

Virolainen NE, Pikkemaat MG, Elferink JWA, Karp MT (2008) Rapid detection of tetracyclines and their 4-epimer derivatives from poultry meet with Bioluminescent biosensor bacteria. J Agric Food Chem 56:11065–11070